**CourseMate**

Engaging. Trackable. Affordable.

CourseMate brings course concepts to life with interactive learning, study, extra problem sets, and exam preparation tools that support *STAT*.

INCLUDES:
Integrated eBook, **Interactive teaching and learning tools,** and **Engagement Tracker**, a first-of-its-kind tool that monitors student engagement in the course.

ON THE
WEB

BEHAVIORAL SCIENCES
STAT
Are you in?

ONLINE RESOURCES INCLUDED!

FOR INSTRUCTORS:
• First Day of Class Instructions
• Custom Options through 4LTR+ Program
• Lecture Booster
• Instructor's Manual
• Test Bank
• PowerPoint® Slides
• Instructor Prep Cards
• Engagement Tracker

FOR STUDENTS:
• Interactive eBook
• Auto-Graded Quizzes
• Flashcards
• Games: Crossword Puzzles,
  Beat the Clock, & Quiz Bowl
• PowerPoint® Slides
• Videos
• Student Review Cards

Students sign in at **login.cengagebrain.com**

WADSWORTH
CENGAGE Learning™

STAT for the Behavioral Sciences
Gary W. Heiman

Senior Publisher:
   Linda Schreiber-Ganster

Executive Editor: Jon-David Hague

Acquisitions Editor: Tim Matray

Developmental Editor: Laura Rush,
   B-books, Ltd.

Associate Development Editor:
   Nicolas Albert

Editorial Assistant: Alicia McLaughlin

Product Development Manager,
   4LTR Press: Steven E. Joos

Brand Executive Marketing Manager,
   4LTR Press: Robin Lucas

Senior Marketing Manager: Jessica Egbert

Marketing Coordinator: Anna Anderson

Executive Marketing Communications
   Manager: Talia Wise

Production Director: Amy McGuire,
   B-books, Ltd.

Senior Content Project Manager:
   Christy A. Frame

Media Editor: Lauren Keyes

Senior Print Buyer: Rebecca Cross

Production Service: B-books, Ltd.

Senior Art Director: Vernon Boes

Cover Design: Denise Davidson

Cover Image: © Etienne Girardet/Getty
   Images

Photography Manager: Don Schlotman

Photo Researcher: Charlotte Goldman

Library of Congress Control Number: 2010932466

ISBN-13: 978-1-111-34206-7
ISBN-10: 1-111-34206-7

Wadsworth Cengage Learning
20 Davis Drive
Belmont, CA 94002-3098
USA

Cengage Learning products are represented in Canada by
Nelson Education, Ltd.

For your course and learning solutions, visit **www.cengage.com**
Purchase any of our products at your local college store or at our
preferred online store **www.CengageBrain.com**

Printed in the United States of America
1 2 3 4 5 6 7 13 12 11 10

For my wife, Karen,
a beauty and a joy forever.

stat
Brief Contents

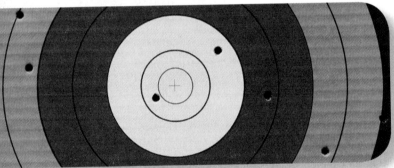

Contents

Contents

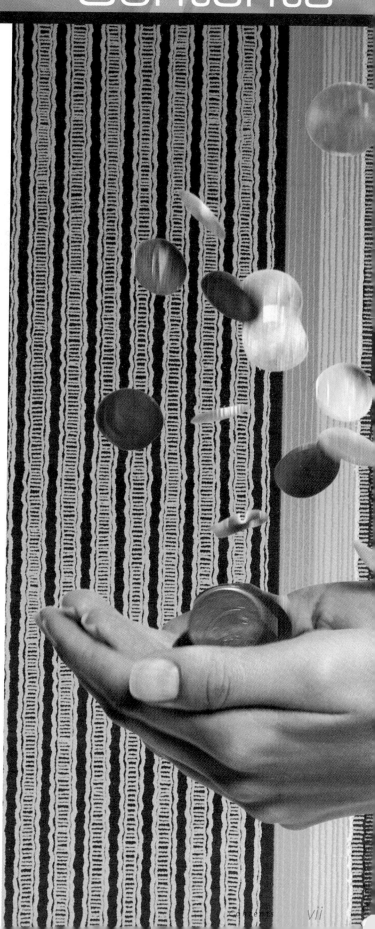

Contents

Contents

Contents

Contents

Introduction
to Statistics and Research

Okay, so you're taking a course in statistics. What does this involve? Well first of all, statistics involve math, but if that makes you a little nervous you can relax: You do not need to be a math wizard to do well in this course. You need to know only how to add, subtract, multiply, and divide—and use a calculator. Also, the term *statistics* is often shorthand for *statistical procedures*, and statisticians have already developed the statistical procedures you'll be learning about. So you won't be solving simultaneous equations, performing proofs and derivations, or doing other mystery math. You will simply learn how to select the statistical procedure—the formula—that is appropriate for a given situation and then compute and interpret the answer. And don't worry, there are not that many to learn, and these fancy sounding "procedures" include such simple things as computing an average or drawing a graph. So really, relax.

As you'll see, statistics are tools that researchers use when performing behavioral research. Therefore, for you to understand statistics, your first step is to understand the basics of research so that you can see how statistics fit in. To get you started, in this chapter we will discuss (1) the logic of research and the purpose of statistics, (2) the two major types of studies that researchers conduct, and (3) the four ways that researchers measure behaviors.

SECTIONS

1.1 WHY IS IT IMPORTANT TO LEARN ABOUT STATISTICS?

Statistics are important because people involved in psychology and other behavioral sciences use statistics and statistical concepts every day. Even if you are not interested in conducting research yourself, statistics are necessary for comprehending other people's research.

What does "understanding statistics" mean? First, you must speak the language. The symbols and terminology you will learn are part of the shorthand "code" used to describe

The purpose of
statistical procedures
is to make sense out
of data.

statistical analyses and to communicate and interpret research results. A major part of learning statistics is actually just learning this code. Once you speak the language, much of the mystery surrounding statistics evaporates, so use flash cards or do whatever else it takes to learn the terminology.

Second, you must understand how researchers use statistics. Behavioral research involves measuring one or more behaviors. This usually results in a large batch of scores—called the data. **The purpose of statistical procedures is to make sense out of data by** *organizing, summarizing, communicating,* **and** *interpreting* **the scores.** You'll see that we organize scores by

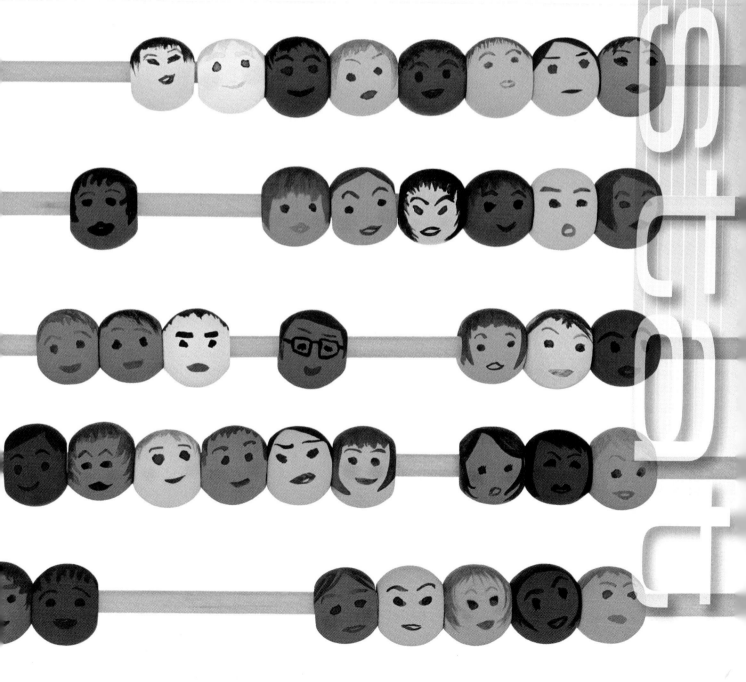

creating tables and graphs, and we summarize scores by computing an average or other single number. We also have rules for how to report a result so that others can easily understand us. You'll also learn new ways to think about numbers, using a logic that allows you to draw conclusions about the events we are studying.

Prepare yourself for the fact that to learn statistics you must practice statistics. Each new discussion incorporates previous procedures and terminology, so review a topic whenever necessary. Some of the formulas may appear difficult, but that's because they're written in code. They too, simply require a little practice. We will discuss each formula and review how to compute it. There are plenty of opportunities to get the necessary practice: at the end of each chapter, there are practice problems; there are also Review and Application Questions on the tear-out Review Cards at the back of this textbook. Seriously work on these questions. Practicing these problems will help you gain a clear understanding of statistics.

Do not, however, get so carried away with the calculations that you forget that statistics are used as a tool that you must learn to apply. Therefore, your goal is to learn when and why to use each procedure and how to interpret its answer.

In this book we'll compute the answers "by hand" so you can see how each is produced. Once you are familiar with the practice, you will probably want to use a computer. One of the most popular statistics programs used by researchers is called *SPSS*. (Recently, the new version 17 was released and the name was changed to *PASW*, but we still refer to it as *SPSS*.) At the end of most chapters, there is a brief section on using SPSS, as well as step-by-step instructions on the Chapter Review Cards. You'll love it—SPSS is *soooo* easy.

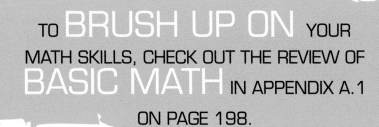

TO BRUSH UP ON YOUR MATH SKILLS, CHECK OUT THE REVIEW OF BASIC MATH IN APPENDIX A.1 ON PAGE 198.

But remember, computer programs only do what you tell them to do. SPSS, and other programs, cannot decide which statistical procedure to compute in a particular situation, nor can it interpret the answer for you. You really must learn *when* to use each statistic and what the answer *means*.

1.2 THE LOGIC OF RESEARCH

The goal of behavioral research is to understand the "laws of nature" that apply to the behaviors of living organisms. That is, researchers assume that specific influences govern every behavior of all members of a particular group. Although any single study is a small step in this process, our goal is to understand every factor that influences the behavior. Thus, when researchers study such things as the mating behavior of sea lions or social interactions between humans, they are ultimately studying the laws of nature.

The reason a study is a small step is because nature is very complex, and so research involves a series of translations that simplify things so that we can examine a specific influence on a specific behavior in a specific situation. Then, using our findings, we *generalize* back to the broader behaviors and laws we began with. For example, here's an idea for a simple study. Say that we think a law of nature is that people must study information in order to learn it. We translate this into the more specific *hypothesis* that "the more you study

But what does it mean?

statistics, the better you'll learn them." Next, we will translate the hypothesis into a situation where we can observe and measure specific people who study specific material in different amounts, to see if they *do* learn differently. Based on what we observe, we have evidence for working back to the general law regarding studying and learning. Part of this translation process involves samples and populations.

Samples and Populations

When researchers talk of a behavior occurring in nature, they say it occurs in the population. A **population** is the entire group of individuals to which a law of nature applies (whether all humans, all men, all four-year-old English-speaking children, etc.). For our example, the population might be all college students who take statistics. A population usually contains all possible members—past, present, and future—so we usually consider it to be infinitely large.

However, to study an *infinite* population would take roughly forever! Instead, we study a sample from the population. A **sample** is a relatively small subset of a population that is intended to represent, or stand in for, the population. Thus, we might study the students in your statistics class as a sample representing the population of all college students studying statistics. The individuals measured in a sample are called the **participants** and it is their scores that constitute our data.

Although researchers ultimately discuss the behavior of individuals, in statistics we often go directly to their scores. Thus, we will talk about the population of scores as if we have already measured the behavior of everyone in the population in a particular situation. Likewise, we will talk about a sample of scores, implying that we have already measured our participants. Thus, a population is the complete group of scores that would be found for everyone in a particular situation, and a sample is a subset of those scores that we actually measure in that situation.

The logic behind samples and populations is this: We use the scores in a sample to *infer*—to estimate—the scores we would expect to find in the population if we could measure it. Then by translating the scores in the sample back into the behaviors they reflect, we can infer the behavior of the population. By describing the behavior of the population, we *are* describing how nature works, because the population *is* the entire group to which the law of nature applies. Thus, if we observe that greater studying leads to better learning for the sample of students in your statistics class, we will infer that similar scores and behaviors would be found in the population of all statistics students. This provides evidence that, in nature, more studying does lead to better learning.

Notice that the above assumes that a sample is *representative* of the population. We discuss this issue in later chapters, but put simply, the individuals in a representative sample *accurately* reflect the individuals that are found in the population. Then our inferences about the scores and behaviors found in the population will also be accurate. Thus, if your class is representative of all college students, then the scores the class obtains are a good example of the scores that everyone in the population would obtain.

On the other hand, any sample can be *unrepresentative* and then it *inaccurately* reflects the population. Maybe your class contains very strange, atypical students who are not at all like those in the population. If so, then their behaviors and scores will mislead us about those of the typical statistics student. Therefore, as you'll see, researchers always consider the possibility that a conclusion about the population—about nature—might be incorrect because it might be based on an unrepresentative sample.

population The large group of all possible scores that would be obtained if the behavior of every individual of interest in a particular situation could be measured

sample A relatively small subset of a population, intended to represent the population; a subset of the complete group of scores found in any particular situation

participants The individuals who are measured in a sample

Understanding Variables

We measure aspects of the situation that we think influence a behavior, and we measure aspects of the behavior itself. The aspects of the situation or behavior that we measure are called variables. A **variable** is anything that can produce two or more different scores. A few of the variables in behavioral research include your age, race, gender, intelligence, and personality type; your salary or the type of job you have; how aggressive you are; and how accurately you can recall an event.

Notice that some variables indicate an amount or quantity: Your height, for example, indicates the *amount* of height you have. Other variables, however, do not indicate an amount, but rather indicate a quality or category. For example, a person's gender is a qualitative variable.

For our research on studying and learning statistics, say that to measure "studying" we select the variable of the number of hours that students spent studying for a particular statistics test. To measure "learning," we select the variable of their performance on the test. After measuring participants' scores on these variables, we examine the *relationship* between them.

variable Anything that, when measured, can produce two or more different scores

relationship A pattern between two variables whereby a change in one variable is accompanied by a consistent change in the other

Understanding Relationships

If nature relates those mental activities we call *studying* to those mental activities we call

learning, then different amounts of learning should occur with different amounts of studying. In other words, there should be a *relationship* between studying and learning. A **relationship** is a pattern in which, as the scores on one variable change, the scores on the other variable change in a consistent manner. In our example, we predict the relationship in which the longer you study, the higher your test grade will be.

Say that we ask some students how long they studied for a test and their subsequent grades on the test. We obtain the data in Table 1.1. To see the relationship, first look at those people who studied for 1 hour and see their grade. Then look at those who studied 2 hours, and see that they had a different grade from those studying 1 hour. And so on. These scores form a relationship because as the study time scores change (increase), the test grades also change in a consistent fashion (also increase). Further, when study time scores do not change (for example, Gary and Bo both studied for 1 hour), their grades also do not change (they both received Fs). We often use the term *association* when talking about relationships: Here, low study times are associated with low test grades and high study times are associated with high test grades.

Because we see a relationship in these sample data, we have evidence that in nature, studying and learning

Table 1.1

Scores Showing a Relationship between the Variables of Study Time and Test Grades

Student	Study Time in Hours	Test Grades
Gary	1	F
Bo	1	F
Sue	2	D
Tony	2	D
Sidney	3	C
Ann	4	B
Rose	4	B
Lou	5	A

FYI: The data presented in this book are fictional. Any resemblance to real data is purely a coincidence.

REMEMBER

In a relationship, *as the scores on one variable change, the scores on the other variable change in a consistent manner.*

do operate as we think: The amount someone studies does seem to make a difference in test grades. In the same way, whenever a law of nature ties behaviors or events together, then we'll see that particular scores from one variable are associated with particular scores from another variable so that a relationship is formed. Therefore, most research is designed to investigate relationships, because relationships are the tell-tale signs of a law at work.

A major use of statistical procedures is to examine the scores in a relationship and the pattern they form. The simplest relationships fit one of two patterns. Let's call one variable X and the other Y. Then, sometimes the relationship fits the description "the more you X, the more you Y." Examples of this include the following: the more you study, the higher your grade; the more alcohol you drink, the more you fall down; the more often you speed, the more traffic tickets you receive; and even that old saying "the bigger they are, the harder they fall."

At other times, the relationship fits the description "the more you X, the *less* you Y." Examples of this include the following: the more you study, the fewer errors you make; the more alcohol you drink, the less coordinated you are; the more you "cut" classes, the lower your grades, and even that old saying "the more you practice statistics, the less difficult they are."

Relationships may also form more complicated patterns where, for example, more X at first leads to more Y, but beyond a certain point even more X leads to *less* Y. For example, the more you exercise the better you feel, until you reach a certain point beyond which more exercise leads to feeling less well, due to pain and exhaustion.

Although the above examples involve quantitative variables, relationships can also involve qualitative variables. For example, men typically are taller than women. If you think of male and female as "scores" on the variable of gender, then this is a relationship because as gender scores change (going from male to female), height

scores decrease. We can study any combination of qualitative and quantitative variables in a relationship.

The Consistency of a Relationship Table 1.1 shows a perfectly consistent association between hours of study time and test grades: All those who studied the same amount received the same grade. In a *perfectly consistent relationship*, a score on one variable is always paired with one and only one score on the other variable. In the real world, however, not everyone who studies for the same time will receive the same test grade. (Life is not fair.) A relationship can be present even if there is only some *degree* of consistency. Then, as the scores on one variable change, the scores on the other variable *tend* to change in a consistent fashion.

For example, Table 1.2 presents a less consistent relationship between the number of hours studied and the number of *errors* made on the test. Notice that the variables are also labeled X and Y. When looking at a relationship, get in the habit of asking, "As the X scores *increase*, do the Y scores change in a consistent fashion?" Answer this by again looking at one study-time score (at one X score) and seeing the error scores (the Y scores) that are paired with it. Then look at the next X score and see the Y scores paired with it. Two aspects of the data in Table 1.2 produce a less consistent relationship: First, not everyone who studies for a particular time receives the same error score (e.g., 12, 13, and 14 errors are all paired with 1 hour). Second, sometimes a particular error score is paired with *different* studying scores (e.g., 11 errors occur with both 1 and 2 hours of

Table 1.2

Scores Showing a Relationship between Study Time and Number of Errors on Test

Student	X Hours of Study	Y Errors on Test
Amy	1	12
Karen	1	13
Joe	1	11
Cleo	2	11
Jack	2	10
Maria	2	9
Terry	3	9
Manny	3	10
Chris	4	9
Sam	4	8
Gary	5	7

A relationship is present (though not perfectly consistent) if there tends to be a different group of scores Y associated with each X score. A relationship is not present when virtually the same batch of Y scores are paired with every X score.

study). These aspects cause overlapping groups of different error scores to occur at each study time, so the overall pattern is harder to see. In fact, the greater the differences among the group of *Y* scores at an *X* and the more the *Y* scores overlap between groups, the less consistent the relationship will be. Nonetheless, we still see the pattern where more studying *tends* to be associated with lower error scores, so a relationship is present. Essentially, one batch of error scores occurs at one study-time score, but a *different* batch of error scores tends to occur at the next study-time score.

Notice that the less consistent relationship above still supports our original hypothesis about how nature operates: we see that, at least to some degree, nature does relate studying and test errors. Thus, we will always examine the relationship in our data, no matter how

Less studying may lead to more errors ...

consistent it is. A particular study can produce anywhere between a perfectly consistent relationship and no relationship, and so in Chapter 10 we will discuss in depth how to describe and interpret the consistency of a particular relationship. (As you'll see, the degree of consistency in a relationship is called its *strength,* and a less consistent relationship is a *weaker* relationship.) Until then it is enough for you to simply know what a relationship is.

When No Relationship Is Present At the other extreme, sometimes the scores from two variables do not form a relationship. For example, say that we had obtained the data shown in Table 1.3.

Here, no relationship is present because the error scores paired with 1 hour are essentially the same as the error scores paired with 2 hours, and so on. Thus, virtually the same (but not identical) batch of error scores shows up at each study time, so no pattern of increasing or decreasing errors is present. These data show that how long people study does not make a consistent difference in their error scores. Therefore, this result would not provide evidence that in nature, studying and learning operate as we think.

Table 1.3

Scores Showing No Relationship between Hours of Study Time and Number of Errors on Test

Student	*X* Hours of Study	*Y* Errors on Test
Amy	1	12
Karen	1	10
Joe	1	8
Cleo	2	11
Jack	2	10
Maria	2	9
Terry	3	12
Manny	3	9
Chris	3	10
Sam	4	11
Jane	4	10
Gary	4	8

1.3 APPLYING DESCRIPTIVE AND INFERENTIAL STATISTICS

Statistics help us make sense out of data, and now you can see that "making sense" means to understand the scores and the relationship they form. However, because we are always talking about samples and populations, we separate statistical procedures into those that apply to samples and those that apply to populations.

Descriptive statistics are procedures for organizing and summarizing *sample* data. The answers from such procedures are often a single number that *describes* important information about the scores. (When you see *descriptive,* think *describe.*) A sample's average, for example, is an important descriptive statistic, because in one number we summarize all scores in the sample. Descriptive statistics are also used to describe the relationship in sample data. For our study-time research, for example, we'd want to know whether a relationship is present, how consistently errors decrease with increased study time, and so on. (We'll discuss the common descriptive procedures in the next few chapters.)

After describing the sample, we want to use that information to estimate or *infer* the data we would find if we could measure the entire population. However, we cannot automatically assume that the scores and the relationship that we see in the sample is what we would see in the population: Remember, the sample might be unrepresentative, so that it misleads us about the population. Therefore, first we apply additional statistical procedures. **Inferential statistics** are procedures for drawing inferences about the scores and relationship that would be found in the population. Essentially, inferential procedures help us to decide whether to believe that our sample accurately represents the population. If it does, then, for example, we would use the class average as an estimate of the average score we'd find in the population of students. Or, we would use the relationship in our sample to estimate how, for everyone, greater learning tends to occur with greater studying. (We discuss inferential procedures in the second half of this book.)

Statistics versus Parameters

Researchers use the following system so that we know when we are describing a sample and when we are describing a population. A number that describes an aspect of the scores in a *sample* is called a **statistic**. Thus, a statistic is an answer obtained from a descriptive procedure. We compute different statistics to describe different aspects of the data, and the symbol for each is a different letter from the *English* alphabet. On the other hand, a number that describes an aspect of the scores in the *population* is called a **parameter**. Thus, a parameter is obtained when applying inferential procedures. The symbols for the different parameters are letters from the *Greek* alphabet.

Thus, for example, the average in your statistics class is a sample average, a descriptive *statistic* that is symbolized by a letter from the English alphabet. If we then estimate the average in the population, we are estimating a *parameter,*

---

descriptive statistics Procedures for organizing and summarizing data so that the important characteristics can be described and communicated

inferential statistics Procedures for determining whether sample data represent a particular relationship in the population

statistic A number from a descriptive procedure that describes a sample of scores; symbolized by a letter from the English alphabet

parameter A number obtained using inferential procedures that describes a population of scores; symbolized by a letter from the Greek alphabet

design The way in which a study is laid out

experiment A research procedure in which one variable is actively changed or manipulated and the scores on another variable are measured to determine whether there is a relationship

independent variable In an experiment, a variable that is changed or manipulated by the experimenter; a variable hypothesized to cause a change in the dependent variable

PARAMETERS USE
GREEK LETTERS;

STATISTICS USE
ENGLISH LETTERS.

and the symbol for a population average is a letter from the Greek alphabet.

After performing the appropriate descriptive and inferential procedures, we stop being a "statistician" and return to being a behavioral scientist: We interpret the results in terms of the underlying behaviors, psychological principles, sociological influences, and so on, that they reflect. This completes the circle, because by describing the behavior of everyone in the population in a given situation, we *are* describing how a law of nature operates.

1.4 UNDERSTANDING EXPERIMENTS AND CORRELATIONAL STUDIES

In research we can examine a relationship using a variety of different kinds of studies. In other words, we use different *designs*. The **design** of a study is how it is laid out—how many samples are examined, how participants are selected and tested, and so on. A study's design is important because different designs require different descriptive and inferential procedures. Recall that your goal is to learn *when* to use each statistical procedure and, in part, that means learning the

particular procedures that are appropriate for a particular design. (On the first tear-out card in the back of your book is a decision tree for selecting procedures, which you should refer to as you build your knowledge of statistics.)

To begin, recognize that we have two major types of designs because we have two general ways of demonstrating a relationship: using experiments or using correlational studies.

Experiments

In an **experiment,** the researcher actively changes or manipulates one variable and then measures participants' scores on another variable to see if a relationship is *produced*. For example, say that we study amount of study time and test errors in an experiment. We decide to compare 1, 2, 3, and 4 hours of study time, so we select four samples of students. We have one sample study for 1 hour, administer the statistics test, and count the number of errors each participant makes. We have another sample study for 2 hours, administer the test, and count their errors, and so on. Then we determine if we have produced the relationship where, as we increase study time, error scores tend to decrease. You must understand the components of an experiment and learn their names.

The Independent Variable An **independent variable** is the variable that is changed or manipulated by the experimenter. We manipulate this variable because we assume that doing so will *cause* the behavior and scores on the other variable to change. Thus, in our example above, amount of study time is our independent variable: we manipulate study time because doing this should cause participants' error scores to change in the predicted way. (To prove that this variable is actually the cause is a very difficult task that we'll save for an advanced discussion. In the meantime, be cautious when using the word *cause*.) You can remember *independent* because this variable occurs *independently* of participants' wishes (we'll have some participants study for 4 hours whether they want to or not).

Technically, a true independent variable is manipulated by doing something *to* participants. However, there are many variables that an experimenter cannot manipulate in this way. For example, we might hypothesize that growing older causes a change in some behavior, but we can't *make* some people be 20 years old and make others be 60 years old. Instead, we would manipulate the variable by selecting one sample of 20-year-olds and one sample of 60-year-olds. We will also call this type of variable an independent variable (although technically it is called a *quasi-independent variable*). Statistically, we treat all independent variables the same.

Thus, the experimenter is always in control of the independent variable, either by determining what is done to each sample or by determining a characteristic of the individuals in each sample. Therefore, a participant's "score" on the independent variable is determined by the experimenter: Above, students in the sample that studied 1 hour have a score of 1 on the study-time variable; people in the 20-year-old sample have a score of 20 on the age variable.

Conditions of the Independent Variable An independent variable is the *overall* variable a researcher manipulates, which is potentially composed of many different amounts or categories. From these the researcher selects the *conditions*. A **condition** is the specific amount or category of the independent variable that creates the specific situation under which participants are studied. Thus, although our independent variable is amount of study time—which could be any amount—our conditions involve 1, 2, 3, or 4 hours of study. Likewise, if we compare 20-year-olds to 60-year-olds, then 20 and 60 are each a condition of the independent variable of age.

The Dependent Variable The **dependent variable** is the variable that measures a behavior or attribute of participants that we expect will be influenced by the independent variable. Therefore, we measure participants' scores on the dependent variable in each condition. You can remember *dependent* because whether a score is high or low presumably *depends* on a participant's reaction to the condition. (This variable reflects the behavior that is "caused" in the relationship.)

Thus, in our studying experiment, test errors is our dependent variable because these scores depend on how participants respond to their particular study time. Or, in a different experiment, if we compare the activity levels of 20- and 60-year-olds, then participants' activity level is the dependent variable because presumably it depends on their age. *Note:* The dependent variable is also called the "dependent measure" and we obtain "dependent scores."

Drawing Conclusions from an Experiment

As we change the conditions of the independent variable, participants' scores on the dependent variable should also change in a consistent fashion. To see this relationship, a useful way to diagram an experiment is shown in Table 1.4. Each column in the diagram is a condition of the independent variable (here amount of study time). The numbers in a column are the scores on the dependent variable from participants who were tested under that condition (here each score is the number of test errors).

Remember that a condition determines participants' scores on the independent variable. Thus, participants in the 1-hour condition each have a score of "1" on the independent variable, those under 2 hours have a score of "2," and so on. Thus, the diagram communicates pairs of scores consisting of 1-13, 1-12, 1-11; then 2-9, 2-8, 2-7, etc. Now look for the relationship as we did previously: First look at the error scores paired with 1 hour, then at the error scores paired with 2 hours, and so on. The pattern here forms a relationship where, as study-time scores increase, error scores tend to decrease. Essentially, participants in the 1-hour condition produce one batch of error scores, those in the 2-hour condition produce a different, lower batch of error scores, and so on.

condition An amount or category of the independent variable that creates the specific situation under which participants' scores on the dependent variable are measured

dependent variable In an experiment, the behavior or attribute of participants that is expected to be influenced by the independent variable

VS.

Table 1.4

Diagram of an Experiment Involving the Independent Variable of Number of Hours Spent Studying and the Dependent Variable of Number of Errors Made on a Statistics Test

Dependent Variable: Number of Errors Made on a Statistics Test →	Independent Variable: Number of Hours Spent Studying			
	Condition 1: 1 Hour	Condition 2: 2 Hours	Condition 3: 3 Hours	Condition 4: 4 hours
	13	9	7	5
	12	8	6	3
	11	7	5	2

We use this diagram because it facilitates applying our statistics. For example, it makes sense to compute the average error score in each condition (each column). Notice, however, that we apply statistics to the *dependent* variable. We do not know what scores participants will produce, so these are the scores that we need help in making sense of (especially in a more realistic study where we might have 100 different scores in each column). We do not compute anything about the independent variable because we know all about it (e.g., above we have no reason to compute the average of 1, 2, 3, and 4 hours). Rather, the conditions simply form the groups of dependent scores that we then examine.

Thus, we will use specific descriptive procedures to summarize the sample's scores and the relationship found in an experiment. Then, to infer that we'd see a similar relationship if we tested the entire population, we have specific *inferential* procedures for experiments. Finally, we will translate the relationship back to the original hypothesis about studying and learning that we began with, so that we can add to our understanding of nature.

Correlational Studies

Not all research is an experiment. Sometimes we do not manipulate or change either variable and instead conduct a correlational study. In a **correlational study,** the researcher measures participants' scores on two variables and then determines whether a relationship is present. Thus, in an experiment the researcher attempts to *make* a relationship happen, while in a correlational study the researcher is a pas-

correlational study A procedure in which subjects' scores on two variables are measured, without manipulation of either variable, to determine whether they form a relationship

sive observer who looks to see if a relationship *exists*. For example, we used a correlational approach previously when we simply asked some students how long they studied for a test and what their test grade was. Or, we would have a correlational design if we asked people their career choices and measured their personality, asking "Is career choice related to personality type?" (As we'll see, correlational studies examine the "correlation" between variables, which is another way of saying they examine the relationship.)

As usual, we want to first describe and understand the relationship we've observed in the sample, and correlational designs have their own descriptive statistical procedures for doing this. Here we do not know the scores that participants will produce for either variable, so the starting point for making sense of them is often to compute the average score on *each* variable. Then, to describe the relationship we would find if we could study the entire population, we have specific correlational inferential procedures. Then, as with an experiment, we will

REMEMBER

In a correlational study, the researcher simply measures participants' scores on two variables to determine if a relationship exists.

translate the relationship back to the original hypothesis about studying and learning that we began with so that we can add to our understanding of nature.

1.5 THE CHARACTERISTICS OF SCORES

We have one more issue to consider when selecting the descriptive or inferential procedure to use in a particular experiment or correlational study. Although we always measure one or more variables, the numbers that comprise the scores can have different underlying mathematical characteristics. The particular characteristics of our scores determine which procedures we should use, because the kinds of math we can perform depends on the kinds of numbers we have. Therefore, always pay attention to two important characteristics of your scores: the scale of *measurement* involved and whether the scale is *continuous* or *discrete*.

The Four Types of Measurement Scales

Numbers mean different things in different contexts. The meaning of a 1 on a license plate is different from that of a 1 in a race, which is different still from the meaning of a 1 in a hockey score. The kind of information that scores convey depends on the *scale of measurement* that is used in measuring the variable. There are four types of measurement scales: *nominal, ordinal, interval,* and *ratio.*

With a **nominal scale,** we do not measure an amount, but rather we categorize or classify individuals. For example, to "measure" your gender we will classify you as either male or female, so we are using a nominal scale. Numbers are used simply for identification (so for nominal, think *name*). For example, we might assign a "1" to males and a "2" to females, but these numbers are assigned arbitrarily—they don't reflect an amount, and we could use any other numbers. Thus, the key here

is that nominal scores indicate only that one individual is *qualitatively* different from another. So, the numbers on football uniforms or your identification number are also nominal scales. In research, we have nominal variables when studying different types of schizophrenia or different therapies. These variables can occur in any design, so for example, in a correlational study, we might measure the political affiliation of participants using a nominal scale by assigning a 5 to democrats, a 10 to republicans, and so on. Then we might also measure participants' income, to determine whether as party affiliation "scores" change, income scores also change. Or, if an experiment compares the job satisfaction scores of workers in several different occupations, the independent variable is the nominal variable of type of occupation.

A different approach is to use an **ordinal scale.** Here the scores indicate rank order—anything that is akin to 1st, 2nd, 3rd . . . is ordinal. (*Ordinal* sounds like *ordered*.) In our studying example, we'd have an ordinal scale if we assigned a 1 to students who scored best on the test, a 2 to those in second place, and so on. Then we'd ask, "As study times change, do students' ranks also tend to change?" Or, if an experiment compares 1st graders to 2nd graders, then this independent variable involves an ordinal scale. The key here is that ordinal scores indicate only a relative amount—identifying who scored relatively high or low. Also, there is no score of 0, and the same amount does not separate every pair of adjacent scores: 1st may be only slightly ahead of 2nd, but 2nd may be miles away from 3rd. Therefore, other examples of ordinal variables include clothing size (e.g., small, medium, large), college year (e.g., freshman or sophomore), and letter grades (e.g., A or B).

A third approach is to use an **interval scale.** Here each score indicates an actual quantity, and an equal amount separates any adjacent scores. (For interval scores, remember *equal* intervals between them.) However, although interval scales do include the number 0, it is not a *true zero*—it does not mean *none* of the variable

nominal scale A measurement scale in which each score is used simply for identification and does not indicate an amount

ordinal scale A measurement scale in which scores indicate rank order

interval scale A measurement scale in which each score indicates an actual amount and there is an equal unit of measurement between consecutive scores, but in which zero is simply another point on the scale (not zero amount)

is present. Therefore, the key is that you can have less than this amount, so an interval scale allows negative numbers. For example, temperature (in Celsius or Fahrenheit) involves an interval scale: because 0° does not mean that zero heat is present, you can have even less heat at −1°. In research, interval scales are common with intelligence or personality tests: A score of zero does not mean zero intelligence or zero personality. Or, in our studying research we might determine the average test score and then assign students a zero if they are average, a +1, +2, etc., for the amount they are above average, and a −1, −2, etc., for the amount they are below average. Then we'd see if more positive scores tend to occur with higher study times. Or, if we create conditions based on whether participants are in a positive, negative, or neutral mood, then this independent variable reflects an interval scale.

The final approach is to use a **ratio scale.** Here, like interval scores, each score measures an actual quantity, and an equal amount separates adjacent scores. However, 0 truly means that none of the variable is present. Therefore, the key is that you cannot have negative numbers, because you cannot have less than nothing. Also, only with a true zero can we make "ratio" statements, such as "4 is twice as much as 2." (So for *ratio*, think *ratio!*) We used ratio scales in our previous examples when measuring the number of errors and the number of hours studied. Likewise, if we compare the conditions of having people on diets consisting of either 1,000, 1,500, or 2,000 calories a day, then this independent variable involves a ratio scale. Other examples of ratio variables include the level of income in a household, the amount of time required to complete a task, or the number of items in a list to be recalled by participants.

We can study relationships that involve any combination of the above scales.

Continuous versus Discrete Scales

In addition, any measurement scale may be either continuous or discrete. A **continuous scale** allows for fractional amounts, so decimals make sense. That is, measurement of the variable *continues* between the whole number amounts, and there is no limit to how small a fraction we may examine. Thus, the variable of age is continuous because it is perfectly intelligent to say that someone is 19.6879 years old. On the other hand, some variables involve a **discrete scale,** which can be measured only in fixed amounts, which cannot be broken into smaller amounts. Usually the amounts are labeled using whole numbers, so decimals do not make sense. For example, being male or female, or being in 1st grade versus 2nd grade are discrete variables, because you can be in one group or you can be in the other group, but you can't be in between. Some variables may be labeled using fractions, as with shoe sizes, but they are still discrete because they cannot be broken into smaller units.

Usually researchers assume that nominal or ordinal variables are discrete, and that interval or ratio variables are at least *theoretically* continuous. For example, intelligence tests are designed to produce whole-number scores, so you cannot have an IQ of 95.6. But theoretically an IQ of 95.6 makes sense, so intelligence is a theoretically continuous (interval) variable. Likewise, it sounds strange if the government reports that the average family has 2.4 children, because this is a discrete (ratio) variable and no one has .4 of a child. However, it makes sense to treat this as theoretically continuous, because we can interpret what it means if the average this year is 2.4, but last year it was 2.8.

ratio scale A measurement scale in which each score indicates an actual amount, there is an equal unit of measurement, and there is a true zero

continuous scale A measurement scale that allows for fractional amounts of the variable being measured

discrete scale A measurement scale that allows for measurement in fixed amounts, which cannot be broken into smaller amounts; usually amounts are labeled using whole numbers

REMEMBER

Whether a variable is continuous *or* discrete *and whether it is measured using a* nominal, ordinal, interval, *or* ratio *scale are factors that determine which statistical procedure to apply.*

1. What is the goal of behavioral research?

2. What is a variable?

3. Describe and contrast experiments and correlational studies.

4. Compare and contrast descriptive and inferential statistics.

5. What is a representative sample?

6. Define the four major types of measurement scales.

7. Describe when a scale is continuous and when it is discrete.

8. What are the differences between statistics and parameters in their use and symbols?

9. Of the three sets of data that follow, which sample shows a perfectly consistent relationship?

Sample A		Sample B		Sample C	
X	Y	X	Y	X	Y
1	5	1	5	1	5
1	5	1	6	1	6
1	5	2	7	2	5
2	7	2	8	2	6
2	7	2	9	3	6
3	9	3	10	3	5
3	9	3	11	4	6

10. For the data sets presented in problem 9, which sample shows no relationship?

11. For each of the following research projects, indicate whether a researcher would be more likely to study the relationship by conducting an experiment or a correlational study:

   a. A comparison of the effects of different amounts of caffeine consumed in one hour on speed of completing a complex motor task

   b. An investigation of the relationship between number of extracurricular activities and GPA

   c. An examination of the relationship between the number of pairs of sneakers a person owns and the person's athletic success

   d. A comparison of the effects of three types of perfume on perceived sexual attractiveness

   e. An investigation of the relationship between GPA and the ability to pay off school loans

   f. A comparison of the effects of different amounts of beer consumed on a person's mood

12. In each of the following experiments, identify the independent variable, the conditions of the independent variable, and the dependent variable:

   a. A researcher studies whether participants' self-esteem is influenced by whether they have completed an easy or a difficult problem-solving task.

   b. A researcher compares young, middle-aged, and senior citizen adults with respect to how much confidence they have in their mental abilities.

   c. A researcher investigates whether people estimate the duration of a time period differently when they view three, five, or seven pictures per minute during the period.

   d. A researcher examines whether length of exposure to a movie containing violent scenes (60 minutes versus 120 minutes) produces differences in subsequent aggressive behavior.

13. In each of the following, identify whether the data are implicitly based on a sample or a population:

   a. Nine out of ten dentists surveyed recommend "Sugarmint" brand chewing gum.

   b. The IRS announced today that 23% of all reported household incomes are below the poverty line.

   c. The average height of professional basketball players in the NBA is 6 feet, 9 inches.

   d. Based on a survey of major cities, the national crime rate increases during the summer months.

14. For each of the following statements, determine what the sample is and what the population is. Also, determine whether the sample would be considered representative.

   a. 200 freshmen at State College were selected at random from among the school's 1,000 freshmen students and asked to indicate their opinions as to how prepared they felt to attend this school.

   b. All 40 teachers at North High School were asked to complete a survey on methods of instruction used in the state's schools.

   c. A psychology student was asked to report on the difficulty of questions asked in the recent state licensing exam.

   d. In a survey of counselors in the United States, 2,000 counselors were selected at random from a national list of 8,000 and asked to rate their job-related anxiety level.

15. In the following chart, complete the cells opposite each variable:

Variable	Type of Measurement Scale	Continuous or Discrete
Nationality		
Hand pressure		
Baseball team rank		
Letter grade on a test		
Pregnancy test		
Checkbook balance		

Creating and Using Frequency Distributions

So we're off into the world of descriptive statistics. Recall that the goal is to make sense of the scores by organizing and summarizing them. One important procedure for doing this is to create tables and graphs, because they show the scores you've obtained and they make it easier to see the relationship between two variables that is hidden in the data. Before we examine the relationship between two variables, however, we first summarize the scores on *each* variable alone. Therefore, this chapter will discuss the common ways to describe scores from one variable by using a *frequency distribution.* You'll see (1) how to show a frequency distribution in a table or graph, (2) the common patterns found in frequency distributions, and (3) how to use a frequency distribution to compute additional information about scores.

SECTIONS

2.1 Some New Symbols and Terminology

2.2 Understanding Frequency Distributions

2.3 Types of Frequency Distributions

2.4 Relative Frequency and the Normal Curve

2.5 Understanding Percentile

2.1 SOME NEW SYMBOLS AND TERMINOLOGY

Looking Back

- From Chapter 1, understand that descriptive statistics are used to describe and summarize the characteristics of data.

- From Chapter 1, know what nominal, ordinal, interval, and ratio scales of measurement are and what *continuous* and *discrete* mean.

The scores we initially measure in a study are called the **raw scores.** Descriptive statistics help us boil down raw scores into an interpretable, "digestible" form. There are several ways to do this, but the starting point is to count the number of times each score occurred. The number of times a score occurs in a set of

raw scores The scores initially measured in a study

data is the score's **frequency.** If we examine the frequencies of every score in the data, we create a *frequency distribution.* The term *distribution* is the general name researchers have for any organized set of data. In a **frequency distribution,** the scores are organized based on each score's frequency. (Actually researchers have several ways to describe frequency, so technically, when we *simply* count the frequency of each score, we are creating a *simple frequency distribution.*)

The symbol for a score's frequency is the lowercase *f*. To find *f* for a score, count how many times that score occurs. If three participants scored 66, then 66 occurred three times, so the frequency of 66 is 3 and so $f = 3$. Creating a frequency distribution involves counting the frequency of every score in the data.

In most statistical procedures, we also count the total number of scores we have. The symbol for the total number

frequency (*f*) The number of times each score occurs within a set of data; also called simple frequency

frequency distribution A distribution of scores, organized to show the number of times each score occurs in a set of data

The frequency *of a score is symbolized by* f. *The total number of scores in the data is symbolized by* N.

of scores in a set of data is the uppercase N. Thus, $N = 43$ means that we have 43 scores. Note that N is not the number of *different* scores, so even if all 43 scores in a sample are the same score, N still equals 43.

2.2 UNDERSTANDING FREQUENCY DISTRIBUTIONS

The first step when trying to understand any set of scores is to ask the most obvious question, "What are the scores that were obtained?" In fact, buried in any data are two important things to know: Which scores occurred, and how often did each occur? These questions are answered simultaneously by looking at the frequency of each score. Thus, frequency distributions are important because they provide a simple and clear way to show the scores in a set of data. Because of this, they are always the first step when beginning to understand the scores from a study. Further, they are also a building block for upcoming statistical procedures.

One way to see a frequency distribution is in a table.

Presenting Frequency in a Table

Let's begin with the following raw scores. (They might measure one of the variables from a correlational study, or they might be dependent scores from an experiment.)

| 14 | 14 | 13 | 15 | 11 | 15 | 13 | 10 | 12 |
| 13 | 14 | 13 | 14 | 15 | 17 | 14 | 14 | 15 |

In this disorganized arrangement it is difficult to make sense of these scores. Watch what happens, though, when we arrange them into the frequency table in Table 2.1.

Researchers have several rules of thumb for making a frequency table. Start with a score column and an f column. The score column has the highest score in the data at the *top* of the column. Below that are all *possible* whole-number scores in decreasing order, down to the lowest score that occurred. Here, our highest score is 17, the lowest score is 10, and although no one obtained a score of 16, we still include it. In the f column opposite each score is the score's frequency: In the sample there is one 17, zero 16s, four 15s, and so on.

Not only can we see the frequency of each score, we can also determine the combined frequency of several scores by adding together their individual fs. For example, the score of 13 has an f of 4 and the score of 14 has an f of 6, so their combined frequency is 10.

Notice that, although 8 scores are in the score column, N is *not* 8. We had 18 scores in the original sample, so N is 18. You can see this by adding together all of the individual frequencies in the f column: The 1 person scoring 17 plus the 4 people scoring 15 and so on adds up to the 18 people in the sample. In a frequency distribution, the sum of the frequencies always equals N.

Table 2.1

Simple Frequency Distribution Table
The left-hand column identifies each score, and the right-hand column contains the frequency with which the score occurred.

Score	f
17	1
16	0
15	4
14	6
13	4
12	1
11	1
10	1
	Total: 18 = N

Graphing a Frequency Distribution

When researchers talk of a frequency distribution, they often imply a *graph* that shows the frequencies of each score. To produce the graph, we place the scores on the X axis and frequency on the Y axis. (A review of basic graphing is in Appendix A.1.) To graph a frequency distribution, place the *scores* on the X axis, regardless of whether the scores reflect actual amounts from an interval or ratio scale, relative rankings from an ordinal scale, or categories from a nominal scale. Label the Y axis using frequency.

We have several ways to draw the graph of a frequency distribution, depending on the scale of measurement that the raw scores reflect. We may create a *bar graph*, a *histogram*, or a *polygon*.

Creating Bar Graphs

We graph a frequency distribution of nominal or ordinal scores by creating a bar graph. A **bar graph** has a vertical bar centered over each X score and the height of the bar corresponds to the score's frequency. Notably, *adjacent bars do not touch*.

Figure 2.1 shows the frequency tables and bar graphs of two samples. The upper table and graph is from a survey in which we counted the number of participants in each category of the nominal variable of political party affiliation. The X axis is labeled using the "scores" of political party, and because this is a nominal variable they can be arranged in any order. In the frequency table, we see that 6 people were Republicans, so we draw a bar at a height of 6 above "Rep.," and so on.

The lower table and graph is from a survey in which we counted the number of participants having a particular military rank (an ordinal variable). The ranks are arranged on the X axis from lowest to highest. Again, the height of each bar is the "score's" frequency.

The reason we create bar graphs with nominal and ordinal scales is that both are *discrete* scales: You can be in one group or the next, but not in-between. The space between the bars in a bar graph also indicates this. On the other hand, recall that interval and ratio scales are usually assumed to be at least theoretically *continuous*: They allow fractional amounts that continue between the whole numbers. To communicate this, these scales are graphed by creating either a *histogram* or a *polygon*.

> **bar graph** A graph in which a free-standing vertical bar is centered over each score on the X axis; used with nominal or ordinal scores

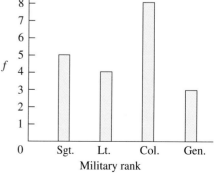

REMEMBER

A graph of a frequency distribution always shows the scores on the X axis and their frequency on the Y axis.

Figure 2.1

Frequency Bar Graphs for Nominal and Ordinal Data
The height of each bar indicates the frequency of the corresponding score on the X axis.

Nominal Variable of Political Affiliation

Party	f
Communist	1
Socialist	3
Democrat	8
Republican	6

Ordinal Variable of Military Rank

Rank	f
General	3
Colonel	8
Lieutenant	4
Sergeant	5

histogram A graph similar to a bar graph but with adjacent bars touching; used to plot the frequency distribution of a small range of interval or ratio scores

frequency polygon A graph that shows interval or ratio scores (*X* axis) and their frequencies (*Y* axis), using data points connected by straight lines

data point A dot plotted on a graph to represent a pair of *X* and *Y* scores

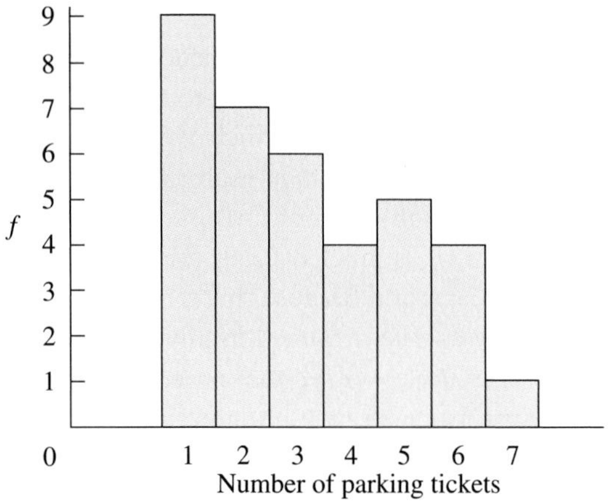

REMEMBER

In a histogram *the adjacent bars touch; in a* bar graph *they do not.*

Creating Histograms

A **histogram** is similar to a bar graph except that *in a histogram the adjacent bars touch*. For example, say that we measured the ratio variable of number of parking tickets that participants received, obtaining the data in Figure 2.2. Again the height of each bar indicates the corresponding score's frequency. Because the adjacent bars touch, there is no gap between the scores on the *X* axis. This communicates that the *X* variable is continuous, with no gaps in our measurements because we can measure it in fractional amounts.

Creating Frequency Polygons

Construct a **frequency polygon** by placing a "dot" above each score on the *X* axis at the height that corresponds to the appropriate frequency on the *Y* axis. Then connect the dots using straight lines. For example, Figure 2.3 shows the previous parking ticket data plotted as a frequency polygon. For a score of 1 the *f* is 9, so we place a dot there. Above a score of 2 the dot is at 7, and so on. Notice, however, that we also include on the *X* axis the next score above the highest score in the data and the next score below the lowest score (in Figure 2.3, scores of 0 and 8 are included). These added scores have a frequency of 0, so the curve touches the *X* axis. In this way we create a complete geometric figure—a polygon—with the *X* axis as its

base. (As with a histogram, we again communicate that we have a continuous variable, because there is a continuous line above all whole and fractional *X* scores.)

Note: A "dot" plotted on any graph is called a **data point**. So, for example, we placed a data point over an *X* of 4 parking tickets at the height on *Y* for a frequency of 4.

Histograms versus Polygons and Grouped Distributions

Technically, we may create either a histogram or polygon when we have a relatively *small* number of different scores occurring on the *X* axis (although there is a bias in research publications to use histograms). Usually, however, we do not create a histogram for a *large* number of different scores. Say we wanted to plot the frequency of between 1 and 50 parking tickets. The 50 bars would need to be very skinny,

Figure 2.2
Histogram Showing the Frequency of Parking Tickets in a Sample

Score	f
7	1
6	4
5	5
4	4
3	6
2	7
1	9

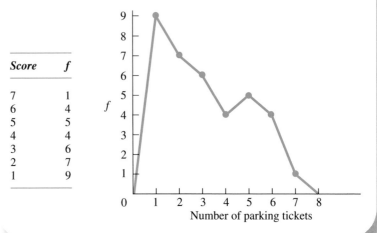

Figure 2.3
Frequency Polygon Showing the Frequency of Parking Tickets in a Sample

Score	f
7	1
6	4
5	5
4	4
3	6
2	7
1	9

so the graph would be difficult to read. We have no rule for how many bars are too many, but when a histogram is unworkable, create a polygon to plot the frequency of each *individual* score.

An alternative approach you may encounter is to reduce the number of points plotted in a figure by first *grouping* the scores. In a **grouped distribution** we combine individual scores into small groups and then report the total frequency (or other description) for each group. Thus, for example,

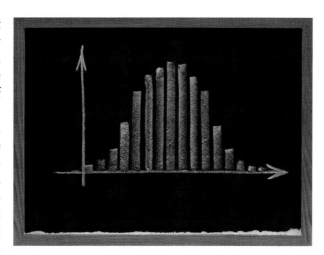

in some data we might group the scores 0, 1, 2, 3, and 4 into the "0–4" group. Then we would add the *f* for the score of 0 with the *f* for the score of 1, and so on, to obtain the frequency of all scores between 0 and 4. Likewise we would combine the scores between 5 and 9 into another group. Then we report the total *f* for each group. The middle score of each group is used to represent the group, so for example, in a graph we'd have the X of 2 stand for the 0–4 group while an X of 7 would be the 5–9 group. This technique can be used to make bar graphs, histograms, or polygons more manageable, as well as to reduce the size of a table.

2.3 TYPES OF FREQUENCY DISTRIBUTIONS

Researchers often encounter polygons that have the same particular shape, so we have names for the most common ones. Each shape comes from an idealized distribution of a population. By far the most important frequency distribution is the *normal distribution*. (This is the big one, folks.)

The Normal Distribution

Figure 2.4 shows the polygon of the ideal normal distribution. (Let's say these are test scores from a population.) Although specific mathematical properties define this polygon, in general it is a bell-shaped curve. But don't call it a bell curve (that's so pedestrian!). Call it a **normal curve** or a **normal distribution,** or say that the scores are *normally distributed.*

Because this polygon represents an infinite population, it is slightly different than for a sample. First, we cannot count the *f* of each score, so no numbers occur on the Y axis. Simply remember that frequencies increase as we proceed higher up the Y axis. Second, the polygon

grouped distribution A distribution created by combining individual scores into small groups and then reporting the total frequency (or other description) of each group

normal curve The symmetrical, bell-shaped curve produced by graphing a normal distribution

normal distribution A set of scores in which the middle score has the highest frequency and, proceeding toward higher or lower scores, the frequencies at first decrease slightly but then decrease drastically, with the highest and lowest scores having very low frequency

Figure 2.4
The Ideal Normal Curve
Scores farther above and below the middle scores occur with progressively lower frequencies.

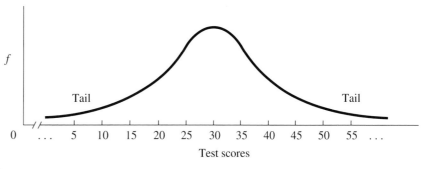

tail of the distribution The far-left or far-right portion of a frequency polygon, containing the relatively low-frequency, extreme scores

negatively skewed distribution A frequency polygon with low-frequency, extreme low scores but without corresponding low-frequency, extreme high ones, so that its only pronounced tail is in the direction of the lower scores

Figure 2.5

Idealized Skewed Distributions

The direction in which the distinctive tail slopes indicates whether the skew is positive or negative.

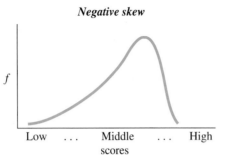

Negative skew

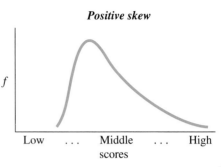

Positive skew

is a smooth curved line. The population contains so many different whole and decimal scores that the individual data points form the curved line. Nonetheless, to see the frequency of a score, locate the score on the X axis and then move upward until you reach the line forming the polygon. Then, moving horizontally, determine whether the frequency of the score is relatively high or low.

As you can see, in a normal distribution, the score with the highest frequency is the middle score (in Figure 2.4 it is at the score of 30.) The normal curve is *symmetrical,* meaning that the left half below the middle score is a mirror image of the right half above the middle score. As you proceed away from the middle, the frequencies decrease, with the highest and lowest scores having relatively very low frequency. However, no matter how low or high a score might be, the curve never actually touches the X axis. This is because in an infinite population theoretically any score might occur sometime, so the frequencies approach—but never reach—zero.

Note: In the language of statistics, the far left and right portions of a normal curve containing the relatively low-frequency, extreme high or low scores are each called the **tail of the distribution.** In Figure 2.4 the tails are roughly below the score of 15 and above the score of 45.

The reason that the normal curve is important is because it is a very common distribution in behavioral research. For most of the variables that we study, most of the individuals score at or close to the middle score, with progressively fewer individuals scoring at the more extreme higher or lower scores. Because of this, the normal curve is also very common in our upcoming

statistical procedures. Therefore, before you proceed, be sure that you can read the normal curve. Can you see in Figure 2.4 that the most frequent scores are between 25 and 35? Do you see that a score of 15 has a relatively low frequency and a score of 45 has the same low frequency? Do you see that there are relatively few scores in the tail above 50 or in the tail below 10? Do you see that the farther into a tail that a score lies, the less frequently the score occurs?

Skewed Distributions

Not all variables form normal distributions. One of the most common *non-normal* distributions is a skewed distribution. A skewed distribution is similar to a normal distribution except that it is not symmetrical and has only *one* pronounced tail. As shown in Figure 2.5 above, a distribution may be either *negatively skewed* or *positively skewed,* and the skew is where the tail is.

A **negatively skewed distribution** contains extreme low scores that have a low frequency but does not contain low-frequency, extreme high scores. The polygon on the left in Figure 2.5 shows an idealized negatively skewed distribution. This pattern might be found, for example, by measuring the running speed of professional football players. Most would tend to run at higher speeds, but a relatively few linemen lumber in at the slower speeds. (To remember *negatively skewed*, remember that the pronounced tail is over the lower scores, sloping toward zero, where the *negative* scores would be.)

On the other hand, a **positively skewed distribution** contains extreme high scores that have low frequency but does not contain low-frequency, extreme low scores. The polygon on the right in Figure 2.5 shows a positively skewed distribution. This pattern is often found, for example, when measuring participants' "reaction time" to a stimulus. Usually, scores will tend to be rather low (fast), but every once in a while a person will "fall asleep at the switch," requiring a large amount of time that produces a high score. (To remember *positively skewed*, remember that the tail slopes away from zero, where the higher, *positive* scores are located.)

Bimodal Distributions

An idealized bimodal distribution is shown in Figure 2.6. A **bimodal distribution** is a symmetrical distribution containing two distinct humps, each reflecting relatively high-frequency scores. At the center of each hump is one score that occurs more frequently than the surrounding scores, and technically the two center scores have the same frequency. Such a distribution would occur with test scores, for example, if most students scored around 60 or 80, with fewer students failing or scoring in the 70s or 90s.

How to Label Distributions

You need to know the names of the previous distributions because descriptive statistics describe the important characteristics of data, and one very important characteristic is the shape of the frequency distribution. First, the shape allows us to understand the data. If, for example, I tell you that my data form a normal distribution, you can mentally envision the distribution and thus instantly understand how my participants generally performed. Also, the shape is important in determining which statistical procedures to employ. Many of our statistics are applied only when we have a normal distribution, while others are for non-normal distributions. *Therefore, the first step when examining any data is to identify the shape of the frequency distribution that is present.*

Data in the real world, however, never form the perfect curves we've discussed. Instead, the scores will form a bumpy, rough approximation to the ideal distribution. Data never form the perfect normal curve, and at best only come close to that shape. However, rather than drawing a different, approximately normal curve in every study, we simplify the task by using the ideal normal curve we saw previously as our one "model" of any distribution that generally has this shape. This gives us one reasonably accurate way of envisioning the various, approximately normal distributions that researchers encounter. The same is true for the other shapes we've seen.

We apply the names of the previous distributions to samples as a way of summarizing and communicating their general shape. Figure 2.7 on the next page shows several examples, as well as the corresponding labels we might use. (Notice that we even apply these names to histograms or bar graphs.) We assume that in the population the additional scores and their frequencies would "fill in" the sample curve, smoothing it out to be closer to the ideal curve.

REMEMBER

Whether a skewed *distribution is* negative *or* positive *corresponds to whether the distinct tail slopes toward or away from zero.*

positively skewed distribution A frequency polygon with low-frequency, extreme high scores but without corresponding low-frequency, extreme low ones, so that its only pronounced tail is in the direction of the higher scores

bimodal distribution A symmetrical frequency polygon with two distinct humps where there are relatively high-frequency scores and with center scores that have the same frequency

Figure 2.6
Idealized Bimodal Distribution

Bimodal

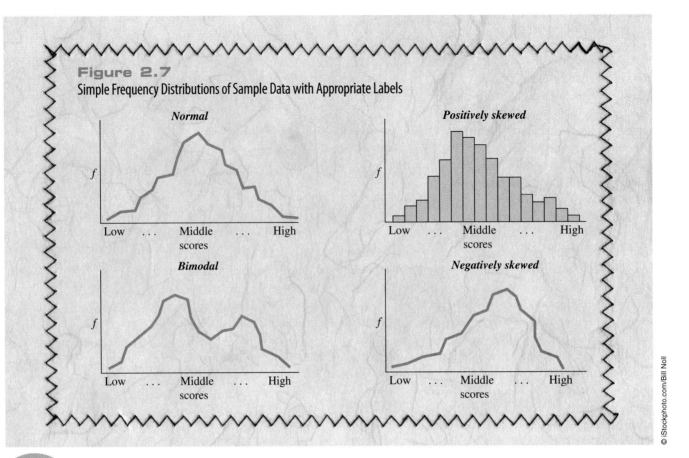

Figure 2.7

Simple Frequency Distributions of Sample Data with Appropriate Labels

Normal

Positively skewed

Bimodal

Negatively skewed

2.4 RELATIVE FREQUENCY AND THE NORMAL CURVE

We will return to frequency distributions—especially the normal curve—throughout the remainder of this course. However, counting the frequency of scores is not the only thing we do. You'll see that an important procedure is to describe scores based on another type of information called *relative frequency*.

relative frequency
The proportion of time a score occurs in a distribution, which is equal to the proportion of the total number of scores that the score's simple frequency represents

Understanding Relative Frequency

Relative frequency is the proportion of the time that a score occurs in a sample. A *proportion* is a decimal number between 0 and 1 that indicates a fraction of the total. Thus, we use relative frequency to indicate what fraction of the sample is made up by the times that a particular score occurs. In other words, we determine the proportion of N that is made up by the *f* of a score.

So that you understand relative frequency, let's first calculate it using a formula, although later we'll use a different approach.

COMPUTING A SCORE'S
RELATIVE FREQUENCY

$$\text{Relative frequency} = \frac{f}{N}$$

This says that to compute a score's relative frequency, divide its frequency (*f*) by the total number of scores (N). For example, if a score occurred 5 times in a sample of 10 scores, then

$$\textit{Relative frequency} = \frac{f}{N} = \frac{5}{10} = .50$$

The score has a relative frequency of .50, meaning that the score occurred .50 of the time in this sample. Or, say that a score occurred 6 times out of an N of 9. Then its relative frequency is 6/9, which is .67. We usually "round off" relative frequency to two decimals. Finally, we might find that several scores have a combined frequency of 10 in a sample of 30 scores: 10/30 equals .33, so these scores together have a relative frequency of .33—they make up .33 of this sample.

We can also work the other direction, beginning with relative frequency and computing the corresponding simple frequency. *To transform relative frequency into simple frequency we multiply the relative frequency times N.* Thus, if a score's relative frequency is .4 when N is 10, then (.40)(10) gives an $f = 4$.

Finally, sometimes we transform relative frequency to percent so that we have the percent of the time that a score occurs. (Remember that officially relative frequency is a proportion.) *To transform relative frequency to a percent, multiply the proportion by 100.* If a score's relative frequency is .50, then we have (.50)(100), so this score occurred 50% of the time. To transform a percent back into relative frequency, divide the percent by 100. The score that is 50% of the sample has a relative frequency of 50%/100 = .50. (For further review of proportions, percents and rounding, consult Appendix A.1.)

When reading research, you may encounter frequency tables that show relative frequency and/or percentages. These are arranged in the same way as the simple frequency table that we saw earlier.

You should know what relative frequency is, but we will not emphasize the above formula. (You're welcome.) Instead, it is important to understand how we use the normal curve to determine a score's relative frequency. This is a core element in later descriptive and inferential statistics.

Finding Relative Frequency Using the Normal Curve

When data are normally distributed, we can use the normal curve to determine relative frequency. To understand this, imagine that you are flying in a helicopter over a parking lot. The X and Y axes are laid out on the ground, and those people who received a particular score are standing in line in front of the marker for their score. The lines of people are packed so tightly together that, from the air, all you see is the tops of many heads. If you painted a line that went

REMEMBER

Relative frequency *indicates the proportion of time (out of N) that a score occurred.*

behind the last person in line at each score, you would have the normal curve shown in Figure 2.8

Think of the normal curve as a solid geometric figure consisting of all participants standing at their scores. The height of the curve above each score reflects the number of people standing at that score. Therefore, reading the graph by moving up from an X score to the line of the curve and then reading the frequency on Y is the same as counting the number of people standing in line at that score. In Figure 2.8, the score of 30 has the highest frequency because literally the longest line of people is standing at this score in the parking lot. Likewise, we could read off the frequencies on the Y axis for several scores, say between 30 and 35. By adding their *f*s together, we'd obtain the frequency of scores between 30 and 35. We'd get the same answer if we counted the people in line above each score and added them together. And if we added together the frequencies for all scores, we'd have the total number of scores (N). This is the same as counting the total number of participants standing in the parking lot.

Using the "parking lot view" you can see how we compute relative frequency. The reason for visualizing the normal curve as people in a parking lot is so that you think of the normal curve as a picture of something solid. The space *under the curve* has *area* that represents individuals and their scores. The entire curve corresponds to everyone in the sample and 100% of the scores. Therefore, any portion of the parking lot—any portion of the space under the curve—corresponds to

Figure 2.8
Parking Lot View of the Ideal Normal Curve
The height of the curve above any score reflects the number of people standing at that score.

f — Tail — Tail

0 . . . 5 10 15 20 25 30 35 40 45 50 55 . . .

Scores

that portion of the sample. For example, in Figure 2.9 a vertical line is drawn through the middle score of 30, and so .50 of the parking lot is to the left of the line. Because the complete parking lot contains all participants, a part that is .50 of it contains 50% of the participants. (We can ignore those relatively few people who are straddling the line.) Participants are standing in the left-hand part of the lot because they received scores of 29, 28, and so on. So in total, 50% of the people obtained scores below 30. Thus, scores below 30 occurred 50% of the time, so the scores below 30 have a combined relative frequency of .50.

We can use the same procedure in any part of the curve. In statistical terms, the total space occupied by people in the parking lot is *the total area under the normal curve*. We take a vertical "slice" of the polygon above certain scores, and the area of this portion of the curve is the space occupied by the people having those scores. We then compare this area to the total area to determine the **proportion of the area under the curve.** Then **the proportion of the area under the normal curve at certain scores corresponds to the relative frequency of those scores.**

Of course, statisticians don't fly around in helicopters, eyeball-

ing parking lots, so here's a different approach: Say that by using a ruler and protractor, we determine that in Figure 2.10 the entire polygon occupies an area of 6 square inches on the page. This total area corresponds to all scores in the sample. Say that the area under the curve between the scores of 30 and 35 covers 2 square inches. This area is due to the number of times these scores occur. Therefore, the scores between 30 and 35 occupy 2 out of the 6 square inches created by all scores, so these scores constitute 2/6, or .33, of the entire distribution. Thus, the scores between 30 and 35 constitute .33 of the distribution, so they have a relative frequency of .33.

Thus, whenever we need to determine the relative frequency of particular scores, we will identify the portion of the distribution containing those scores. Then the proportion of the total area under the curve that is in that portion will equal the relative frequency of the scores in that portion. This technique is especially useful because in Chapter 5 you'll see that statisticians have created a system for easily finding the area under any part of the normal curve, so we can easily determine the relative

Figure 2.9

Normal Curve Showing .50 of the Area under the Normal Curve

The vertical line is through the middle score, so 50% of the distribution is to the left of the line and 50% is to the right of the line.

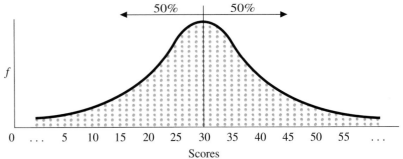

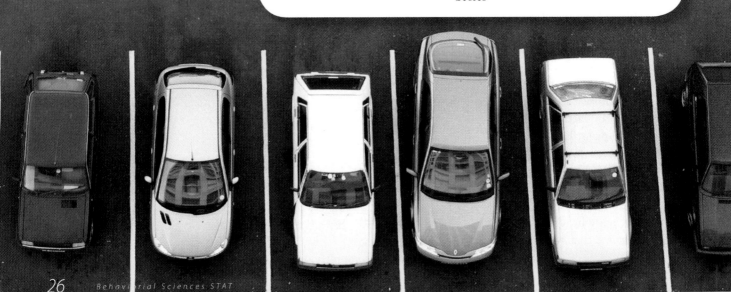

frequency of the scores there. (No, you won't need a ruler and a protractor.) Until then, take special note:

The area under the normal curve corresponds to the times that all scores occur, so a proportion of this area above some of the scores is the proportion of time those scores occur, which is their relative frequency.

2.5 UNDERSTANDING PERCENTILE

One other way to summarize scores is to compute their percentile. Usually a **percentile** is defined as the percent of all scores in the data that are *below* a particular score. Technically, however, a percentile is defined as the percent of the scores that are *at* or below a score, but usually we ignore the relatively few that are at the score. If the score of 40 is at the 50th percentile, we interpret this as indicating that about 50% of the scores are below 40 and 50% of the scores are above 40. Or, if you scored at the 75th percentile, then 75% of the group scored lower than you. If you scored at the 100th percentile, you have the highest score in the data.

A simple method for determining a score's percentile is to again use the area under the normal curve. Percentile describes the scores that are *lower* than a particular score, and on the normal curve, lower scores are to the *left*. Therefore, the percentile for a score corresponds to the percent of the area under the curve that is to the *left* of the score.

For example, Figure 2.11 shows that .50 of the curve is to the left of the score of 30. Because scores to the left of 30 are below it, 50% of the distribution is below 30 (in the parking lot, 50% of the people are standing to the left of the line and all of their scores are less than 30). Thus, the score of 30 is at the 50th percentile. Likewise, say that we find .15 of the distribution is to the left of the score of 20; 20 is at the 15th percentile.

We can also work the other way to find the score at a given percentile. Say that we seek the score at the 85th percentile. We would measure over to the right until 85% of the area under the curve is to the left of a certain point. If, as in Figure 2.11, the score of 45 is at that point, then 45 is at the 85th percentile.

> **percentile** The percentage of all scores in the sample that are below a particular score

Figure 2.10
Finding the Proportion of the Total Area under the Curve
The complete curve occupies 6 square inches, with scores between 30 and 35 occupying 2 square inches.

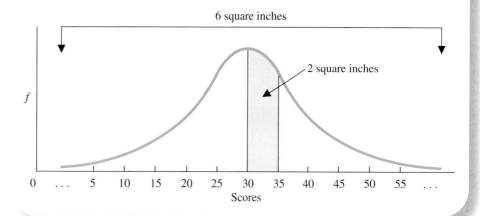

Figure 2.11
Normal Distribution Showing the Area under the Curve to the Left of Selected Scores

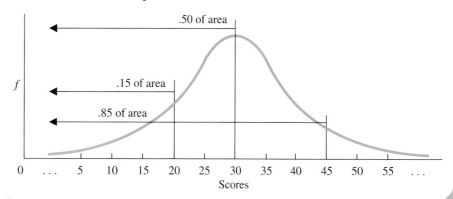

USING SPSS

The SPSS computer program will produce frequency tables and graphs, and you can choose between computing simple frequency, percent, or percentile. On your Chapter Review card, you will find instructions for creating frequency distributions, frequency bar graphs, and frequency bar histograms in SPSS.

From there you can export a table or graph into a word-processing document. However, the program does not always lay out tables using our rules. The source for our rules is the American Psychological Association (APA). We followed "APA format" when creating tables and graphs, and we will later when discussing how to present other statistics. You may need to re-type an SPSS output so that it conforms to APA format.

using what you know

1. What is a frequency distribution?

2. Describe the shapes of normal, positively skewed, negatively skewed, and bimodal distributions when graphed.

3. What does it mean when a score is in one of the tails of the normal distribution?

4. What is the difference between a score's frequency and its relative frequency?

5. A professor observes that a distribution of test scores is negatively skewed. What does this tell the professor about the difficulty of the test? How do you know?

6. In graphing data, when is a bar graph created and when is a histogram or polygon created?

7. Explain why we can compute a score's percentile by finding the proportion of the area under the normal curve to the left of the score and multiplying it by 100.

8. In a test on summarizing scores using frequency distributions and percents, you are given a small sample of raw scores. You are to determine the frequencies, relative frequencies, and percents and put them into a table.
   a. What should you put into the left-hand column of this complete table?
   b. When you add up all the frequencies, what should the total equal?
   c. How would you compute the relative frequencies?
   d. How are the percents calculated?

9. If you are told that your score on an intelligence test is at the 75th percentile, what does this tell you?

10. A scientist asked students to indicate whether a test was easy, somewhat difficult, or very difficult. She wants to graph the frequency distribution of these data. What type of graph should she create? Why?

11. The salaries at a large corporation are known to be positively skewed.
    a. What would this indicate about the pay at this company?
    b. If your salary is in the tail of this distribution, what should you conclude about your salary?

12. The following normal distribution is based on a sample of data. The shaded area represents 13% of the area under the curve.

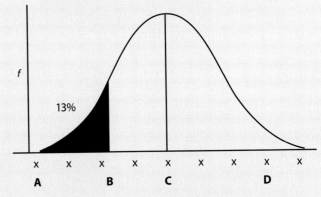

f

13%

x   x   x   x   x   x   x   x   x
A   **B**   **C**   **D**

   a. What is the relative frequency of scores between A and B?
   b. What is the relative frequency of scores between A and C?
   c. What is the relative frequency of scores between B and C?
   d. Rank-order A, B, C, and D to reflect the order of scores from the highest to the lowest frequency.
   e. Rank-order A, B, C, and D to reflect the order of scores from the highest to the lowest score.

13. A study was performed to determine the blood types of a group of 20 astronauts. Create a frequency distribution table for these 20 scores. Then graph the distribution.

Order the blood types as B, AB–, A–, O, and A, with type B as the "highest score."

Subject	Blood Type
1	A
2	O
3	A–
4	O
5	B
6	A–
7	A
8	B
9	AB–
10	A
11	AB–
12	B
13	A
14	A
15	O
16	AB–
17	O
18	O
19	O
20	O

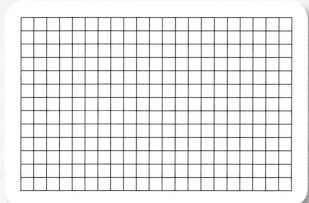

14. If you had to label the shape of the distribution in problem 13, what would you call it?

15. A group of students received the following grades on an exam:

87	83	83	81	84	86	83	86	85	85
86	84	87	86	87	87	86	85	85	86
86	87	85	83	81	87	86	86	86	

Construct a table showing frequency, relative frequency, and percent for these data. Then graph the frequency distribution.

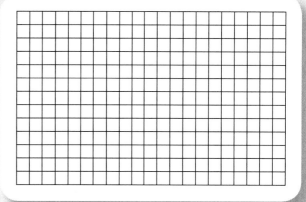

16. a. In a normal distribution 35% of the area under the curve is to the right of the score of 70. At what percentile is 70?

b. The middle score in the distribution is 60. What is the percentile of this score?

c. What is the relative frequency of scores between 60 and 70?

17. A researcher obtained the following data:

Score	f	Relative Frequency	Percent
25	4		
24	5		
23	6		
22	6		
21	8		
20	7		
19	5		
18	4		
17	5		

a. Compute the relative frequency and percent of each score.

b. How do you check that the relative frequencies are correct?

c. How do you check that the percents are correct?

18. a. For the data in problem 17, list the original set of raw scores from which this frequency distribution was generated.

b. Is the distribution positively skewed, negatively skewed, or normal?

Summarizing
Scores with Measures of
Central Tendency

The frequency distributions discussed in Chapter 2 are important because the shape of the distribution is always the first important characteristic of data for us to know. However, graphs and tables are not the most efficient way to summarize a distribution. Instead, we compute individual numbers—statistics—that provide information about the scores. This chapter discusses statistics that describe the important characteristic of data called *central tendency*. The following sections present (1) the concept of central tendency, (2) the three ways to compute central tendency, and (3) how we use them to summarize and interpret data.

SECTIONS

3.1 Some New Symbols and Terminology

3.2 What Is Central Tendency?

3.3 Computing the Mean, Median, and Mode

3.4 Applying the Mean to Research

3.1 SOME NEW SYMBOLS AND TERMINOLOGY

Looking Back

- From Chapter 1, know the logic of statistics and parameters. Understand what independent and dependent variables are and how experiments show a relationship.

- From Chapter 2, recognize normal, skewed, and bimodal distributions and compute percentile using the area under the curve.

Formulas are written to be applied to any set of scores. Usually we use X as the generic symbol for a score. When a formula says to do something to X, it means to do it to all of the scores you are calling X in the sample.

A new symbol you'll see is Σ, the Greek capital letter S, called sigma. Sigma is used with a symbol for scores, especially ΣX. In words, ΣX is pronounced **"sum of X"** and literally means to find the sum of the

sum of X (ΣX) The sum of the scores in a sample

X scores. Thus, ΣX for the scores 5, 6, and 9 is $5 + 6 + 9$, which is 20, so $\Sigma X = 20$. Notice that we do not care whether each X is a different score. If the scores are 4, 4, and 4, then $\Sigma X = 12$.

Also, some statistical answers will contain decimals that we must "round off." The rule for rounding is to *round the final answer to two more decimal places than were in the original raw scores.* Usually we'll have whole-number scores and then the answer contains two decimal places, even if they contain zeros. However, carry more decimal places during your calculations: e.g., if the answer will have two decimals, have at least three decimal places in your calculations. (See Appendix A.1 if you need more details about rounding.)

REMEMBER

A final answer should contain two more decimal places than the original raw scores.

3.2 WHAT IS CENTRAL TENDENCY?

Statistics that measure central tendency are important because they answer a basic question about data: Are the scores in a distribution generally high scores or generally low scores? For example, after taking a test in some class, you first wonder how you did, but then you wonder how the whole class did. Did everyone generally score high, low, or what? You need this information to understand both how the class performed and how you performed relative to everyone else. But it is difficult to do this by looking at individual scores, or even at a frequency distribution. Instead, it is better if you know something like the class average. Likewise, in all research, the first step is to shrink the data into one summary score, called a *measure of central tendency,* that describes the sample as a whole.

To understand central tendency, first change your perspective about what a score indicates. Think of a variable as a continuum (a straight line) and think of a score as indicating a participant's *location* on that variable. For example, if I am 70 inches tall, don't think that I have 70 inches of height. Instead, as in Figure 3.1, my score is at the "address" labeled 70 inches. If my brother is 60 inches tall, then he is located at 60. The idea is not so much that he is 10 inches shorter than I am, but rather that we are separated by a distance of 10 inches. Thus, scores are *locations,* and the difference between any two scores is the *distance* between them.

From this perspective a frequency polygon shows the location of all scores in a distribution. For example, Figure 3.2 shows the height scores from two samples. In our "parking lot" view of a normal curve, participants' scores determine *where* they stand. Further, if we have two distributions containing different scores, then the *distributions* have different locations on the variable.

So when we ask, "Are the scores in a distribution generally high scores or generally low scores?" we are actually asking, "*Where* on the variable is the distribution located?" A **measure of central tendency** is a statistic that indicates the location of a distribution on a variable. Listen to its name: It indicates where the *center* of the distribution *tends* to be located. Thus, it is the point on the variable *around* where most of the scores are located

---

measure of central tendency Statistics that summarize the location of a distribution on a variable by indicating where the center of the distribution tends to be located

---

Figure 3.1
Locations of Individual Scores on the Variable of Height

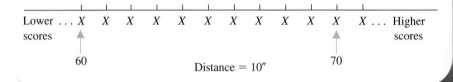

Figure 3.2
Two Sample Polygons on the Variable of Height
Each polygon indicates the locations of the scores and their frequencies.

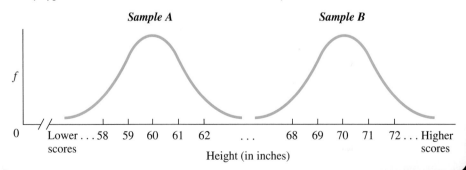

© Corbis/Photolibrary

and it provides an "address" for the distribution. In Sample A in Figure 3.2, most of the scores are in the neighborhood of 59, 60, and 61 inches, so a measure of central tendency will indicate that the distribution is located *around* 60 inches. In Sample B, the distribution is centered at 70 inches.

Notice how descriptive statistics allow us to understand a distribution without looking at every score. If I told you only that one normal distribution is centered at 60 and another is centered around 70, you could envision Figure 3.2 and have a good idea about all of the scores in the data. You'll see other statistics that add to our understanding of a distribution, but measures of central tendency are at the core of summarizing data.

n the following sections we consider the three common ways to measure central tendency: the *mode,* the *median,* and the *mean.*

The Mode
number w/ most occurance in a set of #'s

The **mode** is a score that has the highest frequency in the data. For example, say that we have these scores: 2, 3, 3, 4, 4, 4, 4, 5, 5, 6. The score of 4 is the mode. (There is no conventional symbol for the mode.) The frequency polygon of these scores is on the top in Figure 3.3. It shows that the mode does summarize this distribution because the scores *are* located around 4. Notice that the polygon is roughly a normal curve, with the highest point over the mode. When a polygon has one hump, such as on the normal curve, the distribution is called **unimodal,** indicating that one score qualifies as the mode.

However we may not always have only *one* mode. Consider the scores 2, 3, 4, 5, 5, 5, 6, 7, 8, 9, 9, 9, 10, 11, 12. Here, the two scores of 5 and 9 are tied for the most frequent score. This sample is plotted on the bottom in Figure 3.3. In Chapter 2 such a distribution was called **bimodal** because it has two modes. Identifying the two modes does summarize this distribution, because most of the scores are either around 5 or around 9.

The mode is typically used with scores that reflect a *nominal* scale of measurement (when participants are categorized using a qualitative variable). For example, say that we asked some people their favorite flavor of ice cream and counted the

mode The most frequently occurring score in a sample

unimodal A distribution whose frequency polygon has only one hump and thus has only one score qualifying as the mode

bimodal distribution A frequency polygon with two distinct humps where there are relatively high-frequency scores and with center scores that have the same frequency

Figure 3.3

A Unimodal Distribution (a) and a Bimodal Distribution (b)
Each vertical line marks a highest point on the distribution, thus indicating the most frequent score, which is the mode.

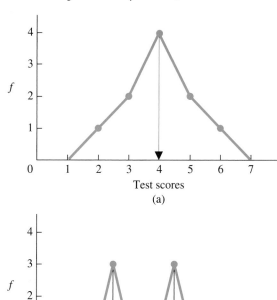

number of people choosing each category. Reporting that the mode was "Goopy Chocolate" does summarize the results, indicating that more people chose this flavor than any other.

There are, however, two potential limitations of the mode. First, the distribution may contain many scores that all have the same highest frequency, and then the mode does not summarize the data. In the most extreme case we might obtain scores such as 4, 4, 5, 5, 6, 6, 7, 7. Here there is no mode.

A second limitation is that the mode does not take into account any scores other than the most frequent score(s), so it ignores much of the information in the data. This can produce a misleading summary. For example, in the skewed distribution containing 7, 7, 7, 20, 20, 21, 22, 22, 23, and 24, the mode is 7. However, most scores are not around 7 but instead are in the low 20s. Thus, the mode may not accurately summarize where *most* scores in a distribution are located.

Because of these limitations, for ordinal, interval, or ratio scores, we usually rely on one of the other measures of central tendency, such as the median.

The Median #directly in middle; diff. than mean

The **median** is simply another name for the score at the 50th percentile. Recall that 50% of a distribution is below the 50th percentile. Thus, if the median is 10, then 50% of the sample scored below 10.

The median presents fewer problems than the mode because (1) a distribution can have only one median and (2) the median will usually be around where most of the scores in a distribution are located. The symbol for the median is **Mdn.**

As you saw in Chapter 2, a score's percentile equals the *proportion of the area under the curve* that is to the left of the score, so the median separates the lower 50% of the distribution from the upper 50%. For example, look at the normal curve in Figure 3.4. Because 50% of the curve is to the left of the line, the score at the line is the 50th percentile, so that score is the median. In the skewed distribution in Figure 3.4, 50% of the curve is also

median (Mdn) The score located at the 50th percentile

Figure 3.4

Location of the Median in a Normal Distribution and in a Skewed Distribution
The vertical line indicates the location of the median, with one-half of the distribution on each side of it.

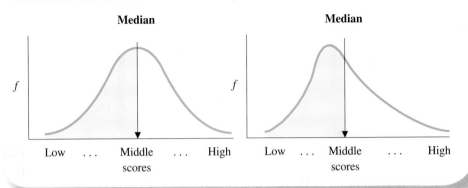

to the left of the vertical line, so the score at the line is the median. In both cases, the score at the line is a reasonably accurate "address" for the entire distribution, with most of the scores around that point.

There are several ways to calculate the median. First, when scores form a perfect normal distribution, the median is also the most frequent score, so it is the same score as the mode. When data are not normally distributed, however, there is no easy way to determine the median. You can, however, *estimate* the median by following these steps:

STEP 1: Arrange the scores from lowest to highest.

STEP 2: Determine N.

STEP 3: If N is an odd number, the median is the score in the middle position.

OR

STEP 4: If N is an even number, the median is the average of the two scores in the middle positions.

For example, for the nine scores 1, 2, 3, 3, 4, 7, 9, 10, 11, the score in the middle position is the fifth score, so the median is the score of 4. On the other hand, for the ten scores 3, 8, 11, 11, 12, 13, 24, 35, 46, 48, the two middle scores are at position 5 (the score of 12) and position 6 (the score of 13). The average of 12 and 13 is 12.5, so the median is approximately 12.5. (To precisely calculate the median, consult an advanced text for the formula. Most computer programs employ this formula, providing the easiest solution.)

We still ignore some information in the data when we compute the median because we count only the frequency of scores in the lower 50% of the distribution without considering their magnitude, and we ignore the scores in the upper 50%. This is actually useful for certain kinds of data. Thus, the median is the preferred measure of central tendency for *ordinal* (rank-ordered) scores. For example, say that a group of students ranked how well a college professor teaches. Reporting that the professor's median ranking was 3 communicates that 50% of the students rated the professor as number 1, 2, or 3. Also, as you'll see, the median is used when interval or ratio scores form a very skewed distribution.

The Mean   *average of all #'s, add #'s together and divide by total amount of #'s*

By far the most common measure of central tendency in behavioral research is the *mean*. The **mean** is defined as the score located at the mathematical center of a distribution, but it is what most people call the *average*. Compute a mean in the same way that you compute an average: Add up the scores and then divide by the number of scores you added. Unlike the mode or the median, the mean considers the magnitude of every score, so it does not ignore any information in the data.

Let's first compute the mean of a sample. Usually we use X to stand for the raw scores in a sample and then the symbol for the *sample mean* is $\overline{X}$. To compute the $\overline{X}$, recall that the symbol that indicates "add up the scores" is ΣX, and the symbol for the number of scores is N, so

For example, in the scores 3, 4, 7, 6:

STEP 1:

Compute ΣX. Add the scores together. Here, $\Sigma X = 4 + 3 + 7 + 6 = 20$.

STEP 2:
Determine N: Here, $N = 4$.

STEP 3:
Divide ΣX *by* N. Here, $\overline{X} = 20/4 = 5$.

Saying that the mean of these scores is 5.00 indicates that the center of this distribution is located at the score of 5.

What is the mathematical center of a distribution? Think of the center as the distribution's *balance point*. For example, on the left side of Figure 3.5 are those scores of 3, 4, 6, and 7. The mean of 5 is at the point that balances the distribution. On the right of Figure 3.5, the mean is the balance point in a different distribution that is not symmetrical. Here, the mean of 4 balances this distribution.

We compute the mean with only interval or ratio data, because it usually does not make sense to compute the average of nominal scores (e.g., finding the average of Democrats and Republicans) or of ordinal scores (e.g., finding the average position in a race). In addition, the distribution should be *symmetrical and unimodal*. In particular, the mean is appropriate with a normal distribution. For example, say we have the scores 1, 2, 3, 3, 4, 4, 4, 5, 5, 6, 7, which form the roughly normal distribution shown in Figure 3.6 on the next page. Here, $\Sigma X = 44$ and $N = 11$, so the mean score—the center—is 4. The reason we use the mean with such data is because the mean is the mathematical center of any distribution:

mean The score located at the mathematical center of a distribution

$\overline{X}$ The symbol used to represent the sample mean

Computing a Sample Mean

$$\overline{X} = \frac{\Sigma X}{N}$$

Figure 3.5
The Mean as the Balance Point of any Distribution

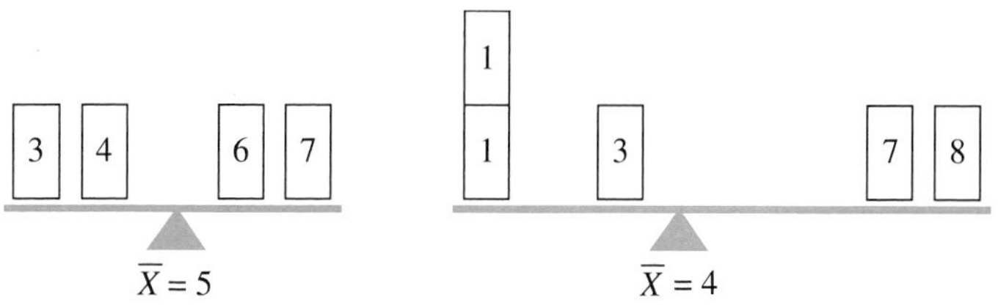

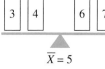

$\overline{X} = 5$ $\overline{X} = 4$

Figure 3.6
Location of the Mean on a Symmetrical, Normal Distribution
The vertical line indicates the location of the mean score, which is the center of the distribution.

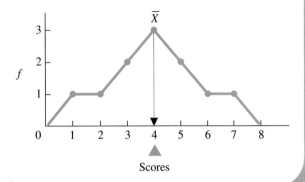

On a normal distribution the center is the point around where most of the scores are located, so the mean is an accurate summary and provides an accurate address for the distribution.

Comparing the Mean, Median, and Mode

In a symmetrical normal distribution, all three measures of central tendency are located at the same score. For example, in Figure 3.6 the mean of 4 also splits the curve in half, so 4 is the median. Also, the mean of 4 has the highest frequency, so 4 is the mode. If a distribution is only roughly normal, then the mean, median, and mode will be close to the same score. However, because the mean uses all information in the data, and because it has special mathematical properties, the mean is the basis for most of the *inferential procedures* we will see. Therefore, when you are summarizing interval or ratio scores, always compute the mean *unless* it clearly provides an inaccurate description of the distribution.

The mean will *inaccurately* describe a highly skewed (nonsymmetrical) distribution. This is because the mean must balance the distribution and to do that the mean must be located toward the extreme tail of the distribution. Then the mean does not describe where *most* of the scores

REMEMBER

With interval or ratio scores the mean is used to summarize normal distributions; the median is used to summarize skewed distributions.

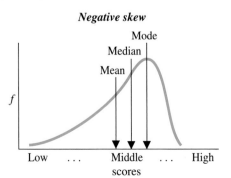

are located. You can see this starting with the symmetrical distribution containing the scores 1, 2, 2, 2, 3. The mean is 2 and this accurately describes the scores. However, including the score of 20 would give the skewed sample, 1, 2, 2, 2, 3, 20. Now the mean is pulled up to 5. But! Most of these scores are not at or near 5. As this illustrates, the mean is at the mathematical center, but in a skewed distribution that center is not where most of the scores are located.

The solution is to use the median to summarize a very skewed distribution. Figure 3.7 shows the relative positions of the mean, median, and mode in skewed distributions. In both graphs the mean is pulled toward the extreme tail and is not where most scores are located. Each distribution is also not centered around its mode. Thus, of the three measures, the median most accurately reflects the central tendency of a skewed distribution.

It is for the above reasons that the government uses the median to summarize such skewed distributions as that of yearly income or the price of houses. For example, the median income in the United States is approximately

Figure 3.7
Measures of Central Tendency for Skewed Distributions
The vertical lines show the relative positions of the mean, median, and mode.

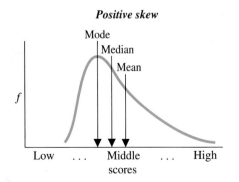

$50,000 a year. But there is a relatively small number of corporate executives, movie stars, professional athletes, and the like who make millions! Averaging in these high incomes would produce a much higher mean. However, because most incomes are not located there, the median is a better summary of this distribution.

Believe it or not, we've now covered the basic measures of central tendency. In sum, the first step in summarizing data is to compute a measure of central tendency to describe the score *around* which the distribution tends to be located.

- Compute the mode with nominal data or with a distinctly bimodal distribution of any type of scores.

- Compute the median with ordinal scores or with a very skewed distribution of interval/ratio scores.

- Compute the mean with a symmetrical, unimodal distribution of interval or ratio scores.

3.4 APPLYING THE MEAN TO RESEARCH

Most often the data in behavioral research are summarized using the mean. This is because most often we measure variables using interval or ratio scores that naturally form a roughly normal distribution. Because the mean is used so extensively, we will delve further into its characteristics and uses in the following sections.

Deviations around the Mean

First, you need to understand why the mean is at the center of a distribution. The answer is because the mean is just as far from the scores above it as it is from the scores below it. That is, the *total distance* that some scores are above the mean equals the *total distance* that the other scores are below the mean.

The distance separating a score from the mean is called the score's **deviation,** indicating the amount the score "deviates" from the mean. A score's deviation is

equal to the score minus the mean, or in symbols, to $X - \overline{X}$. Thus, if the sample mean is 47, a score of 50 deviates by +3 because $50 - 47$ is +3. A score of 40 deviates from the mean of 47 by -7 because $40 - 47 = -7$.

REMEMBER

Always subtract the mean from *the raw score when computing a score's* deviation.

Do not think of deviations as positive and negative numbers. Think of them as indicating a *distance,* which is always a positive number. The sign indicates *direction* from the mean. Thus, a positive deviation indicates that the score is greater than the mean, and a negative deviation indicates that the score is less than the mean. The size of the deviation (regardless of its sign) indicates the distance the score lies from the mean: the *larger* the deviation, the *farther* the score is from the mean. A deviation of 0 indicates that the score equals the mean.

When we determine the deviations of all the scores in a sample, we find the *deviations around the mean.* The **sum of the deviations around the mean** is the sum of all differences between the scores and the mean. And here's why the mean is the mathematical center of a distribution:

The sum of the deviations around the mean always equals zero.

For example, the scores 3, 4, 6, and 7, have a mean of 5. Table 3.1 on the next page shows how to compute the sum of the deviations around the mean for these scores. First, we subtract the mean from each score to obtain its deviation, and then we add the deviations. The sum is zero. In fact, for *any* distribution, having any shape, the sum of the deviations around the mean will equal zero. This is because the sum of the positive deviations

deviation The distance that separates a score from the mean and thus indicates how much the score differs from the mean

sum of the deviations around the mean $[\Sigma(X - \overline{X})]$ The sum of all differences between the scores and the mean

Table 3.1
Computing Deviations around the Mean

X	minus	X̄	equals	Deviation
3	–	5	=	−2
4	–	5	=	−1
6	–	5	=	+1
7	–	5	=	+2
		Sum	=	0

equals the sum of the negative deviations, so the sum of all deviations is zero. In this way the mean is the center of a distribution, because in total the mean is an equal distance from scores above and below it.

Note: We have a symbol for the sum of the deviations. Combining the symbol for a deviation, $(X - \overline{X})$, with the symbol for sum, Σ, gives the sum of the deviations as $\Sigma(X - \overline{X})$. This also says to first subtract the mean from each score to find each deviation, and then add all of the deviations together. Thus, in symbols, $\Sigma(X - \overline{X}) = 0$.

The importance of the sum of the deviations equaling zero is that it makes the mean literally the score *around* which everyone in the sample scored: Some scores are above the mean to the same extent that others are below it. Therefore we think of the mean as the typical score, because it is the one score that more or less describes everyone's score, with the same amounts of more and less. This is why the mean is so useful for summarizing a group of scores.

This characteristic is also why the mean is the best score to use if you want to predict an individual's score. Because the mean is the center score, any errors in our predictions will cancel out over the long run. For example, if the mean on an exam is 80, for any individual we'd estimate a score of 80. If we compare our predictions to what people actually get, sometimes we will be wrong, but our *total error* will equal zero. Say that one student scored 70. We would predict an 80, so we'd be off by −10 because this person's score deviates from the mean by −10. However, the mean is the central score, so at some time another student would score 90. By estimating an 80 again, we'd be off by +10 because this person deviates by +10. Thus, over the entire sample our errors will cancel out to zero because the amounts that some participants score above the mean (having positive deviations) will equal the amount that other participants score below the mean (having negative deviations). A

basic rule in statistics is that if we cannot be perfectly accurate, then the next best thing is to have our errors balance out. Therefore, when we do not know anything else about the scores, we predict that any individual in this situation will score at the mean score. Then, over the long run, our errors—our overestimates and underestimates—will balance out to zero. Only the mean provides this capability.

Summarizing Research

Now you can understand how researchers compute the mean anytime we have a sample of normally distributed interval or ratio scores. Thus, if we've merely observed some participants, we compute the mean number of times they exhibit a particular behavior, or we compute the mean response in a survey. In a *correlational study* we compute the mean score on the X variable and the mean score on the Y variable (symbolized by $\overline{Y}$). Such means are then used to summarize the sample and to predict any individual's score in this situation.

The predominant way to summarize experiments is also to compute means. As an example, say that we conduct a *very* simple study of memory by having participants recall a list in one of three conditions in which the list contains 5, 10, or 15 words. Our dependent variable is the number of words correctly recalled from a list. Table 3.2 shows some idealized recall scores from the participants in each condition (each column). A relationship appears to be present because we see a different batch of higher recall scores occurring with each subsequent list length.

A real experiment would employ a much larger *N*, and so to see the relationship buried in the scores we would compute a mean (or other measure of central tendency) for each condition. When selecting the appropriate measure, remember that the scores are from the *dependent variable,* so compute the mean, median, or mode depending upon (1) the scale of measurement of the dependent variable and (2) for interval or ratio scores, the shape of the distribution. Note: We are primarily concerned with whether the distribution is normal in the population even if our small sample is barely roughly so.

Table 3.2

Number of Words Correctly Recalled from a 5-, 10-, or 15-Item List

Condition 1: 5-Item List	Condition 2: 10-Item List	Condition 3: 15-Item List
4	7	10
3	6	9
2	5	8

Therefore, to select your statistics, read research related to your study and see what researchers say about the population and how they computed central tendency.

In our recall experiment, we compute the mean of each condition (each column), producing Table 3.3. When you are reading research, you will usually see only the means, and not the original raw scores. To interpret each mean, envision the scores that typically would produce it. In our condition 1, for example, a normal distribution producing a mean of 3 would contain scores distributed above and below 3, with most scores close to 3. We then use this information to describe the scores: In condition 1, for example, we'd say participants score around 3, or the typical score is 3, and we'd predict that any participant would have a score of 3 in this situation.

To see the relationship that is present, look at the *pattern* formed by the means: Because a different mean indicates a different batch of raw scores that produced it, a relationship is present when the means change as the conditions change. Table 3.3 shows a relationship because as the conditions change from 5 to 10 to 15 item lists, the changing means show that the scores also change from around 3 to around 6 to around 9, respectively.

Note, however, that not all means must be changing for a relationship to be present. If, for example, our means were 3, 5, and 5, respectively, then at least sometimes we would see a different batch of recall scores occurring for different list lengths, and so a relationship is present. On the other hand, if the means for the three

conditions had been 5, 5, and 5, this would indicate that essentially the same batch of scores occurred regardless of list length, so no relationship is present.

Let's assume that the data in Table 3.3 are *representative* of how the population behaves in this situation. (We must perform *inferential* procedures to check this.) If so, then we have demonstrated that list length makes a *difference* in the mean scores and thus in the individual recall scores. It is important to recognize that demonstrating a difference between the means is the same thing as demonstrating a relationship. In each case we are saying that a different group of scores occurs in each condition. When there is not a difference between the means, no relationship is present. This is important because researchers often imply that they have found a relationship simply by saying that they have found a difference between the means. If they find no difference, they have not found a relationship.

The above logic also applies to the median or mode. A relationship is present if a different median or mode occurs in two or more conditions, because this indicates that a different batch of raw scores is occurring as the conditions change.

REMEMBER

Interpret an experiment by considering the means of the conditions: If some means change as the conditions change, then the raw scores that produced the means are changing, and a relationship is present.

Table 3.3

Means of Conditions in Memory Experiment

Condition 1: 5-Item List	Condition 2: 10-Item List	Condition 3: 15-Item List
$\overline{X} = 3$	$\overline{X} = 6$	$\overline{X} = 9$

Graphing the Results of an Experiment The results of experiments are often reported using graphs. Always label the X axis with the conditions of the independent variable and label the Y axis as the mean (or mode or median) of the dependent scores.

line graph A graph of an experiment when the independent variable is an interval or ratio variable; plotted by connecting the data points with straight lines

bar graph A graph in which a free-standing vertical bar is centered over each score on the *X* axis; used with nominal or ordinal scores

We have two types of graphs we may create. Create either a *line graph* or a *bar graph*, depending on the scale of measurement used to measure the *independent variable*.

Create a line graph when the independent variable is an interval or a ratio variable. In a **line graph** adjacent data points are connected with straight lines. For example, for our previous data with the independent variable of list length (a ratio variable), we create line graph (a) in Figure 3.8. A data point is placed above the 5-item condition opposite the mean of 3 errors, a data point is above the 10-item condition

REMEMBER

If data points form a line that is not horizontal, the Y *scores are changing as* X *changes, and a relationship is present.*

at the mean of 6 errors, and a data point is above the 15-item condition at the mean of 9 errors. Then we connect adjacent data points with straight lines. (Recall that interval and ratio scores are assumed to be *continuous*. Therefore, we plot a continuous line above the possible X scores of a 6-item list, a 7-item list, etc.)

The graph conveys the same information as the sample means did in Table 3.3. Look at the overall pattern: If the vertical positions of the data points go up or down as the conditions change, then the means are changing. Different sample means indicate different scores in each condition, so a relationship is present. However, say that instead, in every condition the mean was 5, producing line graph (b) in Figure 3.8. The result is a horizontal line, indicating that the mean score stays the same as the conditions change, so essentially the same recall scores occur in each condition and no relationship is present.

Our other approach is to create a **bar graph** when the independent variable is a nominal or ordinal variable. Place a bar above each condition on the *X* axis to the height on the *Y* axis that corresponds to the mean score for that condition. As usual, adjacent bars do not touch. (Recall that nominal and ordinal scores are assumed to have *discrete* amounts. Therefore, we plot discrete—separate—bars at each point on the X axis.)

For example, say that we conducted an experiment comparing the recall errors made by psychology majors, English majors, and physics majors. This independent variable involves a nominal scale, so we have the bar graph shown in Figure 3.9. Because the tops of the bars do not form a horizontal line, we know that different means and thus different scores are in each condition. We see that individual scores are around 8 for physics majors, around 4 for psychology majors, and around 12 for English majors, so a relationship is present.

Note: In a different experiment we might have measured a nominal or an ordinal *dependent* variable. In that case we would plot the mode or median on the Y axis for each condition. Then, again depending on the characteristics of the independent variable, we would create either a line or bar graph.

Figure 3.8
Line Graphs Showing (a) the Relationship between Mean Words Recalled and List Length and (b) No Relationship

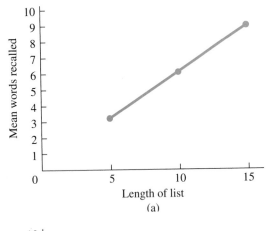

(a)

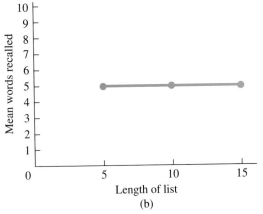

(b)

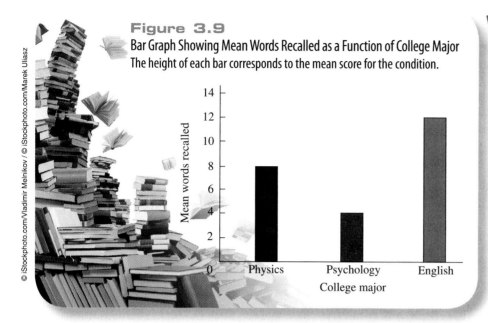

Figure 3.9
Bar Graph Showing Mean Words Recalled as a Function of College Major
The height of each bar corresponds to the mean score for the condition.

μ The symbol used to represent the population mean

Describing the Population Mean

Recall that ultimately we seek to describe the population of scores found in a given situation. Populations are unwieldy, so we also summarize them using measures of central tendency. Usually we have normally distributed interval or ratio scores, so usually we describe the population mean. The mean of a population is symbolized by the Greek letter μ (pronounced "mew"). Thus, to indicate that the population mean is 143, we'd say μ = 143. We use μ simply to show that we're talking about a population, as opposed to a sample, but a population mean has the same characteristics as a sample mean: μ is the average of the scores in the population, it is the center of the distribution, and the sum of the deviations around μ equals zero. Thus, μ is the score around which everyone in the population scored, it is the typical score, and it is the score we predict for any individual in the population.

How do we determine μ? If all scores in the population are known, then we compute μ using the same formula used to compute the sample mean, so $\mu = \Sigma X/N$. Usually, however, a population is infinitely large, so instead, we perform that inferential process we've discussed previously, using the mean of a sample to estimate μ. Thus, if a sample's mean in a particular situation is 99, then, assuming the sample accurately represents the

population, we estimate that μ in this situation is also 99.

Likewise, ultimately we wish to describe any experiment in terms of the scores that would be found if we tested the entire population in each condition. For example, assume that the data from our previous list-length study is representative. Because the mean in the 5-item condition was 3, we expect that everyone should score around 3 in this situation, so we estimate that if the population recalled a 5-item list, the μ would be 3. Similarly, we infer that if the population recalled a 10-item list, μ would equal our condition's mean of 6, and if the population recalled a 15-item list, μ would be 9.

However, instead of using our previous graphs to visualize this relationship in the population, we'll take a different approach. We assume that we know two things: By estimating each μ, we know *where* on the dependent variable each population would be located. Also, by assuming that recall scores are normally distributed, we know the *shape* of each distribution. Thus, we can envision the populations of recall scores we expect for each condition as the frequency polygons shown in Figure 3.10. (These are frequency distributions, so the dependent (recall) scores are on the X axis.) The figure shows a relationship because, as the conditions of the independent variable change, scores on the dependent variable change so that we see a different population of

Figure 3.10
Locations of Populations of Recall Scores as a Function of List Length
Each distribution contains the recall scores we would expect to find if the population were tested under a condition.

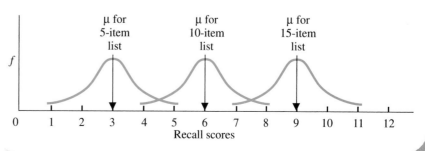

scores for each condition. Essentially, for every 5 items added to a list, the distributions slide to the right, going from scores around 3 to around 6 to around 9.

Conversely, say that we had found no relationship where, for example, every $\overline{X}$ was 3. Then, we'd envision the *one, same* normal distribution located at $\mu = 3$ for all three conditions.

Notice that by envisioning the relationship in the population, we have the scores for describing *everyone's* behavior, so we *are* describing how nature operates in this situation. In fact, as we've discussed here, in every study we (1) use the sample means (or other measures) to describe the relationship in the sample, (2) perform our inferential procedures, and (3) use the sample data to envision the relationship found in the population—in nature. At that point we have basically achieved the goal of our research and we are finished with our statistical analyses.

USING SPSS

We can use SPSS to compute all three measures of central tendency for *one* sample of data (and other statistics we will discuss in Chapter 4 are also produced). Your Chapter Review cards have instuctions on how to do this. However, we do not need to repeatedly run this routine to compute the means for the different conditions in an experiment. Instead, we will compute the means from all conditions at once as part of performing the experiment's inferential procedures that we'll discuss later.

using what you know

1. a. What do measures of central tendency tell us about a distribution of scores?
   b. What are the three measures of central tendency?
   c. What two aspects of the data determine which measure of central tendency to use?

2. Why is the mean the best measure of central tendency for a normal distribution?

3. Why is the mean an inappropriate measure of central tendency in a very skewed distribution?

4. What two limitations can arise when using the mode to describe central tendency?

5. When the mean, the median, and the mode of a distribution have the same value, what does it tell you about the distribution?

6. If the median of a distribution is considerably greater than the mean, what do you know about the shape of the distribution?

7. The following distribution shows the locations of five scores:

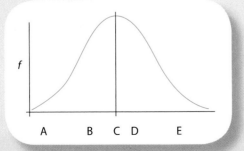

a. Match the deviation scores −7, +1, 0, −2, and +5 with their locations.

   A = _____   B = _____   C = _____
   D = _____   E = _____

b. Rank-order the deviation scores to show the order of the raw scores from highest to lowest.

c. Rank-order the deviation scores to show the order of the raw scores from highest *frequency* to lowest *frequency*.

8. a. What does $\Sigma(X - \overline{X})$ mean?
   b. What two steps are performed in finding $\Sigma(X - \overline{X})$?
   c. What value does $\Sigma(X - \overline{X})$ *always* equal?

9. Why do we use the mean to predict a person's score on a variable when we have no other information?

10. a. What is μ?
    b. What are four characteristics of μ?

11. What assumption must be made in order to use a sample mean to estimate μ?

12. A researcher collected several sets of data. For each, indicate which measure of central tendency she should compute.
    a. The following personality scores:
       0, 2, 3, 3, 8, 4, 9, 6, 7, 5, 6
    b. The following age scores:
       10, 15, 18, 15, 14, 13, 42, 15, 12, 14, 42

c. The following college years:

freshman, senior, junior, junior, freshman, freshman, junior, sophomore, junior

d. The following political affiliations:

Dem., Dem., Rep., Dem., Soc., Com., Com., Soc., Dem., Rep.

13. For each of the four sets of data in problem 12, indicate whether a line graph or a bar graph would be appropriate and why.

14. a. When graphing the results of an experiment, what is plotted on the Y axis?

b. What determines the measure of central tendency you should plot?

15. For each of the graphs pictured, what do you conclude about

a. the $\bar{X}$ in each condition?

b. the scores in each condition?

c. the population mean in each condition?

d. the relationship in the population?

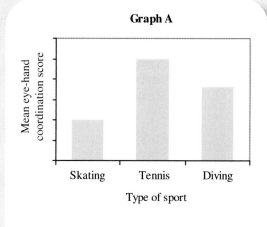

Graph A

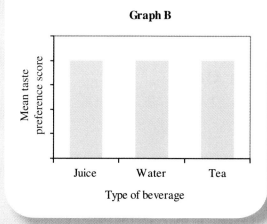

Graph B

16. Does Graph C indicate a relationship? Why or why not?

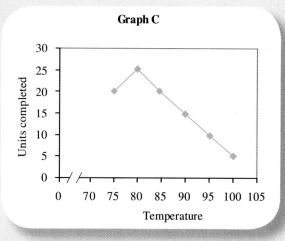

Graph C

17. Does Graph D indicate a relationship? Why or why not?

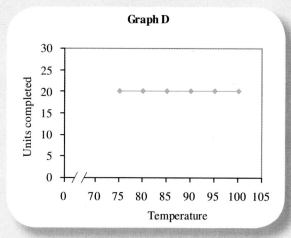

Graph D

18. For the following data, compute (a) the mode, (b) the median, and (c) the mean:

| 76 | 74 | 73 | 78 | 76 | 74 | 74 | 74 | 81 | 90 |
| 84 | 87 | 83 | 76 | 75 | 74 | 77 | 83 | 89 | 77 |

19. What do the measures of central tendency computed in problem 18 tell you about the shape of the distribution of data?

20. Which measure of central tendency most accurately summarizes the distribution of scores in problem 18?

21. Which measure of central tendency best describes the most frequent score for the distribution in problem 18?

22. In a normal distribution of scores, four participants obtained the following deviation scores: –1, +2, +3, and –.15.

a. Which participant obtained the highest raw score?

b. Which participant obtained the lowest raw score?

c. Rank-order the deviation scores in terms of their frequency, putting the score with the highest frequency first.

Summarizing
Scores with Measures
of Variability

You have seen that the first steps in dealing with data are to consider the shape of the distribution and compute the mean (or other measure of central tendency). This information simplifies the distribution and allows you to envision its general properties. But not everyone will behave in the same way, so we may see many different scores. Therefore, to completely describe data you must also determine whether there are large or small differences among the scores. This chapter discusses the statistics for describing the differences among scores, which are called *measures of variability*. In the following sections we discuss (1) the concept of variability, (2) how to compute statistics that describe variability in a sample, and (3) how to estimate the variability in the corresponding population.

4.1 UNDERSTANDING VARIABILITY

Computing a measure of variability is important because it answers the question "How large are the differences among the scores?" Without it, a measure of central tendency provides an incomplete description. For example, look at Table 4.1. Each sample has a mean of 6, so without looking at the raw scores, you might think they are identical distributions. But, Sample A contains scores that differ greatly from each other and from the mean, Sample B contains scores that differ less, and in Sample C there are no differences among the scores.

Looking Back

- From Chapter 2, know how to interpret the normal curve and use the proportion of the area under the curve.

- From Chapter 3, understand what $\bar{X}$ and μ stand for, what a deviation score is, and why the sum of the deviations around the mean is zero.

© eikonas/Alamy

Table 4.1

Three Different Distributions Having the Same Mean Score

Sample A	Sample B	Sample C
0	8	6
2	7	6
6	6	6
10	5	6
12	4	6
$\bar{X} = 6$	$\bar{X} = 6$	$\bar{X} = 6$

Thus, to completely describe a set of data, we need to also calculate statistics called measures of variability. **Measures of variability** describe the extent to which scores in a distribution *differ* from each other. In a sample with more frequent, larger differences among the scores, these statistics will produce larger numbers and we say that the scores (and the underlying behaviors) are more *variable* or show greater *variability*.

Measures of variability communicate three aspects of the data. First, the opposite of variability

measures of variability Statistics that summarize the extent to which scores in a distribution differ from one another

is consistency. Small variability indicates that the scores are consistently close to each other. Larger variability indicates a variety of scores that are inconsistent. Second, the amount of variability implies how accurately a measure of central tendency describes the distribution. Our focus will be on the mean and normal distributions: The more that scores differ from each other, the less accurately they are summarized by the central, mean score. Conversely, the smaller the variability, the closer the scores are to each other and to the mean. Third, we've seen that the difference between two scores can be thought of as the *distance* that separates them. From this perspective, greater differences indicate greater distances between the scores, so measures of variability indicate how *spread out* a distribution is.

Without even looking at the scores in the samples back in Table 4.1, by knowing their variability we'd know that Sample C contains consistent scores that are close to each other, so the mean of 6 accurately describes them. Sample B contains scores that differ more—are more spread out—so they are less consistent and 6 is not so accurate a summary. Sample A contains very inconsistent scores that are spread out far from each other, so 6 poorly describes most of them.

You can see the same aspects when describing or *envisioning* a larger normal distribution. For example, Figure 4.1 shows three normal distributions, which differ because of their variability. Let's use our "parking lot" view. If our statistic indicates relatively small variability, it implies a distribution similar to Distribution A: This is very narrow because most of the people are standing in long lines located close to the mean (with few standing at, say, 40 or 60). Thus, most scores are close to each other, so their differences are small and this is why our statistic indicates small variability. However, if our statistic indicates intermediate variability, then it implies a distribution more like B: This is more spread out because longer lines of people are located at scores

Figure 4.1
Three Variations of the Normal Curve

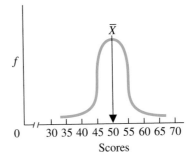

Distribution A

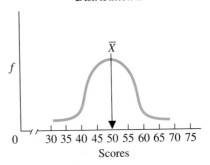

Distribution B

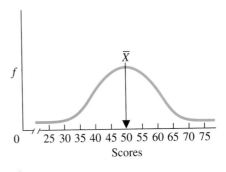

Distribution C

farther above and below the mean (more people stand near 40 and 60). In other words, a greater variety of scores occur here, producing more frequent and larger differences, and this is why our statistic is larger. Finally, when our statistics show very large variability we envision a distribution like C. It is very wide because long lines of people are at scores located farther into the tails (here scores beyond 40 and 60 occur often). Therefore, frequently scores are anywhere between very low and very high, producing many large differences, and this is why the statistic is so large.

Researchers have several ways to measure variability. The following sections discuss the three most common measures of variability: the *range*, the *variance*, and the *standard deviation*.

4.2 THE RANGE

One way to describe variability is to determine how far the lowest score is from the highest score. The descriptive statistic that indicates the distance between the two most extreme scores in a distribution is called the **range.**

COMPUTING THE RANGE

Range = Highest score − Lowest score

Thus, for example, the scores back in Sample A (0, 2, 6, 10, 12) have a range of 12 − 0 = 12. The less variable scores in Sample B (4, 5, 6, 7, 8) have a range of 8 − 4 = 4. And the perfectly consistent Sample C (6, 6, 6, 6, 6) has a range of 6 − 6 = 0. Thus, the range does communicate the spread in the data. However, the range is a rather crude measure. Because it involves only the two most extreme scores, the range is based on the least typical and often least frequent scores, while ignoring all other scores. Therefore, we usually use the range as our sole measure of variability only with *nominal* or *ordinal* data. With nominal data, we compute the range by counting the number of categories we're examining: For example, there is more consistency if the participants in a study belong to one of 4 political parties than if they belong to one of 14 parties. With ordinal data, the range is the distance between the lowest and highest rank: If 100 runners finish a race spanning only the 5 positions from 1st through 5th, this is a close race with many ties; if they span 75 positions, the runners are spread out.

4.3 THE VARIANCE AND STANDARD DEVIATION

Most behavioral research involves interval or ratio scores that form a normal distribution. In such situations (when the mean is appropriate) we use two similar measures of variability called the *variance* and the *standard deviation.*

Understand that we *use* the variance and the standard deviation to describe how different the scores are from each other. We *calculate* them, however, by measuring how much the scores differ from the mean. Because the mean is the center of a distribution, when scores are spread out from each other, they are also spread out from the mean. When scores are close to each other, they are close to the mean.

This brings us to an important point. The mean is the score *around* which a distribution is located: *The variance and standard deviation allow us to quantify "around."* For example, if the grades in a statistics class form a normal distribution with a mean of 80, then you know that most of the scores are around 80. But are most scores between 79 and 81 or between 60 and 100? By measuring how spread out scores are from the mean, the variance and standard deviation will define "around."

Mathematically, the distance between a score and the mean is the difference between the score and the mean. This difference is symbolized by $X - \overline{X}$, which is the amount that a score *deviates* from the mean. Because some scores will deviate from the mean by more than others, it makes sense to compute the average amount the scores deviate from the mean. The larger the "average of the deviations," the greater the variability or spread between the scores and the mean.

We cannot, however, simply compute the average of the deviations. To compute an average, we first sum the scores so we would first sum the deviations. In symbols, this is $\Sigma(X - \overline{X})$. Recall, however, that the sum of the deviations always equals zero because the positive deviations cancel out the negative deviations. Therefore, the average of the deviations will always be zero.

Thus, we want a statistic *like* the average of the deviations, so that we know the average amount the scores are spread out around the mean. But, because the average of the deviations is always zero, we calculate slightly more complicated statistics called the variance and standard deviation. *Think* of them, however, as each producing a number that indicates *something like* the average or typical amount that the scores differ from the mean.

Understanding the Sample Variance

If the problem with the average of the deviations is the positive and negative deviations, then a solution is to *square* the deviations. This removes all negatives, so the sum of the squared deviations is not necessarily zero and neither is their average.

By finding the average squared deviation, we compute the variance. The **sample variance** is the average of the squared deviations of scores around the sample mean. The symbol for the sample variance is S_X^2. Always include the squared sign (^{2}). The capital S indicates that we are describing a sample, and the subscript X indicates that it is computed for a sample of X scores.

We have a formula for the sample variance that defines it:

DEFINING FORMULA FOR THE SAMPLE VARIANCE

$$S_X^2 = \frac{\Sigma(X - \overline{X})^2}{N}$$

This formula is important because it shows you the basis for the variance. Later we will see a different, faster formula to use when you are actually computing the variance. But first, to understand the concept, say

that we measure the ages of some children. As shown in Table 4.2, we first compute each deviation, $(X - \overline{X})$, by subtracting the mean (which here is 5) from each score. Next, as shown in the far right column, we square each deviation. Adding the squared deviations gives $\Sigma(X - \overline{X})^2$, which here is 28. The N is 7 and so

$$S_X^2 = \frac{\Sigma(X - \overline{X})^2}{N} = \frac{28}{7} = 4$$

This sample's variance is 4. In other words, the average squared deviation of the age scores around the mean is 4.

The good news is that the variance is a legitimate measure of variability. The bad news, however, is that the variance does not make much sense as the "average deviation." There are two problems. First, squaring the deviations makes them very large, so the variance is unrealistically large. To say that our age scores differ from their mean by an *average* of 4 is silly, because none of the scores actually deviates from the mean by this much. The second problem is that variance is rather bizarre because it measures in squared units. We measured ages, so the variance indicates that the scores deviate from the mean by 4 *squared* years (whatever that means!).

So, it is difficult to interpret the variance as the "average" deviation. The variance is not a waste of time, however, because it is used extensively in statistics. Also, variance does communicate the *relative* variability of scores. If one sample has $S_X^2 = 1$ and another has $S_X^2 = 3$, you know that the first sample is less variable (more consistent) and more accurately described by its mean. Further, looking back at Figure 4.1, for the smaller variance you might envision the distribution as like Distribution A, while for the larger variance, you would envision one more like Distribution B or C.

Thus, think of variance as a number that generally communicates how variable the scores are: The larger the variance, the more the scores are spread out. The measure of variability that more directly communicates the "average of the deviations" is the *standard deviation*.

Understanding the Sample Standard Deviation

The sample variance is always an unrealistically large number because we square each deviation. A way to solve this problem is to take the square root of the variance. The answer is called the standard deviation. The **sample standard deviation** is the square root of the sample variance (the square root of the average squared

Table 4.2
Calculation of Variance Using the Definitional Formula

Age Score	−	$\overline{X}$	=	$(X - \overline{X})$	$(X - \overline{X})^2$
2	−	5	=	−3	9
3	−	5	=	−2	4
4	−	5	=	−1	1
5	−	5	=	0	0
6	−	5	=	1	1
7	−	5	=	2	4
8	−	5	=	3	9
N = 7					$\Sigma(X - \overline{X})^2 = 28$

deviation of scores around the mean). Conversely, squaring the standard deviation produces the variance.

To create the formula that defines the standard deviation, we simply add the symbol for the square root to the previous defining formula for variance.

DEFINING FORMULA FOR THE SAMPLE STANDARD DEVIATION

$$S_X = \sqrt{\frac{\Sigma(X - \bar{X})^2}{N}}$$

Notice that the symbol for the sample standard deviation is S_X which is the square root of the symbol for the sample variance.

To compute S_X we first compute everything inside the square root sign to get the variance, as we did in Table 4.2. In the age scores the variance (S_X^2) was 4. We take the square root of the variance to find the standard deviation:

$$S_X = \sqrt{4}$$

so

$$S_X = 2$$

The standard deviation of the age scores is 2.

The standard deviation is as close as we come to the "average of the deviations," and we have three related ways to interpret it. First, we interpret our S_X of 2 and as indicating that the age scores differ from the mean by something like an "average" of 2. Further, the standard deviation uses the same units as the raw scores, so the scores differ from the mean age by an "average" of 2 *years*.

Second, the standard deviation allows us to gauge how consistently close together the scores are and, correspondingly, how accurately they are summarized by the mean. If S_X is relatively large, then we know that a large proportion of scores are relatively far from the mean (so envision a relatively wider distribution). If S_X is smaller, then more often scores are close to the mean with relatively few far from the mean (so envision a relatively narrower distribution).

And third, the standard deviation indicates how much the scores below the mean deviate from it and how much the scores above the mean

deviate from it, so the standard deviation indicates how much the scores are spread out *around* the mean. To see this, we can further summarize a distribution by describing the scores that lie at "plus one standard deviation from the mean" ($+1S_X$) and at "minus one standard deviation from the mean" ($-1S_X$). For example, for our age scores of 2, 3, 4, 5, 6, 7, and 8, $\bar{X} = 5$ and $S_X = 2$. The score that is $+1S_X$ from the mean is the score at $5 + 2$, or 7. The score that is $-1S_X$ from the mean is the score at $5 - 2$, or 3. Looking at the individual scores, you can see that it is accurate to say that the majority of the scores are between 3 and 7.

In fact, the standard deviation is mathematically related to the normal curve, so that describing a distribution using the scores that are between $-1S_X$ and $+1S_X$ is especially useful. Recall from Chapter 2 that the *proportion of the area under the normal curve* corresponds to the *relative frequency* of scores. On any normal distribution, approximately .34 of the area and therefore .34 of the scores are between the mean and the score that is $+1S_X$ above the mean. Likewise, approximately .34 of the area—and .34 of the scores—are between the mean and the score that is $-1S_X$ below the mean. For example, say that in the statistics class with a mean of 80, the S_X is 5. The score at $80 - 5$ (at $-1S_X$) is 75, and the score at $80 + 5$ (at $+1S_X$) is 85. Figure 4.2 on the next page shows about where these scores are located. Thus, 34% of the scores are between 75 and 80, and 34% of the scores are between 80 and 85. Altogether, approximately 68% of the scores are always between the scores at $+1S_X$ and $-1S_X$ from the mean, so about 68% of the statistics class has scores between 75 and 85. Conversely, about 32% of the scores are outside this range, with about 16% below 75 and 16% above 85. Thus, saying that most scores are between 75 and 85 is an accurate summary because the majority of scores (68%) are here.

The characteristic bell-shape of *any* normal distribution always places 68% of the distribution between the scores that are $+1S_X$ and $-1S_X$ from the mean. Look back at Figure 4.1 once more. In Distribution A most scores are relatively close to the mean. This will produce a small S_X which is, let's say, 5. Because all scores are relatively close to the mean, 68% of them will be in the small area between 45 ($50 - 5$) and 55 ($50 + 5$.) However, Distribution B is more

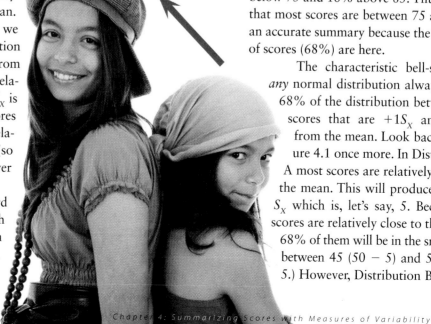

sum of squared Xs (ΣX^2) Calculated by squaring each score in a sample and adding the squared scores

squared sum of X $[(\Sigma X)^2]$ Calculated by adding all scores and then squaring their sum

spread out producing a larger S_X (say it's 7). Because the distribution is wider, we need to go farther above and below the mean to capture the middle 68% of the scores, so now it falls in the wider area between 43 and 57. Finally, Distribution C is the most spread out and will give the largest S_X (let's say it's 12.) Because the distribution is so wide, we need to go far above or below the mean to between 38 and 62 to capture the middle 68%.

In summary, here is how to describe a distribution. If you know the data form a normal distribution, you can envision its general shape. If you know the mean, you know where the center of the distribution is and what the typical score is. And if you know the standard deviation, you know whether the distribution is relatively wide or narrow, you know the "average" amount that scores deviate from the mean, and you know between which two scores the middle 68% of the distribution lies.

4.4 COMPUTING THE SAMPLE VARIANCE AND SAMPLE STANDARD DEVIATION

The previous defining formulas for the variance and standard deviation are important because they show that the core computation is to measure how far the raw scores are from the mean. However, by reworking the defining formulas, we have less obvious but faster *computing* formulas. They involve two new symbols that you must master for later statistics, too. These symbols are:

1 The Sum of the Squared Xs: The symbol ΣX^2 indicates to find the *sum of the squared Xs*. To do so, you first square each X (each raw score) and then sum—add up—the squared Xs. Thus, to find ΣX^2 for the scores 2, 2, and 3, add $2^2 + 2^2 + 3^2$, which becomes $4 + 4 + 9$, which equals 17.

2 The **Squared Sum of X**: The symbol $(\Sigma X)^2$ indicates to find the *squared sum of* X. To do so, work inside the parentheses first, so find the sum of the X scores. Then square that sum. Thus, to find $(\Sigma X)^2$ for the scores 2, 2, and 3, you have $(2 + 2 + 3)^2$, which is $(7)^2$, which is 49.

© iStockphoto.com/BlackJack3D

Figure 4.2

Normal Distribution Showing Scores at Plus or Minus One Standard Deviation

With $S_X = 5$, the score of 75 is at $-1S_X$ and the score of 85 is at $+1S_X$. The percentages are the approximate percentages of the scores falling into each portion of the distribution.

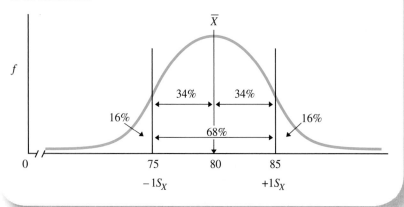

Computing the Sample Variance

The computing formula for variance is derived from its defining formula by replacing the symbol for the mean with its formula and then reducing the components.

COMPUTING FORMULA FOR
THE SAMPLE VARIANCE

$$S_X^2 = \frac{\Sigma X^2 - \frac{(\Sigma X)^2}{N}}{N}$$

The formula says to first find the sum of the Xs, (ΣX), square that sum, and divide the squared sum by N. Then subtract that result from the sum of the squared Xs (ΣX^2). Finally, divide that quantity by N.

For example, we can arrange the previous age scores as shown in Table 4.3.

STEP 1: *Find* ΣX, ΣX^2, *and* N. Here, ΣX is 35, ΣX^2 is 203, and N is 7. Putting these quantities into the formula, we have

$$S_X^2 = \frac{\Sigma X^2 - \frac{(\Sigma X)^2}{N}}{N} = \frac{203 - \frac{(35)^2}{7}}{7}$$

STEP 2: *Compute the squared sum of X.* Here the squared sum of X is 35^2, which is 1225, so

$$S_X^2 = \frac{203 - \frac{1225}{7}}{7}$$

STEP 3: *Divide the* $(\Sigma X)^2$ *by* N. Here 1225 divided by 7 equals 175, so

$$S_X^2 = \frac{203 - 175}{7}$$

STEP 4: *Subtract in the numerator.* Because 203 minus 175 equals 28, we have

$$S_X^2 = \frac{28}{7}$$

STEP 5: *Divide.* After dividing 28 by 7 we have

$$S_X^2 = 4$$

Again, the sample variance for these age scores is 4, and it is interpreted as we discussed previously.

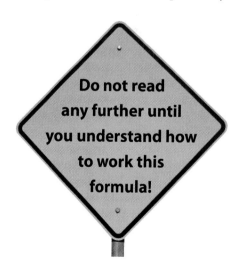

Do not read any further until you understand how to work this formula!

Table 4.3
Calculation of Variance Using the Computational Formula

X Score	X²
2	4
3	9
4	16
5	25
6	36
7	49
8	64
$\Sigma X = 35$	$\Sigma X^2 = 203$

Computing the Sample Standard Deviation

The computing formula for the standard deviation merely adds the square root symbol to the previous formula for the variance.

population standard deviation (σ_X) The square root of the population variance, or the square root of the average squared deviation of scores around the population mean

population variance (σ_X^2) The average squared deviation of scores around the population mean

biased estimators The formula for the variance or standard deviation involving a final division by N, used to describe a sample, but which tends to underestimate the population variability

COMPUTING FORMULA FOR THE SAMPLE STANDARD DEVIATION

$$S_X = \sqrt{\dfrac{\Sigma X^2 - \dfrac{(\Sigma X)^2}{N}}{N}}$$

To demonstrate this formula, we'll use the same age scores from in Table 4.3.

 **1:**

Find ΣX, ΣX^2, and N. Here, ΣX is 35, ΣX^2 is 203, and N is 7. Thus:

$$S_X = \sqrt{\dfrac{203 - \dfrac{(35)^2}{7}}{7}}$$

Follow Steps 2–5 described previously for computing the variance. Inside the square root symbol will be the variance, which is 4, so

$$S_X = \sqrt{4}$$

6: *Compute the square root.*

$$S_X = 2$$

Again the standard deviation of these age scores is 2; interpret it as discussed previously.

4.5 THE POPULATION VARIANCE AND THE POPULATION STANDARD DEVIATION

Recall that our ultimate goal is to describe the population of scores. Sometimes researchers have access to a population and then they directly calculate the actual population variance and standard deviation. The symbol for the *true* or actual **population standard** deviation is σ_X. (The σ is the lowercase Greek letter *s*, or sigma.) Because the squared standard deviation is the variance, the symbol for the true **population variance** is σ_X^2. (In each case the subscript X indicates a population of X scores.) The defining formulas for σ_X and σ_X^2 are similar to those we saw for a sample:

POPULATION STANDARD DEVIATION	POPULATION VARIANCE
$\sigma_X = \sqrt{\dfrac{\Sigma(X - \mu)^2}{N}}$	$\sigma_X^2 = \dfrac{\Sigma(X - \mu)^2}{N}$

The only novelty here is that we are computing how far each score deviates from the population mean, μ. Otherwise, the population standard deviation and variance tell us the same things about the population that we saw previously for a sample: Both are ways of measuring the variability in the scores, indicating how much the scores are spread out. Further, we can interpret the population standard deviation as the "average" deviation of the scores around μ, with 68% of the scores in the population falling between the scores that are at $+1\sigma_X$ and $-1\sigma_X$ from μ.

Usually you will not have a population of scores available, so there is no need for us to create computing formulas. However, you will encounter situations where researchers already know about a population, and so this variance or standard deviation is *given* to you. Therefore, know these symbols and use the previous formulas to understand what each represents.

Thus, we've seen how to describe the variability in a known sample (using S_X^2 or S_X) and how to describe the variability in a known population (using σ_X^2 or σ_X). However, we must discuss one other situation: Although the goal of research is usually to describe the population, often we will not know *all* of the scores in the population. In such situations, we will use our sample data to *infer* or *estimate* the variability in the population.

Estimating the Population Variance and Population Standard Deviation

We use the variability in a sample to estimate the variability we'd find if we could measure the population. However, we do *not* use the previous formulas for the sample variance and sample standard deviation as the basis for this estimate. Those formulas are used only when describing the variability of a *sample*. In statistical terminology, the formulas for S_X^2 and S_X are the **biased estimators**: When used to estimate the population, they are biased toward *underestimating* the true population parameters. Such a bias is a problem, because if we cannot be accurate, we at least want our under- and overestimates to cancel out over the long run. With the biased estimators, the underestimates and overestimates will not cancel out. Although the sample variance and sample standard deviation accurately describe a sample, they are too often too small to use as estimates of the population.

The reason that S_X and S_X^2 underestimate is because their formulas are not designed for estimating the population. When calculating them we measure variability using the mean as our reference point, and so the sum of the deviations must equal zero. Because of this, however, not all of the deviations in the sample are "free" to reflect the variability found in the population. For example, say that the mean of five scores is 6, and that four of the scores are 1, 5, 7, and 9. The sum of their deviations is −2. The final score can only be 8, because we must have a deviation of +2 so that the sum of all deviations is zero. Given the sample mean and the deviations of the other scores, the deviation for the score of 8 is determined by those of the other scores. Therefore, only the deviations produced by the other four scores reflect the variability found in the population. The same would be true for any four of the five scores. Thus, out of the N scores in any sample, only $N - 1$ of them (the N of the sample minus 1) actually reflect the variability in the population.

The problem with the biased estimators (S_X and S_X^2) is that these formulas divide by N. Because we divide by too large a number, the answer tends to be too small. Instead, we should divide by $N - 1$. By doing so, we compute the **unbiased estimators** of the population variance and population standard deviation.

DEFINING FORMULAS FOR THE UNBIASED ESTIMATORS OF THE POPULATION VARIANCE AND STANDARD DEVIATION

Estimated Population Variance

$$s_X^2 = \frac{\Sigma(X - \bar{X})^2}{N - 1}$$

Estimated Population Standard Deviation

$$s_X = \sqrt{\frac{\Sigma(X - \bar{X})^2}{N - 1}}$$

The first formula above is for the **estimated population variance.** Notice that it involves the same basic computation we saw in our sample variance: we are finding the amount each score deviates from the mean, which will then form our estimate of how much the scores in the population deviate from μ. The only novelty is that in computing the "average" of the squared deviations, we divide by $N - 1$ instead of by N.

The second formula above is the defining formula for the **estimated population standard deviation.** As with our sample statistics, we have simply added the square root

unbiased estimators The formula for the variance or standard deviation involving a final division by $N - 1$; calculated using sample data to estimate the population variability

estimated population variance (s_X^2) The unbiased estimate of the population variance calculated from sample data using $N - 1$

estimated population standard deviation (s_X) The unbiased estimate of the population standard deviation calculated from sample data using $N - 1$

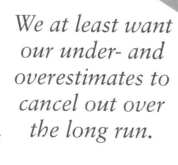
© Photodisc/Getty Images

We at least want our under- and overestimates to cancel out over the long run.

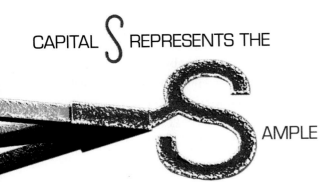

CAPITAL S REPRESENTS THE S AMPLE

lowercase s represents the population estimate

symbol to the formula for the variance: the estimated standard deviation is the square root of the estimated variance.

Because we have new formulas that produce new statistics, we also have new symbols. *The symbol for the unbiased estimated population variance is the lowercase s_X^2.* The square root of the variance is the standard deviation, so *the symbol for the unbiased estimated population standard deviation is s_X.* To keep your symbols straight, remember that the symbols for the *sample* involve the capital or big S, and in those formulas you divide by the "big" value of N. The symbols for *estimates* of the population involve the lowercase or small s, and here you divide by the smaller quantity $N - 1$. Also, think of s_X^2 and s_X as the inferential versions, because the *only* time you use them is to *infer* the variance or standard deviation of the population based on a sample. Think of S_X^2 and S_X as the descriptive variance and standard deviation, because they are used to *describe* the sample.

Computing the Estimated Population Variance and Standard Deviation

The only difference between the computing formula for the estimated population variance and the previous computing formula for the sample variance is that here the final division is by $N - 1$.

COMPUTING FORMULA FOR THE
ESTIMATED POPULATION VARIANCE

$$s_X^2 = \frac{\Sigma X^2 - \dfrac{(\Sigma X)^2}{N}}{N - 1}$$

Notice that in the numerator we still divide by N.

For example, previously we had the age scores of 3, 5, 2, 6, 7, 4, 8. To estimate the population variance, follow the same steps as before. First, find ΣX and ΣX^2; here, $\Sigma X = 35$ and $\Sigma X^2 = 203$. Also, $N = 7$ so $N - 1 = 6$. Putting these quantities into the above formula gives

$$s_X^2 = \frac{\Sigma X^2 - \dfrac{(\Sigma X)^2}{N}}{N - 1} = \frac{203 - \dfrac{(35)^2}{7}}{6}$$

Work through this formula the same way you did for the sample variance: 35^2 is 1225, and 1225 divided by 7 equals 175, so

$$s_X^2 = \frac{203 - 175}{6}$$

Next, 203 minus 175 equals 28, so

$$s_X^2 = \frac{28}{6}$$

and the final answer is

$$s_X^2 = 4.67$$

This answer is slightly larger than the sample variance for these scores, which was $S_X^2 = 4$. Although 4 accurately describes the sample variance, we estimate that the variance in the corresponding population is 4.67. In other words, if we could measure all scores in the population and then compute the true population variance, we would expect σ_X^2 to be 4.67.

The formula for the estimated population standard deviation merely adds the square root sign to the above formula for the variance.

COMPUTING FORMULA FOR THE
ESTIMATED POPULATION STANDARD
DEVIATION

$$s_X = \sqrt{\frac{\Sigma X^2 - \dfrac{(\Sigma X)^2}{N}}{N - 1}}$$

Above, the estimated population variance for our age scores was $s_X^2 = 4.67$. Therefore, s_X is $\sqrt{4.67}$, which is 2.16. If we could compute the standard deviation using the entire population of scores, we would expect σ_X to equal 2.16.

Interpreting the Estimated Population Variance and Standard Deviation

Interpret the estimated population variance and standard deviation in the same way as S_X^2 and S_X, except that here we describe how much we *expect* the scores to be spread out in the population, how consistent or inconsistent we *expect* the scores to be, and how accurately we *expect* the population to be summarized by μ. We can also determine the scores at $+1s_X$ and $-1s_X$ from the μ so that we know between which two scores to *expect* 68% of the population of scores to fall.

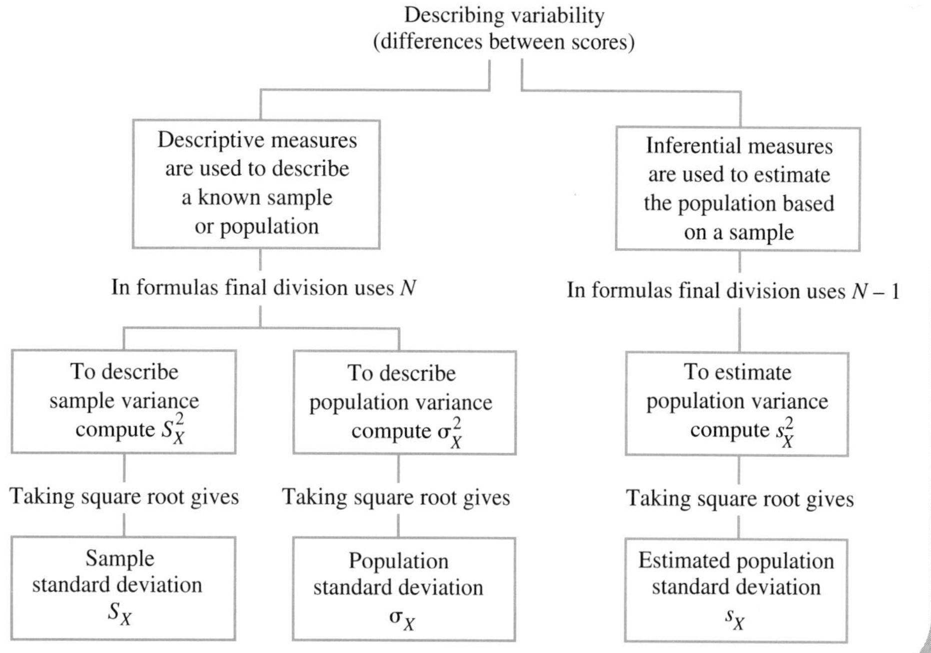

Notice that, assuming a sample is representative, we have reached our ultimate goal of describing the population of scores. For example, based on a statistics class with a mean of 80, we'd infer that the population would score at a μ of 80. The size of s_X (or s_X^2) estimates how spread out the population is, so if s_X turned out to be 6, we'd expect that the "average amount" individual scores deviate from 80 to be about 6. Further, we'd expect about 68% of the population to score between 74 (80 − 6) and 86 (80 + 6.) Because these scores reflect behavior, this description gives us a good idea of how most individuals in the population behave in this situation (which is why we conduct research in the first place).

Figure 4.3

Organizational Chart of Descriptive and Inferential Measures of Variability

Describing variability
(differences between scores)

Descriptive measures are used to describe a known sample or population

Inferential measures are used to estimate the population based on a sample

In formulas final division uses N

In formulas final division uses $N - 1$

To describe sample variance compute S_X^2

To describe population variance compute σ_X^2

To estimate population variance compute s_X^2

Taking square root gives

Taking square root gives

Taking square root gives

Sample standard deviation S_X

Population standard deviation σ_X

Estimated population standard deviation s_X

> How much can I expect the scores to be spread out in the population?

4.6 SUMMARY OF THE VARIANCE AND STANDARD DEVIATION

To keep track of all of the different statistics you've seen, remember that *variability* refers to the differences between scores, which we describe by computing the *variance* and *standard deviation*. In each, we are finding the difference between each score and the mean and then calculating something like the average deviation.

Organize your thinking about the particular measures of variability using Figure 4.3. Any standard deviation is merely the square root of the corresponding variance. For either measure, compute the *descriptive* versions when the scores are available: When describing how far scores are spread out from $\overline{X}$, we use the sample variance (S_X^2) and the sample standard deviation (S_X). When describing how far the scores are spread out from μ, we use the population variance (σ_X^2) and the population standard deviation (σ_X). When the complete population of scores is unavailable, *infer* the variability of the population based on a sample by computing the *unbiased estimators* (s_X or s_X^2) . These *inferential* formulas require a final division by $N - 1$ instead of by N.

© iStockphoto.com/Barbara Sauder / © iStockphoto.com/CurvaBezier

STATISTICS IN THE RESEARCH LITERATURE: REPORTING MEANS AND VARIABILITY

The standard deviation is most often reported in published research because it more directly communicates how consistently close the individual scores are to the mean. Thus, the mean from a study might describe the number of times participants exhibited a particular behavior, and a small standard deviation indicates they consistently did so. Or, in a survey, the mean might describe the typical opinion held by participants, but a large standard deviation indicates substantial disagreement among them. The same approach is used in experiments, in which we compute the mean and standard deviation in each condition. Then each mean indicates the typical score and the score we predict for anyone in that condition. The standard deviation indicates how consistently close the actual scores are to that mean. Or instead, often researchers report the estimated population standard deviation to estimate the variability of scores if everyone in the population was tested under that condition.

However, as if you haven't seen enough symbols already, research journals that follow the publication guidelines of the American Psychological Association do not use our statistical symbols for the sample mean and sample standard deviation. Instead, the symbol for the mean is M. The symbol for the standard deviation is SD, and unless otherwise specified, you should assume it is the estimated population version. On the other hand, when a report discusses the *true* population parameters, our Greek symbols μ and σ are used.

USING SPSS

SPSS will simultaneously compute the mean, median, mode, range, standard deviation, and variance for a sample of data. The SPSS instructions on your Chapter Review Card show you how. Note that SPSS (and some researchers) refer to variability as *dispersion*, and the program computes only the unbiased estimators of the variance and standard deviation (our s_x^2 and s_x). To obtain the $\overline{X}$ and s_x for each condition of an experiment, you may either (1) enter the data from one condition at a time and run the routine, or (2) compute these statistics for all conditions at once as part of performing the inferential procedures we will discuss later.

These cards help make SPSS really easy!

© Brand X Pictures/Jupiterimages

1. What two characteristics of a distribution are needed in order to accurately describe the distribution?

2. a. What information do measures of variability convey?

   b. Why is a measure of central tendency a less accurate description of a distribution if the measure of variability is large?

3. a. Distinguish among the symbols "S_x," "σ_x," and "s_x."

   b. When are the above symbols accompanied by the squared sign (?)?

4. a. With what type(s) of data should the range be used?

   b. With what type(s) of data should the variance and standard deviation be used?

5. a. What do both the variance and the standard deviation indicate about a distribution of data?

   b. To describe the variability of a distribution, which should usually be employed—the standard deviation or the variance? Why?

6. If a distribution has a large variance, what does this tell us about the consistency of the scores?

7. a. When summarizing the results of an experiment, what two statistics do you usually calculate for each condition?

   b. How do you then know if a relationship is present?

8. For the following set of scores, calculate the range, the sample mean ($\overline{X}$), the sample variance (S_x^2), and the sample standard deviation (S_x).

23	25	17	19	21	20	22	20	24	23
21	19	20	17	25	23	19	20	22	22

9. a. Compute the estimated population variance (s_x^2) and the estimated population standard deviation (s_x) for the sample in problem 8.

   b. Why are the numbers computed here larger than those computed in problem 8?

   c. Why must the estimates be computed this way?

10. For a distribution with a mean of 130 and a standard deviation of 15, approximately 68% of the scores will lie between which two scores?

11. Suppose you want to estimate population parameters from a sample of 30 scores. You decide to use the sample mean to estimate the population mean, and then you estimate the population variance as

$$S_X^2 = \frac{740 - \dfrac{(112)^2}{30}}{30} = 10.729$$

a. What did you do wrong?

b. What should your estimate be?

12. If the variance of 10 scores is computed as follows:

$$\sqrt{\frac{40 - \dfrac{(16)^2}{10}}{10}} = 1.20$$

And the standard deviation is then computed as follows:

$$\frac{40 - \dfrac{(16)^2}{10}}{10} = 1.44$$

a. What is the problem here?

b. What should have alerted you something was wrong?

13. You first compute the sample variance as

$$\frac{392 - \dfrac{(55)^2}{10}}{9} = 9.94$$

Then you compute the estimated population variance as

$$\frac{392 - \dfrac{(55)^2}{10}}{10} = 8.95$$

a. What is the problem here?

b. What should have alerted you something was wrong?

14. The new statistician for the football team at the Institute of Psycho-Ceramics has calculated the following player statistics. Player A's average rushing yards per carry is $\overline{X} = 5$ with a standard deviation of 3; Player B's average rushing yards per carry is $\overline{X} = 5$ with a standard deviation of 10.

a. Which player is the more consistent yardage gainer? Why?

b. Which player is more likely to break loose for a big run? Why?

15. A teacher calculates for the test grades in Class A, $\overline{X} = 32$ and $S_x = 4$; for test grades in Class B, $\overline{X} = 32$ and $S_x = 8$.

a. What score should the teacher predict for students in each class?

b. Which class permits the better predictions? Why?

Describing Data
with *z*-Scores and the Normal Curve

In previous chapters we have summarized an entire distribution of scores. In this chapter we'll take a different approach and discuss the statistic to use when we want to interpret an *individual* score. Here we ask the question "How does any particular score compare to the other scores in a sample or population?" We answer this question by transforming raw scores into *z-scores*. In the following sections, we discuss (1) the logic of *z*-scores and their simple computation, (2) how *z*-scores are used to evaluate individual raw scores, and (3) how the same logic can be used to evaluate sample means.

SECTIONS

5.1 UNDERSTANDING *z*-SCORES

relative standing
A description of a particular score derived from a systematic evaluation of the score using the characteristics of the sample or population in which it occurs

Researchers *transform* raw scores into *z*-scores because we usually *don't* know how to interpret a raw score: We don't know whether, in nature, a score should be considered high or low, good or bad, or what. Instead, the best we can do is to compare a score to the other scores in the distribution, describing the score's relative standing. **Relative standing** reflects

Looking Back

- From Chapter 2, know that relative frequency is the proportion of time scores occur and it corresponds to the proportion of the area under the normal curve; know that a score's percentile equals the percent of the curve to the left of the score.

- From Chapter 3, understand that the mean is the center of a distribution and the larger a raw score's deviation, the farther the score is form the mean.

- From Chapter 4, understand that S_X and σ_X indicate the "average" deviation of scores around $\overline{X}$ and μ, respectively.

© John Lund/Paula Zacharias/Blend Images/Jupiterimages

the systematic evaluation of a score relative to the sample or population in which the score occurs. The way to determine the relative standing of a score is to transform it into a *z*-score. From the *z*-score we'll know whether the individual's underlying raw score was *relatively* good, bad, or in-between.

To see how this is done, say that we conduct a study at Prunepit University in which we measure the attractiveness of a sample of males. The scores form the normal curve shown in Figure 5.1 on the next page. We want to interpret these scores, especially those of three men: Chuck, who scored 35; Sean, who scored 65; and Brandon, who scored 90. You already know that the way to do this is to use a score's location on the distribution to determine its *frequency, relative frequency,* and *percentile*. For example, Chuck's score is far below the mean and has a rather low frequency. Also, the proportion of the area under the curve above his score is small, so his score has a low relative frequency. And because little of the distribution is to the left of (below) his score, he also has a low percentile. On the other hand, Sean is somewhat above the mean, so he is somewhat above the 50th percentile. Also, the height of the curve at his score is large, so his score has a

Figure 5.1

Frequency Distribution of Attractiveness Scores at Prunepit U
Scores for three individuals are identified on the *X* axis.

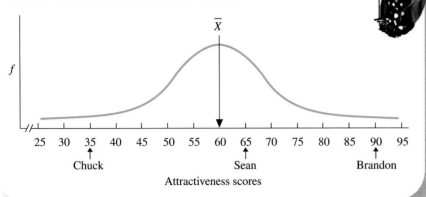

rather high frequency and relative frequency. And then there's Brandon: His score is far above the mean, with a low frequency and relative frequency, and a very high percentile.

The problem with the above descriptions is that they are subjective and imprecise. Instead, to precisely quantify each score's relative standing, we will calculate its *z*-score. This will indicate exactly where on the distribution a score is located so that we can precisely determine the score's frequency, relative frequency, and percentile.

Describing a Score's Relative Location as a z-Score

We began the description of each man's score above by noting whether it is above or below the mean. Likewise, our first calculation is to measure how far a raw score is from the mean by computing the score's *deviation*, which equals $X - \overline{X}$. For example, Brandon's score of 90 deviates from the mean of 60 by $90 - 60 = +30$. A deviation of $+30$ *sounds* as if it might be large, but is it? We need a frame of reference. For the entire distribution, only a few scores deviate by as much as Brandon's score, and *that* makes his an impressively high score.

Thus, a score is impressive if it is far from the mean, and "far" is determined by how frequently other scores deviate from the mean by that amount.

Therefore, to interpret a score's location, we must

z-score The statistic that indicates the distance a score is from its mean when measured in standard deviation units

compare its deviation to the other deviations. As you saw in Chapter 4, the standard deviation is interpreted as the "average deviation." By comparing a score's deviation to the standard deviation, we can describe the score in terms of this average deviation. For example, say that in the attractiveness data the sample standard deviation is 10. Brandon's deviation of $+30$ is equivalent to 3 standard deviations, so Brandon's raw score is located 3 standard deviations above the mean. His raw score is impressive because it is three times as far above the mean as the "average" amount that scores were above the mean.

By transforming Brandon's deviation into standard deviation units, we have computed his *z*-score. A **z-score** indicates the distance a raw score is from the mean when measured in standard deviations. The symbol for a *z*-score is *z*. A *z*-score always has two components: (1) either a positive or a negative sign which indicates whether the raw score is above or below the mean and (2) the absolute value of the *z*-score (ignoring the sign) which indicates how *far* the score is from the mean in standard deviations. So, because Brandon is *above* the mean by 3 standard deviations his *z*-score is $+3$. If he had been below the mean by this amount, he would have $z = -3$.

Brandon's raw score is impressive!

REMEMBER

To interpret any raw score, transform it to a z-score. This indicates how far the score is above or below the mean when measured in standard deviations.

Thus, like any score, a *z*-score is a *location* on a distribution. However, it also simultaneously communicates the *distance* it is from the mean. Therefore, knowing that Brandon scored at $z = +3$ provides us with a frame of reference that we do not have by knowing only that his raw score was 90.

Computing z-Scores in a Sample or Population

We computed Brandon's *z*-score by first subtracting the mean from his raw score and then dividing by the standard deviation, so

TRANSFORMING A RAW SCORE IN A SAMPLE INTO A *z*-SCORE

$$z = \frac{X - \overline{X}}{S_X}$$

(This is both the defining and the computing formula.)

To find Brandon's *z*-score,

STEP 1: *Determine the* $\overline{X}$ *and* S_X. Filling in the formula gives

$$z = \frac{X - \overline{X}}{S_X} = \frac{90 - 60}{10}$$

STEP 2: *Find the deviation in the numerator.* Always subtract $\overline{X}$ from X. Then

$$z = \frac{+30}{10}$$

STEP 3: *Divide.*

$$z = +3$$

Likewise, Sean's raw score was 65, so

$$z = \frac{X - \overline{X}}{S_X} = \frac{65 - 60}{10} = \frac{+5}{10} = +.50$$

Sean's raw score is literally one-half of 1 standard deviation above the mean.

And finally, Chuck's raw score is 35, so

$$z = \frac{X - \overline{X}}{S_X} = \frac{35 - 60}{10} = \frac{-25}{10} = -2.50$$

So Chuck's raw score was 2.5 standard deviations below the mean.

Usually we compute *z*-scores for the raw scores in a sample because usually we are describing scores from data that we've collected. However, if we know a population's mean (μ) and standard deviation (σ_X) we can compute *z*-scores in the population using the same logic as above. Putting the population symbols in the formula gives

TRANSFORMING A RAW SCORE IN A POPULATION INTO A *z*-SCORE

$$z = \frac{X - \mu}{\sigma_X}$$

Now the answer indicates how far a raw score lies from the population mean, measured using the population standard deviation. (*Note:* We do not compute *z*-scores using the estimated population standard deviation, s_X.)

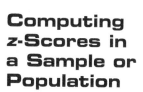

Chuck is Here.

REMEMBER

Always include a positive or negative sign when computing a z-score.

Computing a Raw Score When z Is Known

Sometimes we know a z-score and want to transform it back to the raw score that produced it. Then

TRANSFORMING A z-SCORE IN A SAMPLE INTO A RAW SCORE

$$X = (z)(S_X) + \overline{X}$$

This says to multiply the z-score times the standard deviation and then add the mean. For example, if another man had a z of +1, then to find his attractiveness score

STEP 1: *Determine the $\overline{X}$ and S_X. Ours were $\overline{X} = 60$ and $S_X = 10$, so*

$$X = (+1)(10) + 60$$

STEP 2: *Multiply z times S_X.*

$$X = +10 + 60$$

STEP 3: *Add.*

$$X = 70$$

The raw score of 70 corresponds to a z of +1.

In another case, say that someone had a z-score of −1.30. Then with $\overline{X} = 60$ and $S_X = 10$,

$$X = (-1.30)(10) + 60$$

so

$$X = -13 + 60$$

Adding a negative number is the same as subtracting its positive value, so

$$X = 47$$

The raw score here is 47.

The above logic also applies to finding the raw score for a z from a population, except we use the symbols for the population.

z-distribution
The distribution of z-scores produced by transforming all raw scores in a distribution into z-scores

TRANSFORMING A z-SCORE IN A POPULATION INTO A RAW SCORE

$$X = (z)(\sigma_X) + \mu$$

Here, we multiply the z-score times the population standard deviation and then add μ.

5.2 USING THE z-DISTRIBUTION TO INTERPRET SCORES

Although thinking of the standard deviation as the "average deviation" gives you an intuitive understanding of a z-score, the way to fully and precisely interpret a z-score is to envision the sample or population it is in as a z-distribution. A **z-distribution** is the distribution produced by transforming all raw scores in the data into z-scores. For example, our attractiveness scores produce the z-distribution shown in Figure 5.2.

Notice the two ways the X axis is labeled. This shows that by creating a z-distribution, we only change the way that we identify each score. Saying that Brandon has a z of +3 is merely another way to say that he has a raw score of 90. Because he is still at the same location in the distribution, Brandon's z-score has the

Figure 5.2
z-Distribution of Attractiveness Scores at Prunepit U
The labels on the X axis show first the raw scores and then the z-scores.

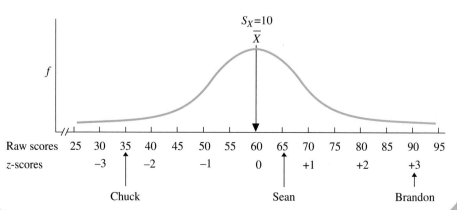

same frequency, relative frequency, and percentile as his raw score.

The advantage of z-scores, however, is that they directly communicate each score's location within the distribution. The z of 0, for example, always corresponds to the mean of the raw scores: In Figure 5.2, a score of 60 is zero distance from the mean of 60. For any other score, the sign indicates the *direction* the score lies in relation to the mean. A "+" indicates that the score is graphed to the right of the mean. A "−" indicates that the score is graphed to the left of the mean. Therefore, z-scores become increasingly larger numbers with a positive sign as we proceed farther to the right of the mean. Such larger z-scores and their *higher* raw scores occur less frequently. Conversely, z-scores become increasingly larger numbers with a negative sign as we proceed farther to the left of the mean. Such larger z-scores and their *lower* raw scores occur less frequently. Note, however, that a negative z-score is not always a "bad" score. For some variables (e.g., test errors or credit card debt) a low raw score is the goal and a larger negative z-score is a better score.

Recognize that, as shown, most of the z-scores are *between* +3 and −3. Although z-scores greater than +3 or −3 are possible, they occur *very* infrequently. (*Note:* the symbol "±" means "plus or minus," so we can restate this by saying that most z-scores are between ±3.)

Figure 5.2 also illustrates three important characteristics of any z-distribution.

1. **A z-distribution always has the same shape as the raw score distribution.** When the underlying raw score distribution is normal, its z-distribution is normal.

2. **The mean of any z-distribution is 0.** Whatever the mean of the raw scores is, it transforms into a z-score of 0. (Also, the average of the positive and negative z-scores is 0.)

3. **The standard deviation of any z-distribution is 1.** Whether the standard deviation of the raw scores is 10 or 100, a score at that distance from the mean is a distance of 1 when transformed to z-scores, so the "average deviation" is now 1. (Also, if we compute S_X using the z-scores in a z-distribution, the answer will be 1.)

REMEMBER

The larger a z-score—whether positive or negative—the farther the corresponding raw score is from the mean, and the less frequently the z-score and raw score occur.

Because of these characteristics, all normal z-distributions are similar, so that a particular z-score will be at the *same* relative location on *every* distribution. Therefore, you can interpret the relative standing of any raw score by envisioning a z-distribution like that in Figure 5.2: If a z-score is close to zero, the raw score is near the mean and is a very frequent score. A z greater than ±1 indicates a raw score that is less frequent. The closer the z is to ±3, the farther into a tail the raw score is, and the closer the raw score is to being one of the few highest or lowest scores in the distribution.

5.3 USING THE z-DISTRIBUTION TO COMPARE DIFFERENT VARIABLES

Above we saw that z-scores allow us to interpret the scores for a variable in *one* distribution. However, because all z-distributions are so similar, a second use of z-scores is to compare scores from *different* variables.

Here's a new example. Say that Althea received a grade of 38 on her statistics quiz and a grade of 45 on her English paper. These scores reflect different kinds of tasks, so it's like comparing apples to oranges. The solution is to transform the raw scores from each class into z-scores. Then we can compare Althea's relative standing in English to her relative standing in statistics, and we are no longer comparing apples and oranges.

NOTE

The *z*-transformation equates or standardizes different distributions, so *z*-scores are often referred to as **standard scores.**

Say that for the statistics quiz, the $\overline{X}$ was 30 and the S_X was 5. Althea's grade of 38 becomes $z = +1.6$. For the English paper, the $\overline{X}$ was 40 and the S_X was 10, so Althea's 45 becomes $z = +.5$. Althea's z of $+1.6$ in statistics is farther above the mean than her z of $+.5$ in English is above the mean, so she performed relatively better in statistics.

Say that another student, Millie, obtained raw scores that produced $z = -2$ in statistics and $z = -1$ in English. Millie did better in English because her z-score of -1 is less distance below the mean.

To see just how comparable the z-scores from these two classes are, we can plot their z-distributions on the same graph. Figure 5.3 shows the result, with the origi-

nal raw scores also plotted. (To read the statistics curve, simply ignore the English curve and vice versa. The English curve is taller because of a higher frequency at each score.) Although the classes have different distributions of raw scores, *the location of each z-score is the same.* For example, any normal distribution is centered over its mean. This center is at $z = 0$, regardless of whether this corresponds to a 30 in statistics or a 40 in English. Also, recall that the score located one standard deviation above the mean places about 34% of the distribution between the score and the mean. Scores that are $+1S_X$ above their respective means are at $z = +1$ regardless of whether this corresponds to a 35 in statistics or a 50 in English. Likewise, the raw scores of 40 in statistics and 60 in English are both 2 standard deviations above their respective means, so both are at the same location called $z = +2$. And so on: When two raw scores are the same distance in standard deviations from their respective mean, they produce the same z-score and are at the same location in the z-distribution.

Using this z-distribution, we can see that Althea scored higher in statistics than in English, whereas Millie scored higher in English than she did in statistics.

5.4 Using the z-Distribution to Compute Relative Frequency

The third important use of z-scores is to determine the relative frequency of specific raw scores. Recall that relative frequency is the proportion of time that a score occurs and that the relative frequency of particular scores equals the proportion of the total area under the curve at those scores. Because a particular z-score is always at the same location on the normal curve, the area under the curve for that z-score is always the

Figure 5.3
Comparison of Distributions for Statistics and English Grades, Plotted on the Same Set of Axes

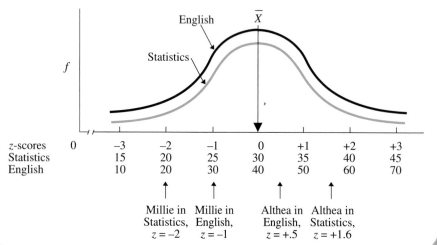

z-scores	0	−3	−2	−1	0	+1	+2	+3
Statistics		15	20	25	30	35	40	45
English		10	20	30	40	50	60	70

Millie in Statistics, $z = -2$; Millie in English, $z = -1$; Althea in English, $z = +.5$; Althea in Statistics, $z = +1.6$

same. *Therefore, the relative frequency of a particular z-score will be the same on all normal z-distributions.* The relative frequency of a *z*-score is also the relative frequency of the corresponding raw scores.

To understand this, look again at Figure 5.3. Although the heights of the two curves differ, the *proportion* under each curve is the same. For example, 50% of each distribution is to the left of its mean. Notice that this is where all of the negative *z*-scores are. In other words, negative *z*-scores occur 50% of the time, so they have a combined relative frequency of .50. (On *any* normal distribution, the negative *z*-scores have a relative frequency of .50.) By knowing the relative frequency of the *z*-scores, we also know the relative frequency of the corresponding raw scores. In Figure 5.3, the statistics students having negative *z*-scores have raw scores between 15 and 30, so the relative frequency of these scores is .50. The English students having negative *z*-scores have raw scores between 10 and 40, so the relative frequency of these scores is .50.

Here's another example. Recall that about 34% of a normal distribution is between the mean and $z = +1$. Thus, in the statistics class, scores between 30 and 35 occur .34 of the time. In English, scores between 40 and 50 occur .34 of the time.

In the same way, we can determine the relative frequencies for any other portion of a distribution, once we envision it as a *z*-distribution. To do so, we employ the *standard normal curve.*

The Standard Normal Curve

Because the relative frequency of a particular *z*-score is always the same for any normal distribution, we don't need to draw a different *z*-distribution for each variable that we measure. Instead, we envision one standard curve that, in fact, is called the standard normal curve. The **standard normal curve** is a perfect normal *z*-distribution that serves as our model of any approximately normal *z*-distribution. The idea is that although most raw scores will produce an approximately normal

z-distribution, we simplify things by operating as if the *z*-distribution fits our one, perfect normal curve. Then we determine the relative frequency of particular *z*-scores on this curve. This relative frequency of the *z*-scores is also the relative frequency of the corresponding raw scores that we would *expect* if our data formed a perfect normal distribution.

You may compute *z*-scores using our previous formulas for finding a *z*-score in a sample or in a population. Either way, usually this provides a reasonably accurate description of our actual distribution, although how accurate we are depends on how closely the data conform to the true normal curve. Therefore, this approach is best when we meet these three criteria: (1) we have a large sample or a population; (2) scores are continuous, interval, or ratio scores; and (3) the scores come close to forming a normal distribution. Recognize, however, that if you have a rather small sample, your expected relative frequencies may contain considerable error.

We determine the relative frequency of any *z*-score by looking at the area under the standard normal curve. Statisticians have already determined the proportion of the area under various parts of the curve, as shown in Figure 5.4. The numbers above the *X* axis indicate the proportion of the total area between the *z*-scores. The numbers below the *X* axis indicate the proportion of the total area between the mean and the *z*-score. (You won't need to memorize them.)

Each proportion is also the relative frequency of the *z*-scores—and raw scores—located in that part of

standard normal curve A theoretical perfect normal curve, which serves as a model of any approximately normal *z*-distribution

Figure 5.4

Proportions of Total Area under the Standard Normal Curve

The curve is symmetrical: 50% of the scores fall below the mean, and 50% fall above the mean.

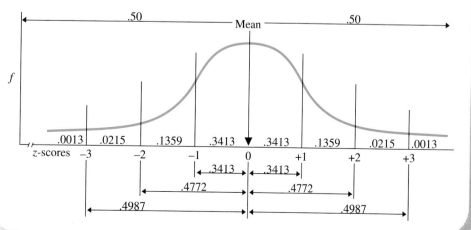

and $z = +2$. Therefore, we also expect .4772, or 47.72%, of all attractiveness scores at Prunepit U to fall between the mean score of 60 and Steve's score of 80. Conversely, .0228 of the area—and scores—are above his score.

We might also ask how *many* people scored between the mean and Steve's score. Then we would convert relative frequency to *simple frequency* by multiplying the N of the sample times the relative frequency. Say that our N was 1000. If we expect .4772 of all scores to fall between the mean and $z = +2$, then $(.4772)(1000) = 477.2$, so we expect about 477 people to have scores between the mean and Steve's score.

We can also determine a score's expected *percentile* (the percent of the scores below—graphed to the left of—a score). As in Figure 5.5, on a normal distribution the mean is the median (the 50th percentile). A positive z-score is above the mean, so Steve's score of $+2$ is above the 50th percentile. In addition, Steve's score is above the 47.72% of the scores that fall between the mean and his score. Thus, in total, 97.72% of all scores are below Steve's score. We usually round off percentile to a whole number, so Steve's raw score of 80 is at the 98th percentile.

the curve. Between a z of 0 and a z of ±1 is .3413 (or 34.13%) of the area; so about 34% of the scores are here. Or, z-scores between $+1$ and $+2$ occur .1359 of the time, which, added to .3413, gives a total of .4772 of the distribution located between the mean and $z = +2$. Or, with .3413 of the scores between the mean and $z = -1$, and .3413 of the scores between the mean and $z = +1$, a total of .6826 or about 68% of the distribution is between zs of -1 and $+1$. And so on. (Notice that z-scores beyond $+3$ or beyond -3 occur only .0013 of the time, which is why the range of z is essentially between ±3.)

In practice we usually apply the standard normal curve by beginning with a particular raw score in mind and then computing its z-score. For example, back in our original attractiveness scores, say that Steve has a raw score of 80. With $\overline{X} = 60$ and $S_X = 10$, we have

$$z = \frac{X - \overline{X}}{S_X} =$$

$$\frac{80 - 60}{10} = +2$$

His z-score is $+2$.

We can envision Steve's location as in Figure 5.5. We might first ask what proportion of scores are expected to fall between the mean and Steve's score. On the standard normal curve, .4772 of the total area falls between the mean

Figure 5.5

Location of Steve's Score on the z-Distribution of Attractiveness Scores
Steve's raw score of 80 is a z-score of $+2$.

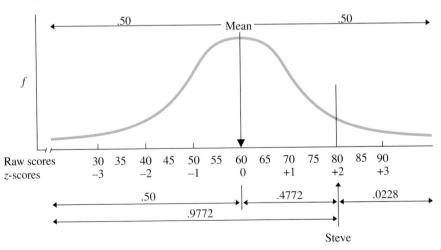

© iStockphoto.com/Audrey Roorda

Finally, we can also work in the opposite direction to find a raw score at a particular relative frequency or percentile. Say that we began by asking above what attractiveness score is .0228 of the distribution. First in terms of z-scores, we see that above a $z = +2$ is .0228 of the distribution. Then to find the raw score that corresponds to this z, we use the formula for transforming a z-score into a raw score: $X = (z)(S_X) + \overline{X}$. We'll find that above a raw score of 80 is .0228 of the distribution.

Using the z-Table

So far our examples have involved whole-number z-scores, although with real data, a z-score may contain decimals. To find the proportion of the total area under the standard normal curve for any two-decimal z-score, look in Table 1 of Appendix B. A portion of this "z-table" is reproduced in Table 5.1.

Say that you seek the area under the curve above or below a $z = +1.63$. First, locate the z in column A, labeled "z," and then move to the right. In column B, labeled "Area between the mean and z," is the proportion of the area under the curve between the mean and the z identified in column A. Here, .4484 of the curve is between the mean and the z of $+1.63$. This is shown in Figure 5.6. Because our z is the positive $+1.63$, we place this area between the mean and the z on the *right-hand* side of the distribution. Next, in column C, labeled "Area beyond z in the tail," is the proportion of the area under the curve that is in the tail beyond the z-score.

Here, .0516 of the curve is in the right-hand tail of the distribution beyond the z of $+1.63$ (also shown in Figure 5.6).

Notice that the z-table shows no positive or negative signs. You must decide whether your z is positive or negative and place the areas in their appropriate locations. Thus, if we had the negative z of -1.63, Column B and C would provide the respective areas shown on the left-hand side of Figure 5.6. If you get confused when using the z-table, look at the normal curve at the top of the table. The different portions indicate the part of the curve described in each column.

To find the z-score that corresponds to a particular proportion, read the columns in the reverse order.

Table 5.1
Sample Portion of the z-Table

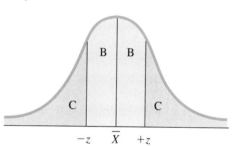

A z	B Area between the mean and z	C Area beyond z in the tail
1.60	.4452	.0548
1.61	.4463	.0537
1.62	.4474	.0526
1.63	.4484	.0516
1.64	.4495	.0505
1.65	.4505	.0495

Figure 5.6
Distribution Showing the Area Under the Curve for $z = -1.63$ and $z = +1.63$

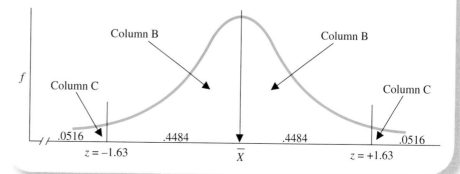

First, find the proportion in column B or C, depending on the area you seek, and then identify the z-score in column A. For example, say that you seek the z-score corresponding to .4484 of the curve between the mean and z. Find .4484 in column B of the table, and then, in column A, the z-score is 1.63.

Use the information from the z-table as we have done previously. For example, say that we examine Johnny's raw score, which happens to transform into the z of +1.63 that we've just discussed in Figure 5.6. If we seek the proportion of scores above his score, then from column C of the z-table we expect that .0516 of the scores are above this score. If we seek the relative frequency of scores between his score and the mean, from column B we expect that .4484 of the scores are between the mean and his raw score. Then we can also compute simple frequency or percentile as discussed previously. Or, if we began by asking what raw score demarcates .4484 or .0516 of the curve, we would first find these proportions in column B or C, respectively, then find the z-score of +1.63 in column A, and then transform the z-score to find the corresponding raw score. Table 5.2 summarizes these procedures.

Sometimes you will need a proportion that is not given in the z-table, or you'll need the proportion corresponding to a three-decimal z-score. In such cases, round to the nearest value in the z-table. (To compute the precise value, consult an advanced textbook to perform "linear interpolation.")

Table 5.2
Summary of Steps When Using the z-Table

If You Seek	First, You Should	Then You
Relative frequency of scores between $\bar{X}$ and X	transform X to z	find area in column B*
Relative frequency of scores beyond X in tail	transform X to z	find area in column C*
X that marks a given relative frequency between X and $\bar{X}$	find relative frequency in column B	transform z to X
X that marks a given relative frequency beyond X in tail	find relative frequency in column C	transform z to X
Percentile of an X above $\bar{X}$	transform X to z	find area in column B and add .50
Percentile of an X below $\bar{X}$	transform X to z	find area in column C
*To find the simple frequency of the scores, multiply relative frequency times N.		

© iStockphoto.com/Snezana Negovanovic

mean. Not only does this allow us to interpret the mean, but it is also the basis for upcoming inferential statistics.

To see how the procedure works, say that we give a part of the Scholastic Aptitude Test (SAT) to a sample of 25 students at Prunepit U. Their mean score is 520. Nationally, the mean of *individual* SAT scores is 500 (and σ_X is 100), so it appears that at least some Prunepit students scored relatively high, pulling the mean to 520.

5.5 USING z-SCORES TO DESCRIBE SAMPLE MEANS

We can also use the logic of z-scores to describe the relative standing of an entire sample. We do this by computing a z-score for a sample

© Robert Warren/The Image Bank/Getty Images

But how do we interpret the performance of the sample as a whole? The problem is the same as when we examined individual raw scores: Without a frame of reference, we don't know whether a particular sample mean is high, low, or in-between.

The solution is to evaluate a sample mean in terms of its relative standing. Previously this meant that we compared a particular raw score to all other scores that occur in this situation. Now we'll compare our sample mean to the other sample means that occur in this situation. Therefore, the first step is to take a small detour and create a distribution showing these means. This distribution is called the *sampling distribution of means*.

Figure 5.7

Sampling Distribution of SAT Means

The X axis shows the different values of $\overline{X}$ obtained when sampling the SAT population.

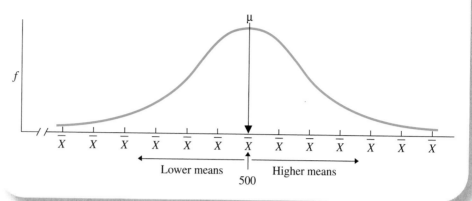

The Sampling Distribution of Means

If the national average of SAT scores is 500, then, in other words, in the population the μ is 500. Because we selected a sample of 25 students and obtained their SAT scores, we essentially drew a sample of 25 scores from this population. To evaluate our sample mean, we first create a distribution showing all other possible means that occur when selecting a sample of 25 scores from the SAT population. To do so, pretend that we record all SAT scores from the population on slips of paper and put them in a large hat. Then we would select a sample with the same size N as ours, compute the sample mean, replace the scores in the hat, draw another 25 scores, compute the mean, and so on. Because the scores selected in each sample would not be identical, not all sample means would be identical. By constructing a frequency polygon of the different values of $\overline{X}$ we obtained, we would create a sampling distribution of means. The **sampling distribution of means** is the frequency distribution of all possible sample means that occur when an infinite number of samples of the same size N are selected from one raw score population. Our SAT sampling distribution of means is shown in Figure 5.7. This is similar to a distribution of raw scores, except that each "score" along the X axis is a sample mean.

In reality we cannot "infinitely" sample a population, but we know that it would look like Figure 5.7 because

of the central limit theorem. The **central limit theorem** is a statistical principle that defines the mean, the shape, and the standard deviation of a sampling distribution.

From the central limit theorem, we know the following:

1. **The mean of the sampling distribution equals the mean of the underlying raw score population from which we create the sampling distribution.** The sampling distribution is the *population* of sample means, so we use the symbol μ. The μ of the sampling distribution is the average sample mean. When the average raw score is a particular score (here, 500), the average sample mean when using those scores will also be that score (again 500).

2. **A sampling distribution is an approximately normal distribution.** Most often each $\overline{X}$ will equal the raw score population's μ. Sometimes, however, a sample will contain more high scores or low scores relative to the population, so the sample mean will be slightly above or below μ. Less frequently, we'll obtain rather strange samples, producing a $\overline{X}$ farther above or below μ.

3. **The standard deviation of the sampling**

sampling distribution of means A frequency distribution showing all possible sample means that occur when samples of a particular size are drawn from a population

central limit theorem A statistical principle that defines the mean, standard deviation, and shape of a theoretical sampling distribution

distribution is mathematically related to the standard deviation of the raw score population. As you'll see in a moment, the variability of the raw scores influences the variability of the sample means.

The importance of the central limit theorem is that we can describe a sampling distribution *without* having to actually infinitely sample a population of raw scores. Therefore, we can create the sampling distribution of means for any raw score population.

Why do we want to see the sampling distribution? We took a small detour, but the original problem was to evaluate our Prunepit mean of 520. Once we envision the distribution back in Figure 5.7, we have a model of the frequency distribution of all sample means that occur when measuring SAT scores. Then we can use this distribution to determine the relative standing of our sample mean. The sampling distribution is a normal distribution, and you already know how to determine the relative standing of any "score" on a normal distribution: We use z-scores. That is, we will determine where a mean of 520 falls on the X-axis of this sampling distribution by finding its distance from the μ of the sampling distribution when measured using the standard deviation of the sampling distribution. This z-score will tell us if our sample mean is one of the more common means that is relatively close to the average sample mean, or whether it is one of the few, higher or lower means that occur in this situation.

To compute the z-score for a sample mean, we need one more piece of information: the "standard deviation" of the sampling distribution.

The Standard Error of the Mean

The standard deviation of the sampling distribution of means is called the **standard error of the mean**. Like a standard deviation, the standard error of the mean can be thought of as the "average" amount that the sample means deviate from the μ of the sampling distribution. That is, in some sampling distributions, the sample means may be very different from one

standard error of the mean ($\sigma_{\bar{X}}$) The standard deviation of the sampling distribution of means

another and deviate greatly from the average sample mean. In other distributions, the means may be very similar and deviate little from μ.

For now, we'll discuss the *true* standard error of the mean, as if we had actually computed it using the entire sampling distribution. Its symbol is $\sigma_{\bar{X}}$. The σ indicates that we are describing a population, but the subscript $\bar{X}$ indicates it is a population of sample means. The central limit theorem tells us that $\sigma_{\bar{X}}$ can be found using this formula:

THE TRUE STANDARD ERROR OF THE MEAN

$$\sigma_{\bar{X}} = \frac{\sigma_X}{\sqrt{N}}$$

Notice that the formula involves σ_X. For now we will discuss those situations where we know about the population of raw scores so that $\sigma_{\bar{X}}$ is given. When we know the true standard deviation of the underlying raw score population, we will know the true "standard deviation" of the sampling distribution.

In the formula, the size of $\sigma_{\bar{X}}$ depends first on the size of σ_X: The more variable the raw scores, the more likely that we'll get a very different set of scores in each sample, and so the more that the sample means will differ and the larger will be $\sigma_{\bar{X}}$. Second, the size of $\sigma_{\bar{X}}$ depends on the size of N: The larger the N, the closer the sample means will be to the population mean, and so $\sigma_{\bar{X}}$ will be smaller.

To compute $\sigma_{\bar{X}}$ for our SAT example:

STEP **1:** *Identify the σ_X of the raw score population and the N used to create the sample.* We used an N = 25, and we know that $\sigma_X = 100$. So

$$\sigma_{\bar{X}} = \frac{\sigma_X}{\sqrt{N}} = \frac{100}{\sqrt{25}}$$

STEP **2:** *Compute the square root of N.* The square root of 25 is 5, so

$$\sigma_{\bar{X}} = \frac{100}{5}$$

STEP 3: *Divide.*

$$\sigma_{\overline{X}} = 20$$

This indicates that in the SAT sampling distribution, the individual sample means differ from the μ of 500 by an "average" of 20 points when the N of each sample is 25.

Now we can calculate a z-score for our sample mean.

Computing a *z*-Score for a Sample Mean

Previously you saw that the formula for transforming a raw score in the population into a z-score was

$$z = \frac{X - \mu}{\sigma_X}$$

We transform a sample mean into a z-score using a similar formula.

TRANSFORMING A SAMPLE MEAN
INTO A z-SCORE

$$z = \frac{\overline{X} - \mu}{\sigma_{\overline{X}}}$$

Don't be confused by this formula: All z-scores show how far a score is from the mean of a distribution, measured in standard deviations of that distribution. Now our "score" is a sample mean, so we find how far the sample mean is from the mean of the sampling distribution (μ), measured in standard error units ($\sigma_{\overline{X}}$).

To compute the z-score for our Prunepit data,

STEP 1: *Compute the standard error of the mean ($\sigma_{\overline{X}}$) as described above, and identify the sample mean and μ of the sampling distribution.*

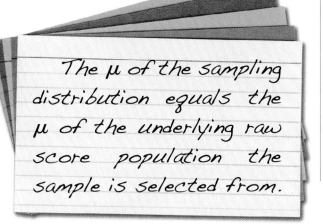

The μ of the sampling distribution equals the μ of the underlying raw score population the sample is selected from.

For the sample from Prunepit U, $\overline{X} = 520$, $\mu = 500$, and $\sigma_{\overline{X}} = 20$, so we have

$$z = \frac{\overline{X} - \mu}{\sigma_{\overline{X}}} = \frac{520 - 500}{20}$$

STEP 2: *Subtract μ from $\overline{X}$.*

$$z = \frac{+20}{20}$$

STEP 3: *Divide.*

$$z = +1$$

Thus, a sample mean of 520 has a z-score of $+1$ on the SAT sampling distribution of means that occurs when N is 25.

Let's combine everything we just did. Say that at ThunderCat U, a sample of 25 SAT scores produced a mean of 440. To find their z-score:

1. First, compute the standard error of the mean ($\sigma_{\overline{X}}$):

$$\sigma_{\overline{X}} = \frac{\sigma_X}{\sqrt{N}} = \frac{100}{\sqrt{25}} = \frac{100}{5} = 20$$

2. Then find z:

$$z = \frac{\overline{X} - \mu}{\sigma_{\overline{X}}} = \frac{440 - 500}{20} = \frac{-60}{20} = -3$$

The ThunderCat sample has a z-score of -3 on the sampling distribution of SAT means.

Describing the Relative Frequency of Sample Means

Everything we said previously about a z-score for an individual score applies to a z-score for a sample mean. The z-score tells us our sample mean's relative location within the sampling distribution, and thus its relative standing among all means that occur in this situation. Because our sample mean from Prunepit U has a z-score of $+1$, we know that it is above the μ of the sampling distribution by an amount equal to the "average" amount that sample means deviate above μ. The sample mean from ThunderCat U, however, has a z-score of -3, so it is one of the lowest SAT means we'd ever expect to obtain.

And here's the nifty part: Because a sampling distribution is always an approximately normal distribution, transforming *all* of the sample means in the sampling distribution into z-scores would produce a normal z-distribution. Recall that the *standard normal curve* is our model of *any* normal z-distribution. Therefore, as we did previously with raw scores, we can use the standard normal curve and z-table to describe the relative frequency of sample means in any part of a sampling distribution.

Figure 5.8 shows the standard normal curve applied to our SAT sampling distribution. This is the same curve, with the same proportions, that we used to describe individual raw scores. Here, however, each proportion is the expected relative frequency of the sample means that occur in this situation. For example, the sample mean 520 from Prunepit U has a z of +1. As shown, and as in column B of the z-table, .3413 of all scores fall between the mean and z of +1 on any normal distribution. Therefore, .3413 of all sample means are expected to fall here, so we expect .3413 of all SAT sample means to be between 500 and 520 (when N is 25).

Or, the mean of 440 from ThunderCat U had a z of −3. If we asked how often we would find means below this score, as shown (from column C of the z-table) only .0013 of the sampling distribution would be expected to be below this z-score. Therefore, we'd expect only .0013 of all SAT means to be below the ThunderCat mean of 440.

We can describe the means from any variable using this procedure. To be honest, researchers do not often compute the z-score for a sample mean solely to determine relative frequency (nor does SPSS include this routine). However, it is extremely important that you understand this procedure, because it is the essence of *all* upcoming inferential statistical procedures (and you'll definitely be doing those!).

Figure 5.8
Proportions of the Standard Normal Curve Applied to the Sampling Distribution of SAT Means

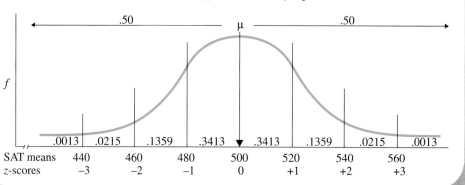

| SAT means | 440 | 460 | 480 | 500 | 520 | 540 | 560 |
| z-scores | −3 | −2 | −1 | 0 | +1 | +2 | +3 |

Summary of Describing a Sample Mean with a z-Score

We can describe a sample mean from any raw score population by following these steps:

 Envision the sampling distribution of means (or better yet, draw it) as a normal distribution with a μ equal to the μ of the underlying raw score population.

 Locate the sample mean on the sampling distribution by computing its z-score.

  a. Using the σ_X of the raw score population and your sample N, compute the standard error of the mean, $\sigma_{\bar{X}}$.

  b. Compute z, finding how far your $\bar{X}$ is from the μ of the sampling distribution, measured in standard error units.

 Use the z-table to determine the relative frequency of scores above or below this z-score, which is the relative frequency of sample means above or below your mean.

REMEMBER

The standard normal curve model and the z-table can be used with any sampling distribution, as well as with any raw score distribution.

USING SPSS

SPSS will transform raw scores into z-scores. You should note, however, that it does this for all the values of a variable and does not output the information in the usual way. Instead, it updates your database with a new variable containing the standardized values for each score in the specified variable(s). You can also use SPSS to generate a plot that will allow you to compare a variable's distribution against that of a normal curve.

1. a. What does a z-score transformation tell you?

    b. What factors determine the absolute value of the z-score?

2. Describe the mean, standard deviation, and shape of the z-distribution.

3. Why are z-scores referred to as "standard scores?"

4. a. What is the standard normal curve?

    b. What is it used for?

    c. What criteria should be met for the standard normal curve to give an accurate description of a sample?

5. Name four (4) uses of z-scores.

6. a. What are the shape, mean, and standard deviation of a sampling distribution of means?

    b. What does the sampling distribution of means reflect?

7. Dr. Jones has administered a test to her students. She calculated an $\overline{X} = 86$ and $S_x = 12$.

    a. What is the z-score of a student with a raw score of 80?

    b. What is the z-score of a student with a raw score of 98?

    c. What is the raw score for a student with a z-score of −1.5?

    d. What is the raw score for a student with a z-score of +1.0?

8. Which z-score corresponds to the smaller raw score in each of the following pairs?

    a. $z = -2.8$ and $z = -1.7$

    b. $z = +1.0$ and $z = +2.3$

    c. $z = -.7$ and $z = +1.5$

    d. $z = 0$ and $z = -1.9$

9. In problem 8, which z-score in each pair has the higher frequency?

10. If a set of scores has a sample mean of 25 and a sample variance of 4, find the following:

    a. the z-score for a raw score of 31

    b. the z-score for a raw score of 18

    c. the raw score for a z-score of −2.5

    d. the raw score for a z-score of +.5

11. Of the z-scores −2.3, +1.0, +.9, and −.6,

    a. which z-score corresponds to the smallest raw score?

    b. which z-score reflects the raw score having the highest frequency?

12. What proportion of the area under the standard normal curve would you expect to be

    a. between $z = -1.2$ and $z = +.6$?

    b. below $z = 1.4$?

    c. below $z = -2.6$?

    d. above $z = -2.0$?

13. Suppose you have two normal distributions containing different scores and different ranges.

    a. If you want to know the relative frequency of scores above $z = +.97$ for each distribution, what should you do?

    b. If you also want to know the expected relative frequency of scores between the $\overline{X}$ and 100 for each distribution, what should you do?

14. Find the relative frequency of scores

    a. between the mean and $z = +1.40$.

    b. below $z = -1.86$.

    c. above $z = +2.68$.

    d. below $z = -2.4$ and above $z = +1.96$.

15. For a distribution in which the mean is 100 and the standard deviation is 12, find the following:

    a. the relative frequency of scores between 76 and the mean

    b. the relative frequency of scores above 112

    c. the percentile of a score of 106

    d. the percentile of a score of 84

16. In problem 15, $N = 700$.

    a. How many participants are expected to score between 76 and 100?

    b. How many are expected to score above 112?

17. In the population, the average score on a test of self-esteem is 50 ($\sigma_x = 5$). You select a sample of 25 students who score a mean of 53.

    a. Sally claims these students don't have very high self-esteem, because they are only 3 points above the population mean. Why can't she make this claim merely by looking at the $\overline{X}$?

    b. What should she do?

    c. What is the z-score for this sample?

    d. What conclusion do you draw about your sample?

18. A researcher investigates the speed with which a sample of 50 participants completes an analogies test. She obtains a sample mean of 136.42 minutes. For the population of people who have taken this test, the mean is 130 and the standard deviation is 18.

    a. What percentage of the time can she expect to obtain a sample mean of 136.42 or above?

    b. Explain why she obtained such an unusual $\overline{X}$.

Using Probability
to Make Decisions about Data

You now know most of the common descriptive statistics used in behavioral research. Therefore, you are ready to begin learning the other type of statistical procedure, called *inferential statistics*. Recall that these procedures are used to draw inferences from sample data about the scores and relationship found in nature—what we call the population. This chapter sets the foundation for these procedures by introducing you to probability. As you'll see, researchers combine their knowledge of probability with the standard normal curve model to make decisions about data. Therefore, although you do not need to be an expert in probability, you do need to understand the basics. In the following sections we'll discuss (1) what probability is, (2) how to determine probability using the normal curve, and (3) how to use probability to draw conclusions about a sample mean.

SECTIONS

6.1 Understanding Probability

6.2 Probability Distributions

6.3 Obtaining Probability from the Standard Normal Curve

6.4 Random Sampling and Sampling Error

6.5 Deciding Whether a Sample Represents a Population

6.1 UNDERSTANDING PROBABILITY

Looking Back

- From Chapter 1, know how to use a sample to draw inferences about the population and the logic behind it.
- From Chapter 5, know how to compute a z-score for a sample mean and how to determine relative frequency from the standard normal curve.

Probability is used to describe random or chance events. By *random* we mean that nature is being fair, with no bias toward one event over another (no rigged roulette wheels or loaded dice). In statistical terminology, the event that occurs is a *sample* that is obtained from the *population* of all events that might occur. Thus, the sample could be drawing a particular playing card from the population of all cards in the deck, or, when tossing a coin, the sequence of heads and tails we see in a sample from the population of possible heads and tails. In research, the sample is the

particular group of individuals selected from the population of individuals we are interested in.

We compute probability only for samples or events that are obtained through random sampling. **Random sampling** is selecting a sample in such a way that all events or individuals in the population have an equal chance of being selected. Thus, in research, random sampling is anything akin to drawing participants' names from a large hat that contains all names in the population. A particular sample occurs or does not occur merely because of the luck of the draw.

But how can we describe an event that occurs only by chance? By paying attention to how *often* the event occurs *over the long run*. If event A happens frequently over the long run, we think it is likely to happen again now, and we say that it has a high probability. If event B happens infrequently, we think that it is unlikely to happen now, and we say that it has a low probability. When we describe how often an event occurs, however, we are making a relative judgment and actually

random sampling
A method of selecting samples so that all members of the population have the same chance of being selected for a sample

probability (p)
The likelihood of an event when a population is randomly sampled; equal to the event's relative frequency in the population

describing the event's *relative frequency* (the proportion of the time that the event occurs). Thus, the **probability** of an event is equal to the event's relative frequency in the population of possible events that can occur. The symbol for probability is **p**.

We use probability to describe the *likelihood*, or our *confidence*, that a particular sample will occur. We assume that an event's past relative frequency in the population will continue over the long run in the future. We translate this into our confidence for a *single* sample by expressing the relative frequency as a probability. For example, I am a rotten typist and I randomly make typos 80% of the time. This means that in the population of my typing, typos occur with a relative frequency of .80. We expect the relative frequency of typos to continue at a rate of .80 in anything else I type. This expected relative frequency is expressed as a probability, so the probability is .80 that I will make a typo when I type the next woid.

Notice that a probability is always expressed as a decimal. So if event A has a relative frequency of zero in a particular situation, then $p = 0$. This means that we do not expect A to occur in this situation because it never does. If A has a relative frequency of .10 in this situation, then it has a probability of .10: Because it occurs only 10% of the time in the population, we have some—but not much—confidence that A will occur in the next sample. On the other hand, if A has a probability of .95, we are confident that it will occur: It occurs 95% of the time in this situation, so we expect it in 95% of our samples. At the most extreme, event A's relative frequency can be 1: It is 100% of the population, so we are positive it will occur in this situation because it always does and we say $p = 1$.

An event cannot happen less than 0% of the time nor more than 100% of the time, so a probability can *never* be less than 0 or greater than 1. Also, all events together constitute 100% of the population. This means that the relative frequencies of all events must add up to 1, and the probabilities must also add up to 1. Thus, if the probability of my making a typo is .80, then because $1 - .80 = .20$, the probability is .20 that a word will be error free.

Finally, understand that except when p equals either 0 or 1, we are never certain that an event will or will not occur. The probability of an event is its relative frequency *over the long run* (in the population). It is up to chance whether a particular sample contains the event. So, even though I make typos 80% of the time, I may go for quite a while without making one. That 20% of the time when I make no typos has to occur sometime. Thus, although the probability is .80 that I will make a typo in each word, it is only over the long run that we expect to see precisely 80% typos.

People who fail to understand that probability implies *over the long run* fall victim to the "gambler's fallacy." For example, after observing my errorless typing for a while, the fallacy is thinking that errors "must" occur now, essentially concluding that errors have become more likely. Or, say we're flipping a coin and get 7 heads in a row. The fallacy is thinking that a head is now less likely to occur, because it's already occurred too often (as if the coin says, "Hold it. That's enough heads for a while!"). The mistake of the gambler's fallacy is failing to recognize that whether an event occurs or not, its probability is not altered, because probability is determined by what happens "over the long run."

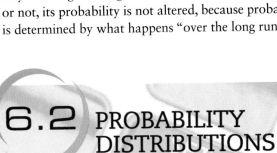

6.2 PROBABILITY DISTRIBUTIONS

To compute the probability of an event, we need only determine its relative frequency in the population. When we know the relative frequency of every event in a population, we have a probability distribution. A

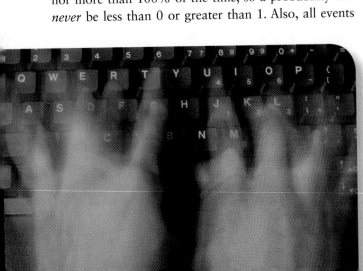

© PNC/Digital Vision/Jupiterimages

probability distribution indicates the probability of all possible events in a population.

One way to create a probability distribution is to observe the relative frequency of events, creating an *empirical probability distribution*. Typically, however, we cannot observe the entire population, so the probability distribution is based on the observed frequencies of events in a sample, which are used to represent the population. For example, say that Dr. Fraud is sometimes very cranky, and his crankiness is random. We observe him on 18 days and he is cranky on 6 of them. Relative frequency equals f/N, so the relative frequency of his crankiness is 6/18, or .33. We expect that he will continue to be cranky 33% of the time. Thus, the probability that he will be cranky today is $p = .33$. Conversely, he was not cranky on 12 of the 18 days, which is 12/18, or .67. Thus, $p = .67$ that he will not be cranky today. Because his cranky days plus his non-cranky days constitute all possible events, we have the complete probability distribution for his crankiness.

Another way to create a probability distribution is to devise a *theoretical probability distribution*, which is based on how we assume nature distributes events in the population. From such a model, we determine the *expected* relative frequency of each event in the population, which is then the probability of each event. For example, consider tossing a coin. We assume that nature has no bias toward heads or tails, so over the long run we expect the relative frequency of heads to be .50 and the relative frequency of tails to be .50. Thus, we have a theoretical probability distribution for coin tosses: The probability of a head on any toss is $p = .50$ and the probability of a tail is $p = .50$.

Or, consider drawing a playing card from a deck of 52 cards. We expect each card to occur at a rate of once out of every 52 draws over the long run. Thus, each card has a relative frequency of 1/52, or .0192, so the probability of drawing any specific card on a single draw is $p = .0192$.

> **probability distribution** The probability of every possible event in a population, derived from the relative frequency of every possible event in that population

And that is the logic of probability: We devise a *probability distribution* based on the expected relative frequency of each event in the population. An event's expected relative frequency equals its probability of occurring in a particular sample.

6.3 OBTAINING PROBABILITY FROM THE STANDARD NORMAL CURVE

Researchers use probability distributions to compute the probability of obtaining particular *scores* in a particular situation: That is, although we randomly select a sample of individuals in a study, we are really interested in the scores they produce. Therefore, we will operate as if we first measured the scores of everyone in the population. Then we will compute the probability of randomly selecting particular scores from that population.

For now we'll assume that the scores are normally distributed, so our probability distribution is based on the *standard normal curve*. Recall from Chapter 5 that with this curve we use z-scores to find the *proportion of the area under the normal curve*. This proportion corresponds to the relative frequency of scores in that part of a distribution. However, now we have seen that the relative frequency of an event *is* its probability. Therefore, the proportion of the total area under the curve for particular scores equals the probability of those scores.

For example, Figure 6.1 on the next page shows the probability distribution for a set of scores. Say that we seek the probability of randomly selecting someone with a raw score between 59 and 69. Raw scores between 59 and 69 transform to z-scores between 0 and $+1$.

As we've seen (and from column B of the z-table in Appendix B), z-scores between the mean and a z of +1 occur .3413 of the time. Thus, the probability is .3413 that we will randomly select one of these z-scores. And, because these z-scores correspond to raw scores between 59 and 69, the probability is also .3413 that we will select someone with a raw score between 59 and 69.

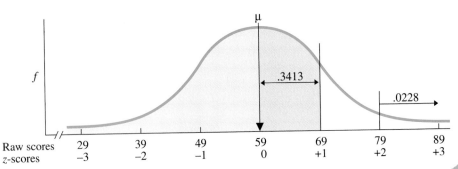

Figure 6.1
z-Distribution Showing the Area for Scores between the Mean and *z* = +1, and Above *z* = +2

Or, say that we seek the probability of selecting a score *greater* than 79. In the right-hand tail of Figure 6.1, a raw score of 79 has a z-score of +2. From column C of the z-table, the relative frequency of scores beyond this z is .0228. Therefore, the probability is .0228 that we will select a participant having a z-score greater than +2 or a raw score greater than 79.

In truth, researchers seldom use this procedure to determine the probability of individual scores. They do, however, use this procedure as part of inferential statistics to determine the probability of sample means.

Determining the Probability of Sample Means

We can compute the probability of obtaining particular sample means by using a *sampling distribution of means,* which is another type of probability distribution. Recall that a sampling distribution is the frequency distribution of all possible sample means that occur when a particular raw score population is sampled an infinite number of times using a particular N. For example, Figure 6.2 shows the sampling distribution of all possible means produced from the population of SAT scores when μ is 500 and N = 25. Recognize that the different values of $\overline{X}$ occur here simply because of the luck of the draw of which scores are in the sample each time. Sometimes a sample mean higher than 500 occurs because, by chance, the sample contains predominantly high scores. At other times, the sample contains predominantly low scores, producing a mean

below 500. Thus, the sampling distribution provides a picture of how often different sample means occur by chance when a sample is selected from the underlying SAT raw score population.

By using the sampling distribution, we can determine the probability that we will obtain particular sample means. First, as in Chapter 5, we compute the mean's z-score. For example, for the SAT means in Figure 6.2 when N is 25, say that $\sigma_X = 100$. First we compute the standard error of the mean using the formula

$$\sigma_{\overline{X}} = \frac{\sigma_X}{\sqrt{N}} = \frac{100}{\sqrt{25}} = \frac{100}{5} = 20$$

Say that we are interested in the means between 500 and 520. Therefore, next we compute the z-score for the mean of 520 using the formula

$$z = \frac{\overline{X} - \mu}{\sigma_{\overline{X}}} = \frac{520 - 500}{20} = \frac{+20}{20} = +1$$

Now, we can determine the probability of obtaining means between the z-scores of 0 and +1. As usual, as in Figure 6.2, the relative frequency of such z-scores is .3413, so the relative frequency of the sample means that produce these z-scores is also .3413. Therefore, the probability is .3413 that we will obtain a sample mean between 500 and 520 from this population. And here's the important part: *The probability of selecting a particular sample mean is the same as the probability of randomly selecting a sample of participants who produce scores that result in that sample mean.* Therefore, we have determined that when we randomly select 25 participants from the population of students who take the SAT, the probability is .3413 that their sample mean will between 500 and 520.

Likewise, say that we seek the probability of obtaining sample means above 540. As in the right-hand tail of Figure 6.2, a mean of 540 has a z-score of +2. As

© iStockphoto.com/Jesus Jauregui

shown, the relative frequency of z-scores above this z is .0228. Therefore, the probability is .0228 that we will select a sample whose SAT scores produce a mean higher than 540.

Or, we might seek the probability of means that are above 540 or below 460. This translates into seeking z-scores beyond a z of ±2 (plus or minus 2). Beyond z = +2 in the right-hand tail is .0228 of the curve, and beyond z = −2 in the left-hand tail is also .0228 of the curve. When we talk about means being in one area of the distribution or the other, mathematically we *add* these areas together. Therefore, a total of .0456 of the curve contains the means that have z-scores beyond ±2. Thus, the probability is .0456 that we'll obtain one of the means that lie there.

Before proceeding, be sure that you understand how z-scores and a sampling distribution indicate the probability of sample means. In particular, look again at Figure 6.2 and see what happens when means have a *larger* z-score that places them *farther* into the tail of the sampling distribution: The height of the curve above the means decreases, indicating that the means occur less often. *Therefore*, the probability of such means occurring *in this situation decreases.* Thus, a sample mean having a larger z-score is less likely to occur when we are dealing with the underlying raw score population. For example, in Figure 6.2, a mean of 560 has a z-score of +3, indicating that we are very unlikely to select a sample of students from the ordinary SAT population that contains such high scores and thus has such a high mean.

REMEMBER

The larger the absolute value of a sample mean's z-score, the less likely the mean is to occur when samples are drawn from the underlying raw score population.

6.4 RANDOM SAMPLING AND SAMPLING ERROR

The procedure for using z-scores to compute the probability of sample means forms the basis for all inferential statistics. The first step in understanding these statistics is to understand why we need them. Recall that in research we want to say that the way a sample behaves is the way that the population behaves. We need inferential statistics because there is no guarantee that the sample accurately reflects the population. In other words, we are never certain that a sample is *representative* of

Figure 6.2
Sampling Distribution of SAT Means When N = 25

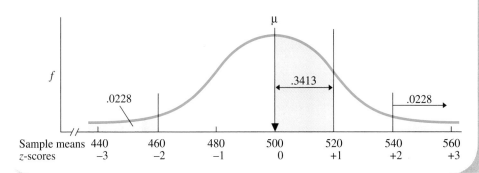

REPRESENTATIVE SAMPLE

POPULATION

© Burazin/Photographer's Choice/Getty Images / © Chromacome/Stockbyte/Jupiterimages

the population. In a **representative sample,** the characteristics of the individuals and scores in the sample accurately reflect the characteristics of individuals and scores found in the population. Put simply, a representative sample is a miniature version of the population in terms of the types of individuals it contains and therefore in terms of the scores they produce. So, if the μ in the SAT population is 500, then the $\overline{X}$ in a representative sample will be 500.

The reason that we select participants using random sampling is so that we will produce a representative sample. A random sample *should* be representative because, by being unselective and random in choosing participants, we allow the characteristics of the population to occur naturally in the sample in the same ways that they occur in the population. For example, say that 20% of the population has an SAT score of 475. Then 20% of a random sample should also score 475, because that's how often individuals with that score are out there. In the same way, random sampling should produce a sample having all of the characteristics of the population.

At least we hope it works that way! A random sample *should* be representative, but nothing forces this to occur. The problem is that *just by the luck of the draw* a sample may be *unrepresentative*, having characteristics that do not match those of the population. And, representativeness

is not all or nothing. Depending on the individuals—and scores—selected, a sample can be somewhat representative, only somewhat matching the population. For example, 20% of the population may score at 475, but simply through the luck of who is selected, we might obtain this score 10% or 30% of the time in our sample. Then the sample will have characteristics that are somewhat different from those of the population, and although μ may be 500, the sample mean will be some other number. In the same way, depending on the scores we happen to select, *any* sample may be unrepresentative of the population from which it is selected, so its sample mean will not equal the population mean it is representing.

The statistical term for communicating that chance produced an unrepresentative sample is to say that the sample reflects sampling error. **Sampling error** occurs when random chance produces a sample statistic (such as $\overline{X}$) that is not equal to the population parameter it represents (such as μ). Sampling error conveys that the reason a sample mean is different from μ is because, by chance, the sample is unrepresentative of the population. In plain English, because of the luck of the draw, a

representative sample A sample whose characteristics accurately reflect those of the population

sampling error The difference, due to random chance, between a sample statistic and the population parameter it represents

REMEMBER

Any sample may poorly represent one population, or it may accurately represent a different population.

sample contains too many high scores or too many low scores relative to the population, so the sample is in *error* to some degree in representing the population.

Here, then, is the central problem for researchers and the reason for inferential statistics: When sampling error produces a sample that is different from the population that it comes from and represents, the sample appears to come from and represent some other population. Thus, although a sample always represents some population, we are never sure *which* population it represents: Through sampling error, the sample may poorly represent one population, or it may accurately represent some other population altogether.

For example, say that we return to Prunepit University and in a random sample obtain a mean SAT score of 550. This is surprising because the ordinary national population of SAT scores has a μ of 500. Therefore, we should have obtained a sample mean of 500 if our sample was perfectly representative of this population. How do we explain a sample mean of 550? On the one hand, maybe we simply have sampling error. Perhaps we obtained a sample of relatively high SAT scores merely because of the luck of the draw of who was selected for the sample. Thus, it's possible that chance produced a sample that is less than perfectly representative, but the population being represented is still that ordinary population where μ is 500. On the other hand, perhaps the sample does not come from or represent the ordinary population of SAT scores: After all, these *are* Prunepit students, so they may belong to a very different population of students, having some other μ. For example, maybe Prunepit students belong to the population where μ is 550, and their sample is perfectly representative of this population.

The solution to this dilemma is to use inferential statistics to make a decision about the population being represented by our sample. The next chapter puts all of this into a research context, but in the following sections we'll examine the basics of deciding whether a sample represents a particular population.

6.5 DECIDING WHETHER A SAMPLE REPRESENTS A POPULATION

We deal with the possibility of sampling error in this way: Because we rely on random sampling, how representative a sample is depends on random chance—the luck of the draw of which individuals and scores are selected. Therefore, using probability, we can determine whether our sample is *likely* to come from and thus represent a particular population. If the sample is *likely* to occur when that population is sampled, then we decide that it *does* represent that population. If our sample is *unlikely* to occur when that population is sampled, then we decide that the sample does *not* represent that population, and instead represents some other population.

Here's a simple example. You obtain a paragraph of someone's typing, but you don't know whose. Is it mine? Does it represent the population of my typing? Say that the paragraph contains zero typos. It's possible that some quirk of chance produced such an unrepresentative sample, but it's not likely: I type errorless words only 20% of the time, so the probability that I could produce an entire errorless *paragraph* is extremely small. Thus, because such a sample is unlikely to come from the population of my typing, you should conclude that the sample represents the population of another, competent typist where such a sample is more likely.

On the other hand, say that there are typos in 75% of the words in the paragraph. This is consistent with what you would expect if the sample represents my typing, but with a little sampling error. Although you expect 80% typos from me over the long run, you should not expect precisely 80% typos in every sample. Rather, a sample with 75% errors seems likely to occur simply by chance when the population of my typing is sampled. Thus, you can accept that this paragraph somewhat poorly represents my typing.

We use the same logic to decide if our Prunepit sample represents the ordinary population of SAT scores where μ is 500: We will determine the probability of obtaining a sample mean of 550 from this population. As you've seen, we determine the probability of a sample mean by computing its *z*-score on the sampling distribution of means. Thus, we first envision the sampling distribution showing the means that occur when we select an infinite number of samples having our N from the ordinary SAT population. This is shown in Figure 6.3.

The next step is to calculate our sample mean's z-score to locate it on this distribution and thus determine its likelihood. In reality, we would not expect a *perfectly* representative sample having a $\overline{X}$ of exactly 500.000… (Think how unlikely that is!) Instead, if our sample represents this population, the sample mean should be *close* to 500. For example, say that the z-score for our sample mean is at location A in Figure 6.3. Read what the frequency distribution indicates by following the dotted line: This mean has a relatively high frequency and thus is very *likely* when drawing a sample from the ordinary SAT population. Anytime you deal with this population, this mean is likely to occur, because often a sample is unrepresentative to this extent, containing slightly higher scores in the sample than occur in the population. Thus, this is exactly the kind of mean we'd expect if our Prunepit sample came from this SAT population. Therefore, using statistical terminology, we say that we *retain* the idea that our sample probably comes from and represents the ordinary SAT population, accepting that the difference between our $\overline{X}$ and μ reflects chance sampling error.

However, say that instead, our sample has a z-score at location B in Figure 6.3: Following the dashed line shows that this is a very infrequent and *unlikely* mean. Thus, we seldom obtain a sample that is *so* unrepresentative of the ordinary SAT population that it will produce this mean. In other words, our sample is unlikely to be representing this population, because this mean almost never happens with this population. Therefore, we say that we *reject* that our sample represents this population, rejecting that the difference between the $\overline{X}$ and μ reflects sampling error. Instead, it makes more

sense to conclude that the sample represents some other raw score population (having some other μ), where this sample mean is more likely.

This logic is used in *all* inferential procedures, so be sure that you understand it:

We will always begin with a known, underlying raw score population that a sample may or may not represent. From the underlying raw score population we envision the sampling distribution of means that would be produced. Then we determine the location of our sample mean on the sampling distribution. The farther into the tail of the sampling distribution the sample mean is, the less likely that the sample comes from and represents the original underlying raw score population that we began with.

© iStockphoto.com/Bulent Ince

Setting Up the Sampling Distribution

To decide if our Prunepit sample represents the ordinary SAT population (with $\mu = 500$), we must perform two tasks: (1) Determine the probability of obtaining our sample from the ordinary SAT population, and (2) decide whether the sample is unlikely to be representing this population. We perform both tasks simultaneously by setting up the sampling distribution.

We formalize the decision process in this way: At some point, a sample mean is so far above or below 500 that it is unbelievable that chance produced such an unrepresentative sample. Any samples *beyond* this point that are farther into the tail are even more unbelievable. To identify this point, as shown in Figure 6.4, we literally draw a line in each tail of the distribution. In statistical terms, the shaded areas beyond the lines make up the *region of rejection*. As shown, very infrequently are samples so poor at representing the SAT population that they have means in the region of rejection.

Figure 6.3

Sampling Distribution of SAT Means Showing Two Possible Locations of Our Sample Mean

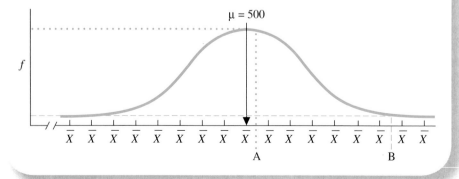

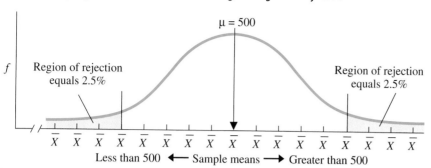

$\mu = 500$

f

Region of rejection
equals 2.5%

Region of rejection
equals 2.5%

$\overline{X}$ $\overline{X}$ $\overline{X}$ $\overline{X}$ $\overline{X}$ $\overline{X}$ $\overline{X}$ $\overline{X}$ $\overline{X}$ $\overline{X}$ $\overline{X}$ $\overline{X}$ $\overline{X}$ $\overline{X}$ $\overline{X}$ $\overline{X}$ $\overline{X}$

Less than 500 ◀— Sample means —▶ Greater than 500

region of rejection
That portion of a sampling distribution containing values considered too unlikely to occur by chance, found in the tail or tails of the distribution

criterion The probability that defines whether a sample is unlikely to have occurred by chance and thus is unrepresentative of a particular population

Thus, the **region of rejection** is the part of a sampling distribution containing means that are so unlikely that we *reject* that they represent the underlying raw score population. Essentially, we "shouldn't" get a sample mean that lies in the region of rejection if we're representing the ordinary SAT population because such means almost never occur with this population. Therefore, if we do get such a mean, we probably aren't representing this population: We reject that our sample represents the underlying raw score population and decide that the sample represents some other population.

Conversely, if our Prunepit mean is not in the region of rejection, then our sample is not unlikely to be representing the ordinary SAT population. In fact, by our definition, samples not in the region of rejection are likely to represent this population, but just not perfectly. In such cases we *retain* the idea that our sample is simply poorly representing this population of SAT scores.

How do we know where to draw the line that starts the region of rejection? By defining our *criterion*. The **criterion** is the probability that defines samples as unlikely to be representing the raw score population. Researchers usually use .05 as their criterion probability. (You'll see in Chapter 7 that

we use such a small probability because then we reduce the likelihood of making certain errors.) Thus, using this criterion, those sample means that together occur only 5% of the time when representing the ordinary SAT population are defined as so unlikely that if we get any one of them, we'll reject that our sample represents this population.

Our criterion determines the size of our region of rejection. In Figure 6.4, the sample means that occur 5% of the time are those that make up the extreme 5% of the sampling distribution. However, we're talking about the means above *or* below 500 that together are a *total* of 5% of the curve. Therefore, we divide the 5% in half so the extreme 2.5% of the sampling distribution will form our region of rejection in each tail.

Now the task is to determine if our sample mean falls into the region of rejection. To do this, we compare the sample's z-score to the *critical value*.

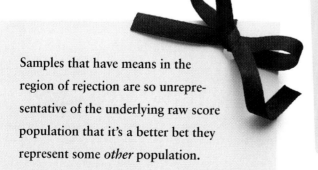

Samples that have means in the region of rejection are so unrepresentative of the underlying raw score population that it's a better bet they represent some *other* population.

REMEMBER

The criterion probability *that defines samples as unlikely—and also determines the size of the* region of rejection—*is usually* p = .05.

REJECTED

© iStockphoto.com/Katarzyna Krawiec / © iStockphoto.com/DNY59

Identifying the Critical Value

At the spot on the sampling distribution where the line marks the beginning of the region of rejection is a specific *z*-score. Because the absolute value of *z*-scores gets larger as we go farther into the tails, if the *z*-score for our sample is larger than the *z*-score at the line, then our sample mean lies *in the* region of rejection. The *z*-score at the line is called the critical value. A **critical value** marks the inner edge of the region of rejection and thus defines the value required for a sample to fall into the region of rejection. Essentially, it is the minimum *z*-score that defines a sample as too unlikely.

How do we determine the critical value? By considering our criterion. With a criterion of .05, the region of rejection in each tail is the extreme .025 of the total area under the curve. From column C in the *z*-table, the extreme .025 lies beyond the *z*-score of 1.96. Therefore, in each tail, the region of rejection begins at 1.96, so ±1.96 is the critical value of *z*. Thus, as shown in Figure 6.5, labeling the inner edges of the region of rejection with ±1.96 completes how you should set up the sampling distribution. (*Note:* In the next chapter, using both tails like this is called a *two-tailed test*.)

We'll use Figure 6.5 to determine whether our Prunepit mean lies in the region of rejection by comparing our sample's *z*-score to the critical value.

A sample mean lies in the region of rejection if its *z*-score is *beyond* the critical value.

Thus, if our Prunepit mean has a *z*-score that is *larger* than ±1.96, then the sample lies *in the* region of rejection. If the *z*-score is *smaller than* or *equal to* the critical value, then the sample is *not* in the region of rejection.

Deciding Whether the Sample Represents a Population

Finally, we can evaluate our sample mean of 550 from Prunepit U. First, we compute the sample's *z*-score on the sampling distribution created from the ordinary SAT population. There, $\sigma_X = 100$ and $N = 25$, so the standard error of the mean is

$$\sigma_{\bar{X}} = \frac{\sigma_X}{\sqrt{N}} = \frac{100}{\sqrt{25}} = 20$$

Then the *z*-score is

$$z = \frac{\bar{X} - \mu}{\sigma_{\bar{X}}} = \frac{550 - 500}{20} = +2.5$$

To complete the procedure, we compare the sample's *z*-score to the critical value to determine where the sample mean is on the sampling distribution. As shown in Figure 6.6, our sample's *z* of +2.5—and the underlying sample mean of 550—lies in the region of rejection. This tells us that a sample mean of 550 is among those means that are extremely unlikely to occur when the sample represents the ordinary population of SAT scores. In other words, very seldom does chance—the luck of the draw—produce such unrepresentative samples from this population, so it is not a good bet that chance produced *our* sample from this population. Therefore, we reject that our sample represents the population of SAT raw scores having a μ of 500.

Notice that we make a definitive, yes-or-no decision. Because our sample is unlikely to represent the SAT raw score population where μ is 500, we decide that no, it does not represent this population.

critical value The score that marks the inner edge of the region of rejection in a sampling distribution; values that fall beyond it lie in the region of rejection

Figure 6.5

Setup of Sampling Distribution of SAT Means Showing Region of Rejection and Critical Values

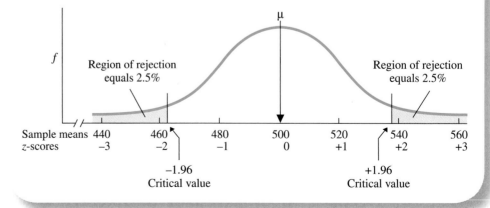

When a sample's z-score is beyond the critical value, reject *that the sample represents the underlying raw score population. When the z-score is not beyond the critical value*, retain *the idea that the sample represents the underlying raw score population.*

We wrap up our conclusions in this way: If the sample doesn't represent the ordinary SAT population, then it must represent some other population. For example, perhaps the Prunepit students obtained the high mean of 550 because they lied about their scores, so they may represent the population of students who lie about the SAT. Regardless, we use the sample mean to estimate the μ of the population that the sample *does* represent. A sample having a mean of 550 is most likely to come from a population having a μ of 550. Therefore, our best estimate is that the Prunepit sample represents a population of SAT scores that has a μ of 550.

On the other hand, say that our sample mean had been 474, resulting in a z-score of $(474 - 500)/20 = -1.30$. Because -1.30 does *not* lie beyond the critical value of ±1.96, this sample mean is *not* in the region of rejection. Look at Figure 6.6, and see that where the sample mean of 474 is located indicates that this is a relatively frequent and thus likely mean. Therefore, we know that chance will often produce such a sample from the ordinary SAT population, so it is a good bet that chance produced *our* sample from this population. Because of this, we can accept that random chance produced a less than perfectly representative sample for us but that it probably represents the ordinary SAT population where μ is 500.

Summary of How to Decide about the Population Being Represented

The basic question answered by all inferential statistical procedures is "Do the sample data represent a particular raw score population?" To answer this:

1. Envision (draw) the sampling distribution of means with a μ equal to the μ of the underlying raw score population.

2. Compute the sample mean and its z-score.

   a. Compute the standard error of the mean, $\sigma_{\bar{X}}$.

   b. Compute z using $\bar{X}$ and the μ of the sampling distribution.

3. **Set up the sampling distribution.** Select the criterion probability (usually .05), locate the region of rejection, and determine the critical value (±1.96 in a two-tailed test).

4. **Compare the sample's z-score to the critical value.** If the sample's z is beyond the critical value, it is in the region of rejection: Reject that the sample represents the underlying raw score population. If the sample's z is not beyond the critical value, do not reject that the sample represents the underlying population.

Other Ways to Set Up the Sampling Distribution

Previously, we placed the region of rejection in both tails of the distribution because we wanted to identify unrepresentative sample means that were either too far

Figure 6.6

Completed Sampling Distribution of SAT Means Showing Location of the Prunepit U Sample Relative to the Critical Value

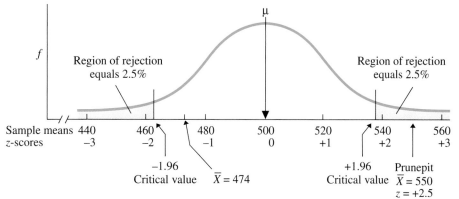

above or too far below 500. Instead, however, we can place the region of rejection in only one tail of the distribution. (In the next chapter you'll see why you would want to use this *one-tailed test.*)

Say that we are interested only in SAT means that are *less* than 500, having negative *z*-scores. Our criterion is still .05, but now we place the *entire* region of rejection in the lower, left-hand tail of the sampling distribution, as shown in part (a) of Figure 6.7. *This produces a different critical value.* The extreme lower 5% of a distribution lies beyond a *z*-score of −1.645. Therefore, the *z*-score for our sample must lie beyond −1.645 for it to be in the region of rejection. If it does, we will again conclude that the sample mean is so unlikely to occur when sampling the SAT raw score population that we'll reject that our sample represents this population. If the *z*-score is anywhere else on the sampling distribution, even far into the upper tail, we will *not* reject that the sample represents the population where μ = 500.

On the other hand, say that we're interested only in sample means *greater* than 500, having positive *z*-scores. Then we place the entire region of rejection in the upper,

right-hand tail of the sampling distribution, as shown in part (b) of Figure 6.7. Now the critical value is *plus* 1.645, so only if our sample's *z*-score is beyond +1.645 does the sample mean lie in the region of rejection. Only then do we reject the idea that our sample represents the underlying raw score population.

Figure 6.7

Setup of SAT Sampling Distribution to Test (a) Negative *z*-Scores and (b) Positive *z*-Scores

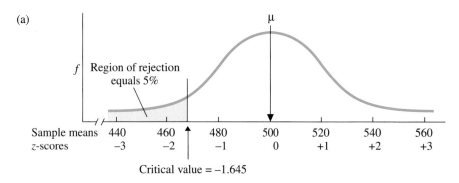

(a)

f Region of rejection equals 5%

Sample means	440	460	480	500	520	540	560
z-scores	−3	−2	−1	0	+1	+2	+3

Critical value = −1.645

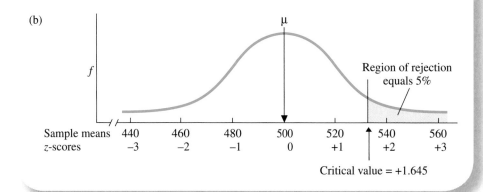

(b)

f Region of rejection equals 5%

Sample means	440	460	480	500	520	540	560
z-scores	−3	−2	−1	0	+1	+2	+3

Critical value = +1.645

using what you know

1. a. What is the probability of an event in a sample based on?

   b. How is probability expressed?

2. a. What is a probability distribution?

   b. What is a theoretical probability distribution?

3. What information is provided by a sampling distribution of means?

4. In testing whether a sample mean is representative of a population,

   a. what is the criterion?

    b. what is the critical value?

   c. what is the region of rejection?

5. a. What decision do we make when we compute a statistic lying beyond the critical value?

   b. What if the statistic does not lie beyond the critical value?

6. Why is a proportion of the area under the standard normal curve equal to a probability?

7. A student keeps a mood diary for a social psychology experiment. Over a 30-day period, he records that he

is happy on 13 days, grumpy on 5 days, sad on 2 days, and anxious on 10 days. Compute the probability that tomorrow he will be (a) happy, (b) sad, (c) anxious, (d) grumpy, (e) in one of these moods.

8. Given a standard deck of 52 playing cards, determine the probability of the following events:

    a. probability of drawing a 7

    b. probability of drawing a club

    c. probability of drawing a 7 or a 10

    d. probability of drawing a 7 or a club

9. Find the probability of each of the following:

    a. a z-score below -2.10

    b. a z-score between $\pm.97$

    c. a z-score below -1.53 or above 2.53

    d. a z-score below 1.37

10. It is known that 86% of persons with insomnia also have reported a problem with fatigue. Dr. Delgado's patient Bob reports no problem with fatigue.

    a. Based on this information, should Dr. Delgado decide Bob has insomnia? Explain your answer.

    b. If further testing reveals Bob *does* have insomnia, explain how the previous decision in the first part of this problem could have occurred.

11. In a sample with a mean of 46 and a standard deviation of 8, what is the probability of randomly selecting each of the following raw scores?

    a. a score above 64

    b. a score between 40 and 50

    c. a score of 48 or below

12. The population from which the sample in problem 11 was randomly drawn has a population mean of 51 and a population standard deviation of 14.

    a. What is the probability of obtaining a random sample of 25 scores with a mean of 46 or less?

    b. Out of 1000 samples, how many would you expect to have a mean of 46 or below?

13. A researcher obtains a sample mean of 66, which produces a z of $+1.45$. With the critical value of ±1.96, this researcher decides to reject the idea that the sample is representative of the underlying raw score population having a μ of 60.

    a. Draw the sampling distribution and indicate the relative locations of $\overline{X}$, μ, the computed z-score, and the two critical values.

    b. Is the conclusion correct? Explain your answer.

14. In a study with a sample mean of 14, a z-score of $+2.0$ is obtained. With a critical value of ±1.96, the research team decides not to reject the idea that the sample represents a μ of 12.

    a. Draw the sampling distribution and indicate the relative locations of $\overline{X}$, μ, the computed z-score, and the two critical values.

    b. What should the team conclude regarding the population where $\mu = 12$?

    c. What population is this sample likely to represent?

15. In problem 13 and problem 14, the sample means do not equal the corresponding μs.

    a. Give two explanations for this occurrence you might have used before examining z.

    b. If both means differ from their respective μs for the same two possible reasons, why are the final conclusions so different?

16. Consider a population for which $\mu = 53$ and $\sigma_x = 15$. Using a criterion of $p = .05$ and both tails of the sampling distribution, which of the following samples ($N = 50$) can be called unrepresentative of the population?

    a. a sample with $\overline{X} = 56$

    b. a sample with $\overline{X} = 47$

17. A test of funniness administered to the population of successful late-night comedians yields a mean of 72 and a standard deviation of 8. Answer the following using a decision criterion of $p = .05$ and the lower tail of the sampling distribution:

    a. Is a random sample of 25 comedians scoring $\overline{X} = 69$ unrepresentative of this population?

    b. If it is, what is the μ of the population it represents?

    c. What type of comedian would be found in this population?

18. The mean number of hours of television watched per week for the population of all two-year-olds is 20.4 hours. The standard deviation of this population is 4.8 hours. A sample of 40 two-year-olds yields a mean of 21.75 hours of television watched per week. Using a decision criterion of $p = .05$ and the upper tail of the sampling distribution, decide whether this sample is representative of the population where $\mu = 20.4$.

19. The population mean μ on a national scholastic achievement test is 100 with a $\sigma_x = 30$. The students in Mr. Smart's class got the following scores:

127	121	123	128	118	126
120	130	128	119	127	125

Using the criterion of .05 in the *upper* tail only, determine if Mr. Smart's class is representative of the population.

20. On a national test of "mental intensity," μ is $20 (\sigma_x = 6.28)$. Students in your class produce the following scores:

25	26	34	14	33	29	22	18
16	13	21	20	22	21	34	30

Using the criterion of .05 and both tails of the sampling distribution, determine if your class is representative of the population.

Overview
of Statistical Hypothesis Testing: The z-Test

In the previous chapter, you learned the basics involved in inferential statistics. Now we'll put these procedures into a research context and present the statistical language and symbols used to describe them. Until further notice, we'll be talking about experiments. This chapter shows (1) how to set up a one-sample experiment, (2) how to perform the "z-test," and (3) how to interpret the results of an inferential procedure.

SECTIONS

7.1 THE ROLE OF INFERENTIAL STATISTICS IN RESEARCH

As you saw in the previous chapter, a random sample may be more or less representative of a population because, just by the luck of the draw, the sample may contain too many high scores or too many low scores relative to the population. Because the sample is not perfectly representative of the population, it reflects *sampling error*, so the sample mean does not equal the population mean.

The possibility of sampling error creates a dilemma when researchers try to infer that a relationship exists in nature. Recall that in an experiment, we hope to see a relationship in which, as we change the conditions of the independent variable, participants' scores on the dependent variable change in a consistent fashion. If the means

Looking Back

- From Chapter 1, know what the conditions of an independent variable are and what the dependent variable is.

- From Chapter 3, understand that a relationship in the population occurs when different means from the conditions represent different μs and thus different distributions of dependent scores.

- From Chapter 6, know that when a sample's z-score falls in the region of rejection, the sample is unlikely to represent the underlying raw score population.

© Pixtal Images/Photolibrary

for our conditions differ, we infer that, in nature, each condition would produce a different population of scores located at a different μ. But! Here is where sampling error comes in. Perhaps we are wrong and the relationship does not exist in nature. Maybe all of the scores actually come from the same population, and the means in our conditions differ simply because of which participants we happened to select for each—because of sampling error. We won't know this, so we will be misled into thinking the relationship does exist. For example, say we compare the creativity scores of some men and women, although we are unaware that in nature men and women do not differ on this variable. Through sampling error, however, we might select some females who are more creative than our males or vice versa. Then sampling error will mislead us into thinking there's a relationship here, although really there is not.

Or, perhaps there is a relationship in the population, but because of sampling error, we see a different relationship in our sample data. For example, say we measure the heights of some men and women and, by chance, obtain a sample of relatively short men and a sample of tall women. If we didn't already know that in the population men are taller, sampling error would mislead us into concluding that women are taller.

inferential statistics Procedures for determining whether sample data represent a particular relationship in the population

parametric statistics Inferential procedures that require certain assumptions about the raw score population represented by the sample; used when we compute the mean of the scores

nonparametric statistics Inferential procedures that do not require stringent assumptions about the raw score population represented by the sample data

In every study it is possible that we are being misled by sampling error so that the relationship we see in our sample data is not the relationship found in nature. This is the reason why, in every study, we apply inferential statistics. **Inferential statistics** are used to decide whether sample data represent a particular relationship in the population. Essentially, we decide if we should *believe* our sample data: Should we believe what the sample data appear to indicate about the relationship in the population, or instead, is it likely that the sample relationship is a coincidence produced by sampling error that misrep-

resents what is found in the population?

The specific inferential procedure employed in a given situation depends upon the *study's design* and on the *scale of measurement* used to measure the *dependent variable*. There are two general categories of inferential statistics. **Parametric statistics** are procedures that require specific assumptions about the raw score populations being rep-

resented. Each procedure has its own assumptions, but there are two assumptions common to all parametric procedures: (1) The population of dependent scores should be at least approximately normally distributed. (2) The scores should be interval or ratio scores. Parametric procedures are used when it is appropriate to compute the mean of the scores. In this and upcoming chapters we'll focus on these procedures.

The other category is **nonparametric statistics,** which are inferential procedures that do not require stringent assumptions about the populations being represented. These procedures are used with nominal or ordinal scores or with skewed interval or ratio distributions. Chapter 13 presents nonparametric procedures.

REMEMBER

Parametric *and* nonparametric inferential statistics *are for deciding if the data reflect a relationship in nature, or if sampling error is misleading us into thinking there is a relationship.*

7.2 SETTING UP INFERENTIAL PROCEDURES

Researchers perform four steps in an experiment: They create the experimental hypotheses, design and conduct the experiment to test the hypotheses, translate the experimental hypotheses into statistical hypotheses, and test the statistical hypotheses.

Creating the Experimental Hypotheses

An experiment always tests two **experimental hypotheses.** They describe the predicted outcome we may or may not find. One hypothesis states that we will demonstrate the predicted relationship (manipulating the independent variable will work as expected). The other hypothesis states that we will not demonstrate the predicted relationship (manipulating the independent variable will not work as expected).

We can predict a relationship in one of two ways. Sometimes we predict a relationship, but we are not sure whether scores will increase or decrease as we change the independent variable. This leads to a "two-tailed" test. A **two-tailed test** is used when we do not predict the direction in which scores will change. Thus, we'd have a two-tailed hypothesis if we thought men and women differed in creativity but were unsure who would score higher. The other approach is a one-tailed test. A **one-tailed test** is used when we predict the *direction* in which scores will change. We may predict that the dependent scores will only increase, or that they will only decrease. Thus, we'd have a one-tailed test if we predicted only that men would be more creative than women or if we predicted only the opposite.

Let's first examine a study involving a two-tailed test. Say we've discovered a substance related to intelligence that we will test with humans in an "IQ pill." The quantity of a substance in the pill is our independent variable and the person's resulting IQ is the dependent variable. We believe this pill will affect IQ, but we are not sure whether it will make people smarter or dumber. Therefore, here are our two-tailed experimental hypotheses:

1. We will demonstrate that the pill works by either increasing or decreasing IQ scores.

2. We will not demonstrate that the pill works, because IQ scores will not change.

Designing a One-Sample Experiment

There are a number of ways that we might design an experiment to test the IQ pill. However, the simplest is as a *one-sample experiment.* We will randomly select one sample of participants and give each person, say, one pill. Then we'll give participants an IQ test. The sample will represent the population of people when they have taken one pill, and the sample $\overline{X}$ will represent that population's μ.

However, *to perform a one-sample experiment, we must already know the population mean under some other condition of the independent variable.* This is because to demonstrate a relationship, we must demonstrate that *different* conditions produce *different* populations having different μs. Therefore, we must compare the population represented by our sample to some other population that receives some other quantity of the pill. One other quantity of the pill is zero: Say that our IQ test has been given to many people over the years who have *not* taken the pill, and this population has a μ of 100. We will compare this population without the pill to the population with the pill represented by our sample. If the population without the pill has a different μ than the population with the pill, then we will have demonstrated a relationship in the population.

Creating the Statistical Hypotheses

So that we can apply statistical procedures, we translate the experimental hypotheses into two *statistical hypotheses.* **Statistical hypotheses** describe the population parameters that the sample data represent if the predicted relationship does or does not exist. The two statistical hypotheses are called the *alternative hypothesis* and the *null hypothesis*.

The Alternative Hypothesis It is easier to create the alternative hypothesis first, because it corresponds to the experimental

experimental hypotheses Two statements made before a study is begun, describing the predicted relationship that may or may not be demonstrated by the study

two-tailed test The test used to evaluate a statistical hypothesis that predicts a relationship but not whether scores will increase or decrease

one-tailed test The test used to evaluate a statistical hypothesis that predicts that scores will only increase or only decrease

statistical hypotheses Two statements that describe the population parameters the sample statistics will represent if the predicted relationship exists or does not exist

hypothesis that the experiment *does* work as predicted. The **alternative hypothesis** describes the population parameters that the sample data represent if the predicted relationship exists. It says that changing the independent variable produces the predicted difference in the populations.

For example, Figure 7.1 shows the populations of IQ scores if the pill *increases* IQ. Without the pill, the population is centered at a score of 100. By giving everyone one pill, however, their score is increased so that the distribution moves to the right, over to the higher scores. We don't know how much scores will increase, but we do know that the μ of the population with the pill will be *greater* than 100, because 100 is the μ of the population without the pill. Recall that these distributions are how we envision a relationship in the population: As the amount of the pill changes, IQ changes so that we have one "batch" of scores for people without the pill and a different batch of scores with one pill.

On the other hand, Figure 7.2 shows the populations if the pill *decreases* IQ. Here, with the pill, the distribution moves to the left, over to the lower scores. We also don't know how much scores might decrease, but we do know that the μ of the population with the pill will be *less than* 100.

The alternative hypothesis communicates all of the above. If the pill works as predicted, then the population with the pill will have a μ that is either greater than or less than 100. In other words, the population mean with the pill will *not equal* 100. The symbol for the alternative hypothesis is H_a. The symbol for not equal is "≠," so our alternative hypothesis is

$$H_a: \mu \neq 100$$

This proposes that our sample mean represents a μ not equal to 100. Because the μ without the pill is 100, H_a implies that there is a relationship in the population. (In a two-tailed test, H_a is always μ ≠ some value.)

alternative hypothesis (H_a)
The hypothesis describing the population parameters that the sample data represent if the predicted relationship does exist

Figure 7.1
Relationship in the Population if the IQ Pill Increases IQ Scores

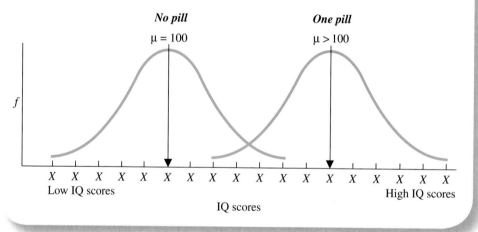

Figure 7.2
Relationship in the Population if the IQ Pill Decreases IQ Scores

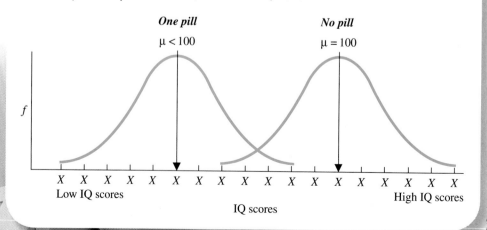

The Null Hypothesis The statistical hypothesis that corresponds to the experimental hypothesis that the independent variable does *not* work as predicted is called the null hypothesis. The **null hypothesis** describes the population parameters that the sample data represent if the predicted relationship does *not* exist. It says that changing the independent variable does *not* produce the predicted differences in dependent scores in the population. And, the null hypothesis says this is true *regardless* of what our data shows. Therefore, we may see the predicted relationship in the sample, but the null hypothesis says this is due to sampling error and that the data are poorly representing that the relationship is *not* found in the population.

If the IQ pill does not work, then it would be as if the pill were not present. The population of IQ scores without the pill has a μ of 100. Therefore, if the pill does not work, then the population of scores will be unchanged and μ will still be 100. Accordingly, if we measured the population with and without the pill, we would have one population of scores, located at the μ of 100, as shown in Figure 7.3.

The null hypothesis communicates the above. The symbol for the null hypothesis is H_0. (The subscript is 0 because *null* means zero, as in zero relationship.) The null hypothesis for the IQ pill study is

$$H_0: \mu = 100$$

This proposes that our sample with the pill comes from and represents the population where μ is 100. Because this is the same population found without the pill, H_0 implies that the predicted relationship does *not* exist in nature. (In a two-tailed test, H_0 is always $\mu =$ some value.)

REMEMBER

The alternative hypothesis (H_a) always says the sample data represent a μ that reflects the predicted relationship. The null hypothesis (H_0) says the sample data represent the μ that's found when the predicted relationship is not present.

The Logic of Statistical Hypothesis Testing

The statistical hypotheses for the IQ pill study are $H_0: \mu = 100$ and $H_a: \mu \neq 100$. Remember, these are hypotheses—guesses—about the population represented by our sample *with* the pill. (We have no uncertainty about what happens without the pill; we *know* the μ there.) Notice that, together, H_0 and H_a include all possibilities, so one or the other must be true. We use inferential procedures to test (choose between) these hypotheses, so these procedures are called "statistical hypothesis testing."

Say that we randomly selected 36 people, gave them the pill, measured their IQ, and found that their mean was 105. We would *like* to say this: People who have not taken this pill have a mean IQ of 100, so if the pill did not work, the sample mean should have been 100. Therefore, a mean of 105 suggests that the pill does work; raising IQ scores about 5 points. If the pill does this for the sample, it should do this for the population. Therefore, we expect that a population receiving the pill would have a μ of 105. Our results appear to support our alternative hypothesis ($H_a: \mu \neq 100$). If we measured everyone in the population with and without the pill, we would have the two distributions shown previously in Figure 7.1, with the population that received the pill located at the μ of 105.

But hold on! We just assumed that our sample is *perfectly* representative of the population it represents. But what if we have sampling error?

null hypothesis (H_0) The hypothesis describing the population parameters that the sample data represent if the predicted relationship does not exist

Figure 7.3
Population of Scores if the IQ Pill Does Not Affect IQ Scores

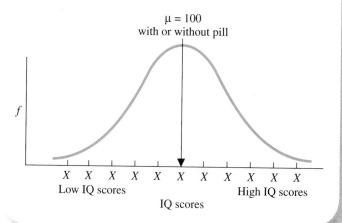

$\mu = 100$
with or without pill

f

X X X X X X X X X X X
Low IQ scores   High IQ scores
IQ scores

z-test The parametric procedure used to test the null hypothesis for a single-sample experiment when the true standard deviation of the raw score population is known

Maybe we obtained a mean of 105 not because the pill works, but because we inaccurately represented the situation where the pill does *not* work. After all, it is unlikely that any sample is *perfectly* representative, so even if our sample represents the population where μ is 100, we don't expect our $\overline{X}$ to equal exactly 100! So, maybe the pill does nothing, but by chance we happened to select participants who *already* had an above-average IQ. Thus, maybe the null hypothesis is correct: Maybe our sample actually represents the population where μ is 100.

Likewise, in any study, we cannot automatically infer that the relationship exists in the population when our sample data show the predicted relationship, because there are always two things that can produce such data: sampling error or a relationship in nature. Essentially, the H_0 says that sampling error produced the sample relationship, so we should not believe there is this relationship in nature. The H_a says that a relationship in nature produced the sample relationship, so that we can believe that nature operates as the sample data suggest.

The only way to prove whether H_0 is true is to give the pill to everyone in the population and see whether μ is 100 or 105. We cannot do that. We can, however, determine how *likely* it is that H_0 is true. That is, using the procedure discussed in the previous chapter, we will determine the likelihood of obtaining a sample mean of 105 from the population that has a μ of 100. If such a mean is too unlikely, then we will *reject* that our sample represents this population, rejecting that H_0 is the correct hypothesis for our study.

All parametric and nonparametric procedures use this logic. To select the correct procedure for a particu-

lar experiment, you should check that the design and dependent scores fit the assumptions of the procedure. The IQ pill study meets the assumptions of the parametric procedure called the *z-test*.

7.3 PERFORMING THE z-TEST

The **z-test** is the procedure for computing a z-score for a sample mean on the sampling distribution of means that we've discussed in previous chapters. The z-test has four assumptions:

1. We have randomly selected one sample.

2. The dependent variable is at least approximately normally distributed in the population and involves an interval or ratio scale.

3. We *know* the mean of the population of raw scores under another condition of the independent variable.

4. We *know* the true standard deviation of the population (σ_X) described by the null hypothesis.

Say that from the research literature, we know that IQ scores meet the requirements of the z-test and that in the population where μ is 100, the standard deviation is 15. Therefore, the next step is to perform the z-test. (*Note:* SPSS does not perform the z-test.)

Setting Up the Sampling Distribution for a Two-Tailed Test

To perform the z-test we must create the sampling distribution of means and identify the region of rejection as we

REMEMBER

The mean of the sampling distribution always equals the μ of the raw score population that H_0 says we are representing.

did in the previous chapter. The sampling distribution is always created from the raw score population that H_0 says our sample represents. Therefore, in our pill study, it is as if, using our N of 36, we infinitely sampled the raw score population without the pill where μ is 100. This will produce a sampling distribution of means with a μ of 100. In any study, the μ of the sampling distribution equals the value of μ given in the null hypothesis.

The finished sampling distribution is shown in Figure 7.4. To get there, we have some new symbols and terms.

1 **Select the alpha.** Recall that the *criterion* is the probability that defines sample means as unlikely to be representing the underlying raw score population. The symbol for the criterion is α, the Greek letter **alpha**. Usually our criterion, our "alpha level," is .05. As we'll see, we may select a smaller probability, but we never use a criterion larger than .05. We'll use .05 for now, so in symbols we say $\alpha = .05$.

2 **Locate the region of rejection.** Recall that we may use one or both tails of the sampling distribution. Which arrangement to use depends on whether we have a two-tailed or one-tailed test. Above, we created a two-tailed hypothesis, predicting that the pill makes people either smarter or dumber, producing an $\overline{X}$ that is either larger than 100 or smaller than 100. Therefore, we have a two-tailed test, with part of the region of rejection in each tail.

3 **Determine the critical value.** We'll abbreviate the critical value of z as z_{crit}. Recall that with $\alpha = .05$, the region of rejection in each tail is 2.5% of the distribution. From the z-table, $z = 1.96$ demarcates this region. Thus, we complete Figure 7.4 by adding that z_{crit} is ±1.96.

Computing z

> **alpha (α)** The Greek letter that symbolizes the criterion probability

Now it's time to compute the z-score for our sample mean. The z-score we compute is "obtained" from the data, so we'll call it z *obtained*, which we abbreviate as z_{obt}. You know how to compute this from previous chapters.

z-TEST FORMULA

$$z_{obt} = \frac{\overline{X} - \mu}{\sigma_{\overline{X}}}$$

The value of μ to put in the formula is the μ of the sampling distribution, which is also the μ of the underlying raw score population that H_0 says the sample represents. The $\overline{X}$ is computed from the scores in the sample. The $\sigma_{\overline{X}}$ is the standard error of the mean, which is computed as

$$\sigma_{\overline{X}} = \frac{\sigma_X}{\sqrt{N}}$$

where N is the N of the sample and σ_X is the true population standard deviation.

For the IQ pill study, the σ_X is 15 and N is 36. Thus,

$$\sigma_{\overline{X}} = \frac{\sigma_X}{\sqrt{N}} = \frac{15}{\sqrt{36}} = \frac{15}{6} = 2.5$$

Then the z-score for our sample mean of 105 is

$$z_{obt} = \frac{\overline{X} - \mu}{\sigma_{\overline{X}}} = \frac{105 - 100}{2.5} = \frac{+5}{2.5} = +2.00$$

The final step is to interpret this z_{obt} by comparing it to z_{crit}.

Figure 7.4
Sampling Distribution of IQ Means for a Two-Tailed Test
A region of rejection is in each tail of the distribution, marked by the critical values of ±1.96.

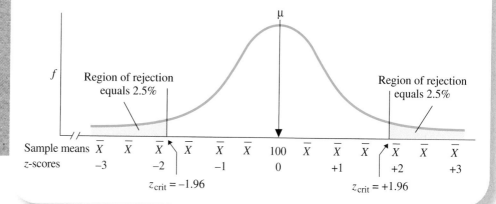

Comparing the Obtained z to the Critical Value

The sampling distribution always describes the situation *when null is true:* Here, it shows all possible means that occur when, as H_0 says, our sample comes from the population where μ is 100 (from the situation where the pill does not work). If we are to believe H_0, the sampling distribution should show that a $\overline{X}$ of 105 is relatively frequent and thus likely in this situation. However, Figure 7.5 shows just the opposite. A z_{obt} of +2.00 tells us that a $\overline{X}$ of 105 seldom occurs by chance when we are representing the population where μ is 100. This makes it difficult to believe that *our* sample mean of 105 occurred by chance from this population: A mean like ours hardly ever occurs in the situation where the pill doesn't work. In fact, because a z_{obt} of +2.00 is beyond the z_{crit} of ±1.96, our sample is in the region of rejection. Therefore, we conclude that our sample is so unlikely to represent the population where $\mu = 100$ that we reject that our sample represents this population. In other words, we reject that our results poorly represent the situation where the pill does not work.

In statistical terms, we say that we have "rejected" the null hypothesis. If we reject H_0, then we are left with H_a, and so we "accept H_a." Here, H_a is $\mu \neq 100$, so we accept that our sample represents a population where μ is not 100.

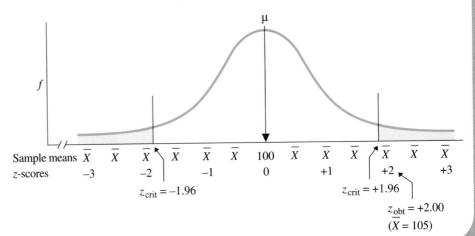

Figure 7.5
Sampling Distribution of IQ Means
The sample mean of 105 is located at $z_{obt} = +2.00$.

$z_{crit} = -1.96$
$z_{crit} = +1.96$
$z_{obt} = +2.00$
$(\overline{X} = 105)$

REMEMBER

When a sample statistic falls beyond the critical value, the statistic lies in the region of rejection, so we reject H_0 and accept H_a.

7.4 INTERPRETING SIGNIFICANT AND NONSIGNIFICANT RESULTS

Once we have made a decision about the statistical hypotheses (H_0 and H_a), we then make a decision about the corresponding original experimental hypothesis. When we reject H_0 we also reject the experimental hypothesis that our independent variable does not work as predicted. Therefore, we will reject that our pill does not work. By accepting H_a we accept that our pill *appears* to work.

If this makes your head spin, it may be because the logic actually involves a double negative. When our sample falls in the region of rejection,

we say "no" to the H_0 that says we're representing the population with $\mu = 100$. But H_0 says this as a way of saying there is *no relationship* involving our pill. By rejecting H_0 we are saying "No" to *no relationship*. This is actually saying "Yes, there *is* a relationship involving our pill," which is what H_a says.

Thus, it appears that we have demonstrated a *relationship in nature* such that the pill would change the population of IQ scores. In fact, we can be more specific: A sample mean of 105 is most likely to represent the population where μ is 105. So without the pill, the population of IQ scores is at a μ of 100, but with the pill we expect scores will rise to produce a population at around a μ of 105.

Finally, we do not know the exact μ of the population represented by our sample. In our pill study, assuming that the pill does increase IQ scores, the population μ would probably not be *exactly* 105. Our sample mean may contain (you guessed it) sampling error! That is, the sample may accurately reflect that the pill increases IQ, but it may not perfectly represent how *much* the pill increases scores. Therefore, we conclude that the μ produced by our pill would probably be *around* 105.

Bearing these qualifications in mind, we interpret the $\overline{X}$ of 105 the way we wanted to several pages back: Apparently, the pill increases IQ scores by about 5 points. But now, because the results are significant, we are confident that we are not being misled by sampling

REJECTING THE NULL HYPOTHESIS IS SAYING "NO" TO THE IDEA THAT THERE IS NO RELATIONSHIP.

Interpreting Significant Results

The way to communicate that we have rejected H_0 and accepted H_a is to use the term *significant*. Significant does *not* mean important or impressive. **Significant** indicates that our results are unlikely to occur if the predicted relationship does not exist in the population. Therefore, we imply that the relationship found in the experiment is "believable," representing a relationship found in nature, and that it does not reflect sampling error from the situation in which the relationship does not exist.

Although we accept that a relationship exists, there are three very important restrictions on how far we can go with this claim. First, we never *prove that H_0 is false.* The sampling distribution in Figure 7.5 shows that a mean of 105 does occur once in a while when we *are* representing the population where μ is 100. Maybe our sample was one of these times. Maybe the pill did not work, and our sample was simply very unrepresentative of this.

Second, *we do not prove that our independent variable caused the scores to change.* In our pill study, although we're confident that our sample represents a population with a μ above 100, we have not proven that it was the pill that produced these scores. Some other hidden variable might have actually caused the higher scores in the sample.

error. Instead, we are confident that we are describing a relationship found in nature. Therefore, after describing this relationship, we return to being behavioral researchers and attempt to explain how nature operates in terms of the variables and behaviors they reflect, describing how the ingredients in the pill affect intelligence, what brain mechanisms are involved, and so on.

Interpreting Nonsignificant Results

Let's say that the IQ pill had instead produced a sample mean of 99. Now the z-score for the sample is

$$z_{\text{obt}} = \frac{\overline{X} - \mu}{\sigma_{\overline{X}}} = \frac{99 - 100}{2.5} = \frac{-1}{2.5} = -.40$$

As in Figure 7.6 on the next page, a z_{obt} of $-.40$ is *not* beyond the z_{crit} of ± 1.96, so the sample mean does not lie in the region of rejection. This indicates that a mean of 99 is likely when sampling the population where $\mu = 100$. Thus, our sample mean was likely to have occurred if we were representing this population. In other words, our mean of 99 was likely—a mean

significant
Describes results that are too unlikely to accept as resulting from sampling error when the predicted relationship does not exist; it indicates rejection of the null hypothesis

nonsignificant
Describes results that are considered likely to result from chance sampling error when the predicted relationship does not exist; it indicates failure to reject the null hypothesis

Figure 7.6
Sampling Distribution of IQ Means
The sample mean of 99 has a z_{obt} of $-.40$.

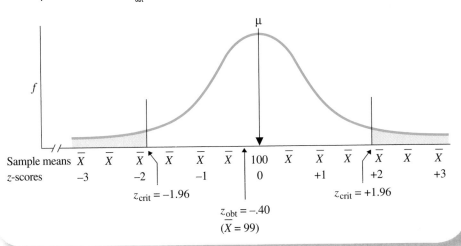

you'd expect—assuming we were representing the situation where the pill does not work. Therefore, we will *not* conclude that the pill works. After all, it makes no sense to claim that the pill works if the results were likely to occur *without* the pill. Thus, our null hypothesis—that our sample represents the population of scores without the pill—is a reasonable hypothesis, so we will not reject it. However, we have not proven that H_0 is true; in such situations, we "fail to reject H_0" or we "retain H_0."

To communicate the above we say that we have *nonsignificant* or not significant results. (Don't say *insignificant*.) **Nonsignificant** indicates that the differences reflected by our results were likely to have occurred through chance sampling error without there being a relationship in nature.

Nonsignificant results do not prove that the independent variable does not work. We have simply failed to find convincing evidence that it *does* work. The only

thing we're sure of is that sampling error *could* have produced our data. Therefore, we still have two hypotheses that are both viable:

1. H_0, that the results do *not* really represent a relationship

2. H_a, that the results *do* represent a relationship.

Maybe the pill does work, but the sample data do not convincingly show this. Or, maybe the pill does not work. We simply don't know. Therefore, when you do not reject H_0, do not say anything about whether the independent variable influences behavior or not. All that you can say is that you have failed to convincingly demonstrate the predicted relationship in the population.

REMEMBER

Nonsignificant *indicates that we have failed to reject* H_0 *because our results are not in the region of rejection and are thus likely to occur when there is not a relationship in nature.*

Nonsignificant *results provide no convincing evidence—one way or the other—as to whether a relationship exists in nature.*

7.5 SUMMARY OF THE z-TEST

Altogether, the preceding discussion can be summarized as follows. For a one-sample experiment that meets the assumptions of the z-test:

1. **Determine the experimental hypotheses and create the statistical hypothesis.** Predict the relationship the study will or will not demonstrate. Then H_0 describes the μ that the $\overline{X}$ represents if the predicted relationship does not exist. H_a describes the μ that the $\overline{X}$ represents if the relationship does exist.

2. **Set up the sampling distribution.** Select α, locate the region of rejection, and determine the critical value.

3. **Compute the $\overline{X}$ and z_{obt}.** First, compute $\sigma_{\overline{X}}$. Then in the formula for z, the value of μ is the μ of the sampling distribution, which is also the μ of the raw score population that H_0 says is being represented.

4. **Compare z_{obt} to z_{crit}.** If z_{obt} lies beyond z_{crit}, then reject H_0, accept H_a, and say the results are "significant." Then describe the relationship. If z_{obt} does not lie beyond z_{crit}, do not reject H_0, and say the results are "nonsignificant." Do not draw any conclusions about the relationship.

7.6 THE ONE-TAILED TEST

Recall that a *one-tailed test* is used when we predict the *direction* in which scores will change. The statistical hypotheses and sampling distribution are different in a one-tailed test.

The One-Tailed Test for Increasing Scores

Say that we had developed a "smart pill." Then the experimental hypotheses are (1) the pill makes people smarter by increasing IQ, or (2) the pill does not make people smarter. For the statistical hypotheses, start with the alternative hypothesis: People without the pill produce a μ of 100, so if the pill makes them smarter, our sample will represent a population with a μ greater than 100. The symbol for greater than is ">"; therefore, H_a: $\mu > 100$. For the null hypothesis, if the pill does not work as predicted, either it will leave IQ scores unchanged or it will decrease them (making people dumber). Then our sample mean represents a μ either equal to 100 or less than 100. The symbol for less than or equal to is "$\leq$," so H_0: $\mu \leq 100$.

We test H_0 by testing whether the sample represents the raw score population in which μ *equals* 100. This is because our sample mean must be above 100 to conclude that our pill makes people smarter. If we then conclude that our mean is too high to represent a μ equal to 100, then we automatically reject that it represents a μ less than 100.

REMEMBER

A one-tailed null hypothesis always includes that μ equals some value. Test H_0 by testing whether the sample data represent the population with that μ.

Thus, as shown in Figure 7.7, the sampling distribution again shows the means that occur if we are representing a $\mu = 100$ (the situation where the pill does nothing to IQ). We again set $\alpha = .05$, but the region of rejection is in only *one tail* of the sampling distribution. You can identify which tail to put it in by identifying the result you must have to claim that your independent variable works as predicted (to support H_a). To say that the pill makes people smarter, the sample mean must be *significant* and *larger* than 100. Means that are significantly larger than 100 are in a region of rejection in the *upper* tail of the sampling distribution. Therefore, the entire region is in the upper tail of the distribution. Then, as in the previous chapter, the region of rejection is 5% of the curve, so z_{crit} is +1.645.

Say that after testing the pill we find $\overline{X} = 106.58$. The sampling distribution is still based on the population with $\mu = 100$ and $\sigma_X = 15$. Say that $N = 36$, so

$$\sigma_{\overline{X}} = \frac{15}{\sqrt{36}} = 2.5$$

Then $z_{obt} = (106.58 - 100)/2.5 = +2.63$. As in Figure 7.7, this z_{obt} is beyond z_{crit}, so it is in the region of rejection. Therefore, the sample mean is unlikely to represent the population having $\mu = 100$, and it's even less likely to represent a population that has a $\mu < 100$. Therefore, we reject the null hypothesis that $\mu \leq 100$, and accept the alternative hypothesis that $\mu > 100$. We conclude that the pill produces a *significant increase* in IQ scores, and estimate that with the pill, μ would equal about 106.58.

If z_{obt} had not been in the region of rejection, we would retain H_0 and have no evidence as to whether the pill makes people smarter or not.

The One-Tailed Test for Decreasing Scores

Say that, instead, we had created a pill to *lower* IQ scores. If the pill works, then the sample mean represents a μ *less than* 100. The symbol for less than is "<", so H_a: $\mu < 100$. However, if the pill does not work, either it will leave scores unchanged, or it will increase scores. Therefore, the sample mean represents a μ greater than or equal to 100. The symbol for greater than or equal to is "≥," so H_0: $\mu \geq 100$. However, we again test H_0 by testing whether $\mu = 100$.

For us to conclude that the pill lowers IQ, our sample mean must be significantly *less* than 100. Therefore, the region of rejection is in the lower tail of the distribution, as in Figure 7.8. With $\alpha = .05$, z_{crit} is now *minus* 1.645.

Figure 7.7
Sampling Distribution of IQ Means for a One-Tailed Test of Whether Scores Increase
The region of rejection is entirely in the upper tail.

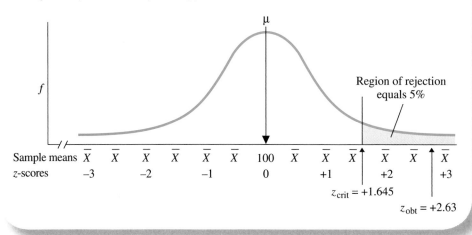

Figure 7.8
Sampling Distribution of IQ Means for a One-Tailed Test of Whether Scores Decrease
The region of rejection is entirely in the lower tail.

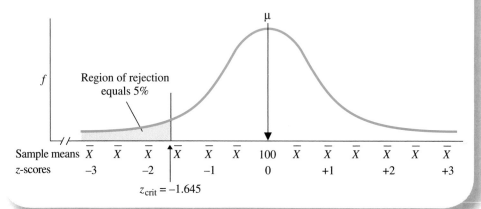

If the sample produces a *negative* z_{obt} beyond -1.645 (for example, $z_{obt} = -1.69$), then we reject the H_0 that the sample mean represents a μ equal to or greater than 100, and accept the H_a that the sample represents a μ less than 100. However, if z_{obt} does not fall in the region of rejection (for example, if $z_{obt} = -1.25$), we do not reject H_0, and we have no evidence as to whether the pill works or not.

7.7 STATISTICS IN THE RESEARCH LITERATURE: REPORTING z

When reading published reports, you'll often see statements such as "the IQ pill produced a *significant difference* in IQ scores," or "it had a *significant effect* on IQ." These are just other ways to say that the results reflect a believable relationship, because they are unlikely to occur through sampling error. However, whether a result is significant depends on the probability used to define "unlikely," so we must also indicate our α. The convention for reporting a result is to indicate the statistic computed, the obtained value, and α. Thus, for a significant z_{obt} of $+2.00$, we report $z = +2.00$, $p < .05$. Notice that instead of indicating that α equals .05, we indicate that the probability (p) is *less than* .05. (We'll discuss the reason for this shortly.) For a nonsignificant z_{obt} of $-.40$, we report $z = -.40$, $p > .05$. Notice that with nonsignificant results, p is *greater* than .05.

7.8 ERRORS IN STATISTICAL DECISION MAKING

There is one other issue to consider when performing hypothesis testing, and it involves potential errors in our decisions. Regardless of whether we conclude that the sample does or does not represent the predicted relationship, we may be wrong.

Type I Errors

Sometimes the variables we investigate are *not* related in nature, and so H_0 is really true. When in this situation, if we obtain data that cause us to reject H_0, then we make an error. A **Type I error** is defined as rejecting H_0 when H_0 is true. This occurs because our sample so poorly represents the situation where the independent variable does not work that we are fooled into concluding that the variable does work. For example, say our pill does not make people smarter, but sampling error produces a mean that falls in the region of rejection. Then we'll mistakenly conclude that the pill does work.

In practice, we never know when we make a Type I error, because only nature knows if the variables are related or not. However, we do know that the theoretical probability of a Type I error equals α, and this is true for a one- or two-tailed test. Our α is the total size of the region of rejection. Think of the sampling distribution as showing the population

> **Type I error**
> Deciding to reject the null hypothesis when the null hypothesis is true (that is, when the predicted relationship does not exist)

of all means we will obtain when H_0 is true if we repeatedly conduct the study over the long run. Our variable never works, but sometimes, because of extreme sampling error, we will obtain a mean that falls into the region of rejection, causing us to make a Type I error. If $\alpha = .05$, then 5% of the time we'll obtain such means, so in the population the relative frequency of Type I errors is .05. Therefore, any time we reject H_0, the theoretical probability that we've made a Type I error is .05.

You either will or will not make the correct decision when H_0 is true. If α is the probability of making a Type I error, then $1 - \alpha$ is the probability of avoiding a Type I error: retaining H_0 when it is true. Thus, if 5% of the time samples are in the region of rejection when H_0 is true, then 95% of the time they are not in the region of rejection when H_0 is true. Therefore, the theoretical probability is .95 that we will retain such an H_0 and thus avoid making a Type I error.

Although the *theoretical* probability of a Type I error equals α, the *actual* probability is slightly less than α. This is because the region of rejection includes the critical value, but to reject H_0, z_{obt} must be *larger* than the critical value. We cannot determine the precise area under the curve for the point located at z_{crit}, so we can't remove it from our 5%. All that we can say is that when α is .05, the region of rejection is slightly less than 5% of the curve. Because the area of the region of rejection is less than α, the probability of a Type I error is also less than α.

In our previous examples when we rejected H_0, the probability that we made a Type I error was less than .05. That is why we reported our significant results using $p < .05$. This is code for "the probability of a Type I error is less than .05." On the other hand, we reported nonsignificant results using $p > .05$. This communicates that we did not call a result significant because to do so would require a region of rejection *greater than* .05 of the curve. But then the probability of a Type I error would be greater than our α of .05 and that's unacceptable.

Sometimes making a Type I error is so dangerous that we want to reduce its probability even further.

REMEMBER

When H_0 is true: Rejecting H_0 is a Type I error, and its probability is α; retaining H_0 is avoiding a Type I error, and its probability is $1 - \alpha$.

Then we usually set alpha at .01, so the probability of making a Type I error is $p < .01$. For example, if the smart pill had some dangerous side effects, we would set $\alpha = .01$ so that we are even less likely to conclude that the pill works when it does not. However, we use the term *significant* in an all-or-nothing fashion: A result is not "more" significant when $\alpha = .01$ than when $\alpha = .05$. If z_{obt} lies *anywhere* in a region of rejection, the result is significant, period! The only difference is that when $\alpha = .01$, the probability that we've made a Type I error is smaller.

Type II Errors and Power

Sometimes the variables we investigate really are related in nature, and so H_0 really is false. When in this situation, if we obtain data that cause us to retain H_0, then we make a Type II error. A **Type II error** is defined as retaining H_0 when H_0 is false (and H_a is true). In practical terms, we fail to identify that an independent variable really does work. This occurs because the sample so poorly represents the situation where the independent variable works that we are fooled into concluding that the variable does not work. For example, say our pill does make people smarter but sampling error produces a mean that does not fall in the region of rejection. Then we'll mistakenly conclude that the pill does not work.

Anytime we discuss Type II errors, it's a given that H_0 is false and H_a is true. However, we never know when we make this error, because we do not know the truth about nature. (We can determine its probability, but the computations for this are beyond the introductory level.) Thus, when we retained H_0 and decided our pill did not work, perhaps we made a Type II error. If we reject H_0 in this situation, then we avoid a Type II error: We've made the correct decision by concluding that the pill works when it does work.

For help in understanding Type I and Type II errors, remember that the type of error you can potentially make is determined by your situation—what nature "says" about whether or not a relationship exists. You can be in only *one* of these situations at a time. Then whether you actually make the error depends on whether you agree or disagree with nature. Thus, four outcomes are possible in any study:

Type I and Type II Errors

When H_0 is true—there is no relationship:

1. Our data cause us to *reject H_0*, so we make a Type I error.
2. Our data cause us to *retain H_0*, so we avoid a Type I error.

When H_0 is false—the relationship exists:

3. Our data cause us to *retain H_0*, so we make a Type II error.
4. Our data cause us to *reject H_0*, so we avoid a Type II error.

Researchers always use a small α, because the primary concern is to minimize the probability of Type I errors: Concluding that an independent variable works when really it does not can cause untold damage. On the other hand, avoiding Type II errors is also important so that we don't miss those independent variables that

do work. This ability has a special name: **Power** is the probability that we will reject H_0 when it is false, correctly concluding that the sample data represent a relationship. In other words, power is the probability of not making a Type II error.

Researchers try to maximize the power of a study by maximizing the chances that the results will be significant. Then we are confident that if the relationship exists in nature, we will not miss it. If we still end up retaining H_0, we are confident that this is because the relationship is not there. Therefore, we are confident in our decision to retain H_0, and, in statistical lingo, we are confident we are avoiding a Type II error. (At the same time, we keep α at .05 or less, so that if we reject H_0 we are confident that we are avoiding a Type I error.)

We have several ways to increase the power of a study. First, it is better to design a study that employs *parametric* procedures, because they are more powerful than *nonparametric* procedures: Because of its theoretical basis, a parametric test is more likely to produce significant results. Second, a one-tailed test is more powerful than a two-tailed test: A z_{obt} is more likely to be beyond the one-tailed z_{crit} of 1.645 than the two-tailed z_{crit} of 1.96, so the one-tailed test is more likely to be significant. Finally, testing a larger N provides greater power: With more scores in a sample (with at least 30), extreme sampling error is less likely, so we are less likely to misrepresent a relationship that is present. (When you study how to design research, you'll learn of other ways to build in power.)

power The probability that we will detect a true relationship and correctly reject a false null hypothesis; the probability of avoiding a Type II error

103

using what you know

1. a. For what purposes are inferential statistical procedures used?
    b. What are the two major types of inferential statistics?

2. a. What two assumptions must be met by the data in order to perform any parametric procedure?
    b. When are nonparametric procedures used?

3. A researcher obtains a sample mean of 82, although the known population μ is 60.
    a. What are the two possible explanations for why the $\overline{X}$ is different from μ?
    b. What must be done to decide which explanation is correct?

4. In problem 3, the population forms a normal distribution and the data are measured on a ratio scale.
    a. Which type of inferential statistic should be used? Why?
    b. If, instead, the data were measured on an ordinal scale, what type of statistic should be used? Why?
    c. If the population is only a roughly normal distribution of ratio scores, what type of statistic should be used?

5. a. What does H_0 describe?
    b. What does H_a describe?
    c. What situation does the H_0 always maintain?

6. a. To which experimental hypothesis does H_0 correspond?
    b. To which experimental hypothesis does H_a correspond?

7. a. What does it mean to predict the direction of a relationship in an experiment?
    b. When do you perform a two-tailed test, and when do you perform a one-tailed test?

8. If we reject H_0, which of the following are true and which are false?
    a. We have proven the independent variable works as predicted.
    b. We have shown H_0 is false.
    c. We have found a sample of N scores is unlikely to produce a particular $\overline{X}$ if the scores are representing a particular population having a certain μ.
    d. We have proven our sample mean represents a μ around a certain value.
    e. We have proven the independent variable causes scores to change.
    f. We have proven the difference between $\overline{X}$ and μ is not due to sampling error.
    g. We have convincing evidence that the predicted relationship exists.

h. The independent variable may not work, and we may have sampling error in representing this.
i. We have nonsignificant results.

9. How should each of the false statements in problem 8 be rephrased so as to be true?

10. If we retain H_0, which of the following are true and which are false?
    a. We have demonstrated that the experiment did not work as predicted.
    b. We have proven that the independent variable does not cause scores to change as predicted.
    c. We are convinced that the independent variable does not work.
    d. We should conclude that there is no relationship in the population.
    e. We have no information about the relationship in the population.
    f. The independent variable may work, but we may have sampling error in representing this.
    g. We have significant results.

11. How should each of the false statements in problem 10 be rephrased so as to be true?

12. a. What is the statistical definition of a Type I error? Why is it important to avoid Type I errors?
    b. What is the statistical definition of a Type II error? Why is it important to avoid Type II errors?

13. What is the meaning of "power"?

14. a. What is the advantage of one-tailed tests over two-tailed tests?
    b. What is the disadvantage?

15. Andy is conducting a study of whether adolescent males who are enrolled in an anger-management class exhibit more aggressive behaviors than other adolescent boys. For the test of aggression Andy has chosen to use, $\mu = 57$ and $\sigma_X = 7$. He collects data on a sample of 25 boys in the anger-management class and obtains $\overline{X} = 60$.
    a. Should Andy do a one-tailed or a two-tailed test? Explain your answer.
    b. State the appropriate H_0 and H_a, given your answer in part a.
    c. Use $\alpha = .05$. What is the value of z_{crit}?
    d. Calculate z_{obt}.
    e. Using symbols, report your findings.
    f. What should Andy conclude?

16. In problem 15:
    a. What is the probability the researcher made a Type I error?
    b. What would the Type I error be in terms of the conclusion about the relationship between the independent and dependent variables?

c. What is the probability the researcher made a Type II error?

d. What would the Type II error be in terms of the conclusion about the relationship between the independent and dependent variables?

17. Denise is interested in whether the physical coordination skills among low-income preschool children are different from those of other children. She knows the population mean for the Preschool Coordination Activity Test (PCAT) is 120 with $\sigma_x = 10$. She tests 80 preschoolers from low-income families and obtains an $\bar{X} = 122$.

a. Should Denise do a one-tailed or a two-tailed test? Explain your answer.

b. State the appropriate H_0 and H_a, given your answer in part a.

c. Use $\alpha = .05$. What is the value of z_{crit}?

d. Calculate z_{obt}.

e. Using symbols, report your findings.

f. What should Denise conclude?

18. A researcher investigates whether children attending day-care centers have a different degree of emotional attachment to their mothers from those not attending day-care centers. On a national test of emotional attachment, children not attending day-care score $\mu = 140$ ($\sigma_x = 12.5$). For a random sample of 20 day-care children, $\bar{X} = 146.07$ ($S_x = 13.3$).

a. What are H_0 and H_a?

b. What is the value of z_{obt}?

c. What is the value of z_{crit}?

d. Report the statistical results in the correct format.

e. In terms of the relationship between the independent and dependent variables, what do you conclude about this study?

f. What do you predict as the average score of any child attending a day-care center?

g. What is the μ for children attending day-care centers?

19. Cherise is working on her Master's thesis and is studying whether seniors age 65 and older take more or fewer prescription drugs when they are treated by a physician specializing in gerontology. From the U.S. Census Bureau, she knows seniors take $\mu = 5$ prescription drugs with $\sigma_x = 3$. She collects data from 20 seniors who are treated by gerontology specialists. The number of prescription drugs each is taking is given below.

4	7	3	0	2	3	5	10	4	6
1	2	0	3	2	4	1	5	2	5

a. Should Cherise do a one-tailed or a two-tailed test? Explain your answer.

b. State the appropriate H_0 and H_a, given your answer in part a.

c. Use $\alpha = .05$. What is the value of z_{crit}?

d. Calculate z_{obt}.

e. Using symbols, report your findings.

f. What should Cherise conclude?

20. Dr. Sharp wants to know if the students in his college have better than average study skills. He knows the norms for the Collegiate Study Skills Test report the $\mu = 80$ and $\sigma_x = 8$. He obtains data on a random sample of 15 students from his college. Using the sample data below and $\alpha = .05$, complete all the steps for a z-test.

78	84	83	75	79	82	80	79
76	84	80	81	77	79	82	

a. Should Dr. Sharp do a one-tailed or a two-tailed test? Explain your answer.

b. State the appropriate H_0 and H_a, given your answer in part a.

c. Use $\alpha = .05$. What is the value of z_{crit}?

d. Calculate z_{obt}.

e. Using symbols, report your findings.

f. What should Dr. Sharp conclude?

21. You hypothesize children will be more relaxed when they are tested with their mothers present. On a standard test, the national average relaxation score of children tested without their mothers present is 44 ($\sigma_x = 6.32$). You test the relaxation of children when their mothers are present. Using the sample data below, complete all the steps for a z-test.

50	52	68	28	66	59	44	36
32	26	42	40	44	42	68	60

a. Should you use a one-tailed or a two-tailed test? Explain your answer.

b. State the appropriate H_0 and H_a, given your answer in part a.

c. Use $\alpha = .05$. What is the value of z_{crit}?

d. Calculate z_{obt}.

e. Using symbols, report your findings.

f. What should you conclude?

Hypothesis Testing
Using the One-Sample *t*-Test

The logic of hypothesis testing discussed in the previous chapter is common to all inferential procedures. The goal now is to learn how different procedures are applied to different research designs. This chapter begins the process by introducing the *t*-test, which is very similar to the previous *z*-test. Therefore, much of this chapter contains more of a variation on a theme than new material. The chapter presents (1) the similarities and differences between the *z*-test and *t*-test, (2) when and how to perform the *t*-test, and (3) a new procedure—called the *confidence interval*—that is used to more precisely estimate μ.

8.1 UNDERSTANDING THE ONE-SAMPLE *t*-TEST

Like the *z*-test, the *t*-test is used for significance testing in a one-sample experiment. In fact, the *t*-test is used slightly more often in behavioral research. This is because the *z*-test required that we *know* the population the standard deviation (σ_X). However, usually researchers do not know such things about the population because they're exploring uncharted areas of behavior. Instead, we must estimate the population variability by using the sample data to compute the *unbiased estimators* (the $N - 1$ formulas) of the population's standard deviation or variance. Then we compute something *like* a *z*-score for our sample mean, but, because the formula is slightly different, it is

Looking Back

- From Chapter 4, know that s_X is the estimated population standard deviation, that s_X^2 is the estimated population variance, and that both involve dividing by $N - 1$.

- From Chapter 7, understand the basics of significance testing.

© Lou Cypher/Fancy/Jupiterimages

called *t*. The **one-sample *t*-test** is the parametric procedure used in a one-sample experiment when the standard deviation of the raw score population is not known.

Here's an example: Say one of those "home-and-gardening/good-housekeeper" magazines has a test of one's housekeeping abilities. The magazine is targeted to women and reports the national average test score for women is 10 (so their μ is 10). It does not report the standard deviation. Our question is, "How do men score on this test?" Therefore, we'll give the test to a sample of men and use their $\overline{X}$ to estimate the μ for the population of all men. Then we can compare the μ for men to the μ of 10 for women. If men score differently from women, then we've found a relationship in which, as gender changes, test scores change.

As usual, we first set up the statistical test.

one-sample t-test The parametric procedure used to test the null hypothesis for a one-sample experiment when the standard deviation of the raw score population must be estimated

1. **The statistical hypotheses:** We're open-minded and look for any kind of difference, so we have a two-tailed test. If men are different from women, then our sample represents a μ for

Use the z-test when σ_X is known; use the t-test when it is not known.

men that will not equal the μ for women of 10, so H_a is $\mu \neq 10$. If men are not different, then their μ will equal that of women, so H_0 is $\mu = 10$.

2. Alpha: We select alpha; .05 sounds good.

3. Check the assumptions: The one-sample t-test requires the following:

   a. You have a one-sample experiment using interval or ratio scores.
   b. The raw score population forms a normal distribution.
   c. The variability of the raw score population is not known and must be estimated from the sample.

Our study meets these assumptions, so we proceed. For simplicity, we test 9 men. (For *power,* you should

never collect so few scores.) Say the sample produces a $\overline{X} = 7.78$. Based on this, we might conclude that the population of men would score around a μ of 7.78. Because females score around a μ of 10, maybe we have demonstrated a relationship between gender and housekeeping scores. On the other hand, as in H_0, maybe gender is not related to test scores and we are being misled by sampling error: Maybe by chance we selected some exceptionally sloppy men for our sample, but in the general population men are no sloppier than women, and so the sample actually represents the one population containing men and women where $\mu = 10$.

To test this null hypothesis, we'll use the same logic we've used previously: H_0 says that the men's mean represents a population where μ is 10. We will compute t_{obt}, which will indicate the location of our sample on the sampling distribution of means when we are sampling from this raw score population. The critical value that marks the region of rejection is t_{crit}. If t_{obt} is beyond t_{crit}, our sample mean lies in the region of rejection, so we'll reject the idea that the sample represents the population where $\mu = 10$.

The only novelty here is that t_{obt} is calculated differently and t_{crit} comes from the t-distribution. Therefore, first we'll see how to compute t_{obt} and then we'll see how to interpret it.

8.2 PERFORMING THE ONE-SAMPLE *t*-TEST

The computation of t_{obt} consists of three steps that parallel the three steps in the z-test. The first step in the z-test was to find the true standard deviation (σ_X) of the raw score population. For the t-test, we can compute the estimated standard deviation (s_X), or, as we'll see, the estimated population variance (s_X^2).

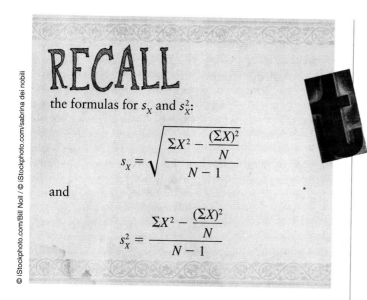

RECALL

the formulas for s_X and s_X^2:

$$s_X = \sqrt{\frac{\sum X^2 - \frac{(\sum X)^2}{N}}{N-1}}$$

and

$$s_X^2 = \frac{\sum X^2 - \frac{(\sum X)^2}{N}}{N-1}$$

The second step of the z-test was to compute the standard error of the mean ($\sigma_{\overline{X}}$), which is the "standard deviation" of the sampling distribution of means. However, because now we estimate the population variability, we compute the **estimated standard error of the mean,** which is an estimate of the "standard deviation" of the sampling distribution of means. The symbol for the estimated standard error of the mean is $s_{\overline{X}}$. (The s stands for an estimate of the population, and the subscript $\overline{X}$ indicates it is a population of means.)

Previously, we computed the standard error using this formula:

$$\sigma_{\overline{X}} = \frac{\sigma_X}{\sqrt{N}}$$

Using the estimated population standard deviation produces this very similar formula:

$$s_{\overline{X}} = \frac{s_X}{\sqrt{N}}$$

Although you may use this formula to compute $s_{\overline{X}}$, a more efficient formula involves the estimated population *variance*.

THE ESTIMATED STANDARD ERROR OF THE MEAN

$$s_{\overline{X}} = \sqrt{\frac{s_X^2}{N}}$$

This formula divides the estimated population variance by the N of our sample and then takes the square root.

The third step in the z-test was to compute z_{obt} using this formula:

$$z_{obt} = \frac{\overline{X} - \mu}{\sigma_{\overline{X}}}$$

The very similar final step in the t-test is to compute t_{obt}:

THE ONE-SAMPLE t-TEST

$$t_{obt} = \frac{\overline{X} - \mu}{s_{\overline{X}}}$$

$\overline{X}$ is the sample mean, μ is the mean of the sampling distribution (which equals the value of μ described in H_0), and $s_{\overline{X}}$ is the estimated standard error of the mean computed above.

For example, imagine our housekeeping study yielded the data in Table 8.1.

STEP 1: *Compute the $\overline{X}$ and the estimated variance (s_X^2) using the sample data.* Here $\overline{X} = 7.78$. The s_X^2 equals

$$s_X^2 = \frac{\sum X^2 - \frac{(\sum X)^2}{N}}{N-1} = \frac{574 - \frac{(70)^2}{9}}{9-1} = 3.695$$

STEP 2: *Compute the estimated standard error of the mean ($s_{\overline{X}}$).*

$$s_{\overline{X}} = \sqrt{\frac{s_X^2}{N}} = \sqrt{\frac{3.695}{9}} = \sqrt{.411} = .64$$

> **estimated standard error of the mean ($s_{\overline{X}}$)** An estimate of the standard deviation of the sampling distribution of means, used in calculating the one-sample t-test

Table 8.1
Housekeeping Scores of Nine Males

Participants	Grades (X)	X^2
1	9	81
2	8	64
3	10	100
4	7	49
5	8	64
6	8	64
7	6	36
8	4	16
9	10	100
$N = 9$	$\sum X = 70$	$\sum X^2 = 574$
	$\overline{X} = 7.78$	

STEP 3: *Compute* t_{obt}.

$$t_{obt} = \frac{\overline{X} - \mu}{s_{\overline{X}}} = \frac{7.78 - 10}{.64} = -3.47$$

Thus, we have computed something similar to a z-score, showing that when measuring how far our men's mean is from the population μ that H_0 says we're representing, our mean has a $t_{obt} = -3.47$.

The *t*-Distribution and *df*

To evaluate a t_{obt} we must compare it to t_{crit}, and for that we examine the t-distribution. Think of the t-distribution in the following way. Once again we infinitely draw samples of the same size N from the raw score population described by H_0. For each sample we compute the $\overline{X}$ and its t_{obt}. Then we plot the frequency distribution of the different means, labeling the X axis with t_{obt} as well. Thus, the **t-distribution** is the distribution of all possible values of t computed for random sample means selected from the raw score population described by H_0.

You can envision the t-distribution as in Figure 8.1. A sample mean equal to μ has a t equal to zero. Means greater than μ have positive values of t, and means less than μ have negative values of t. The larger the absolute value of t, the farther it and the corresponding sample mean are into the tail of the distribution.

But there is one important novelty here: The t-distribution does not fit the perfect standard normal curve (and z-table) the way our previous sampling distributions did. Instead, there are actually *many* versions of the t-distribution, each having a slightly different shape. The shape of a particular distribution depends on the size of the samples that are used when creating it. When using small samples, the t-distribution is only roughly normally distributed. With larger samples, the t-distribution is a progressively closer approximation to the perfect normal curve. This is because with a larger sample, our estimate of the population variance or standard deviation is closer to the true population variance and standard deviation. As we saw, when we have the true population variability, as in the z-test, the sampling distribution *is* a normal curve.

Recall that computing the estimated variance or standard deviation involves dividing by the quantity $N - 1$. It is this number that actually determines the shape of a particular t-distribution. We have a special name for $N - 1$: It is called the **degrees of freedom** and is symbolized as *df*.

> DEGREES OF FREEDOM IN THE ONE-SAMPLE t-TEST
>
> $df = N - 1$,
>
> where N is the number of scores in the sample

In our housekeeping study $N = 9$, so our $df = 8$.

The larger the *df*, the closer the t-distribution comes to forming a normal curve. However, a tremendously

---

t-distribution
The sampling distribution of all possible values of *t* that occur when samples of a particular size are selected from the raw score population described by the null hypothesis

degrees of freedom (df) The number of scores in a sample that reflect the variability in the population; used when estimating the population variability

---

Figure 8.1
Example of a *t*-Distribution of Random Sample Means

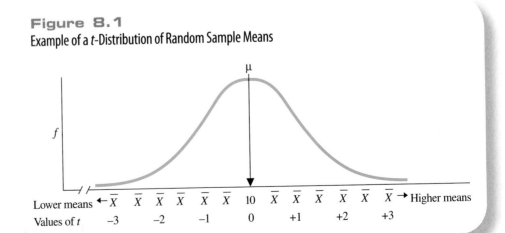

In a one-sample experiment the degrees of freedom—the df—equal N −1 and determine the shape of the t-distribution.

large sample is not required to produce a perfect normal *t*-distribution. When *df* is greater than 120, the *t*-distribution is virtually identical to the standard normal curve, and *t* is the same as *z*. But when *df* is between 1 and 120 (which is often the case in research) a differently shaped *t*-distribution will occur for each *df*.

The fact that there are differently shaped *t*-distributions is important for one reason: Our region of rejection should contain precisely that portion of the curve defined by our α. If α = .05, then we want to mark off precisely 5% of the curve. However, on distributions that are shaped differently, we mark off that 5% at different locations. Because the location of the region of rejection is marked off by the critical value, *with differently shaped t-distributions we will have different critical values.* For example, Figure 8.2 shows two *t*-distributions. Notice the size of the (blue) region of rejection in the tails of Distribution A. Say that this is the extreme 5% of Distribution A and has

the critical value shown. If we also use this t_{crit} on Distribution B, the region of rejection is larger, containing *more* than 5% of the distribution. Conversely, the t_{crit} marking off 5% of Distribution B will mark off *less* than 5% of Distribution A. (The same problem exists for a one-tailed test.)

This issue is important because not only is α the size of the region of rejection, it is also the probability of a Type I error. Unless we use the appropriate t_{crit}, the actual probability of a Type I error will not equal our α and that is not supposed to happen! Thus, there is only one version of the *t*-distribution to use when performing a particular *t*-test: the one that would be created by using the *same df* as in our sample.

Therefore, you no longer automatically have the critical values of 1.96 or 1.645 as in previous chapters. Instead, when your *df* is between 1 and 120, use the *df* to identify the appropriate sampling distribution for your study. The t_{crit} on *that* distribution will accurately mark off the region of rejection so that the probability of a Type I error equals your α. In the housekeeping study with an *N* of 9, we will use the t_{crit} from the *t*-distribution for *df* = 8. In a different study, however, where *N* might be 25, we use the different t_{crit} from the *t*-distribution for *df* = 24. And so on.

REMEMBER: The appropriate t_{crit} for the one-sample t-test comes from the t-distribution that has df equal to N − 1, where N is the number of scores in the sample.

Figure 8.2
Comparison of Two *t*-Distributions Based on Different Sample *N*s

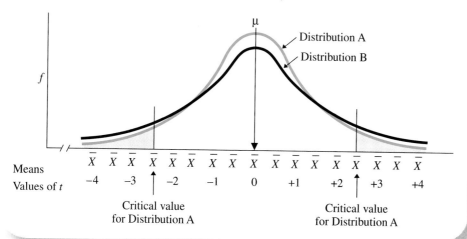

Using the t-Table

We obtain the different values of t_{crit} from Table 2 in Appendix B, entitled "Critical Values of *t*." In this "*t*-table" you'll find separate tables for two-tailed and one-tailed tests. Table 8.2 on the next page contains a portion of the two-tailed table.

To find the appropriate t_{crit}, first locate the appropriate column for your α level (either .05 or .01). Then find the value of t_{crit} in the

Table 8.2
A Portion of the *t*-Table

df	Alpha Level	
	$\alpha = .05$	$\alpha = .01$
1	12.706	63.657
2	4.303	9.925
3	3.182	5.841
4	2.776	4.604
5	2.571	4.032
6	2.447	3.707
7	2.365	3.499
8	2.306	3.355

row at the *df* for your sample. For example, in the housekeeping study, *N* is 9, so *df* is $N - 1 = 8$. For a two-tailed test with $\alpha = .05$ and $df = 8$, t_{crit} is 2.306.

In a different study, say the sample *N* is 61. Therefore, the $df = N - 1 = 60$. Look in Table 2 of Appendix B to find t_{crit}. With $\alpha = .05$, the two-tailed $t_{crit} = 2.000$; the one-tailed $t_{crit} = 1.671$.

The table contains no positive or negative signs. In a two-tailed test you add the "±," and in a one-tailed test you supply the appropriate "+" or "−."

Finally, you will not find a critical value for every *df* between 1 and 120. When the *df* of your sample does not appear in the table, select the two values of t_{crit} for the two *df*s that bracket above and below your *df* (e.g., if $df = 65$, select the critical values at *df*s of 60 and 120). If your t_{obt} is larger than the larger t_{crit}, then your results are significant. If your t_{obt} is smaller than the smaller t_{crit}, then your results are not significant. For the *very* few times your t_{obt} is between the critical values, either use SPSS to perform the *t*-test or consult an advanced book to use "linear interpolation" to compute the precise t_{crit}.

8.3 INTERPRETING THE *t*-TEST

Once you've calculated t_{obt} and identified t_{crit}, you can make a decision about your results. Remember the housekeeping study? We must decide whether or not the men's mean of 7.78 represents the same population of scores that women have. Our t_{obt} was −3.47, and the two-tailed t_{crit} is ± 2.306. Thus, we can envision the sampling distribution in Figure 8.3. Remember, this can be interpreted as the frequency distribution of all means that occur by chance when H_0 is true—here, when men and women belong to the same population with $\mu = 10$. But, our t_{obt} lies beyond t_{crit}, so the results are significant: Our $\overline{X}$ is so unlikely to occur if our men had been representing the same population as found with women, that we reject that we were representing this population—we reject H_0. With $\alpha = .05$, the probability is $p < .05$ that we've made a Type I error (incorrectly rejecting H_0), so we now apply the rules for interpreting significant results as discussed in the previous chapter: Because our $\overline{X}$ is 7.78, we conclude the population of men score at a μ around 7.78, but that women score at a μ of 10. The results demonstrate a relationship in the population between the independent variable (gender) and the dependent variable (housekeeping scores).

If t_{obt} had not fallen beyond t_{crit} (for example, if $t_{obt} = +1.32$), then it would not lie in the region of rejection and would not be significant. Then we would consider whether we had sufficient *power* so that we've avoided

Figure 8.3
Two-Tailed *t*-Distribution for *df* = 8 When H_0 is True and $\mu = 10$

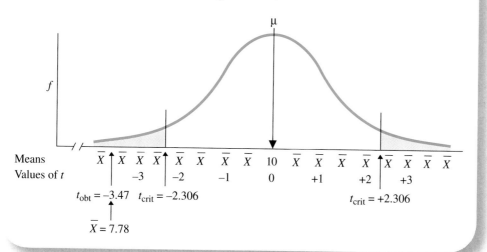

tion with μ greater than 10 (H_a: $\mu > 10$). Our H_0 would be that the sample represents a population with a μ less than or equal to 10 (H_0: $\mu \leq 10$). We compute t_{obt} as shown previously, but we find the one-tailed t_{crit} from the *t*-table for our *df* and α. To decide which tail of the sampling distribution to put the region of rejection in, determine what's needed to support H_a. For our sample to represent a population of *higher* scores, the $\overline{X}$ must be *greater* than 10 and be *significant*. As shown in the left-hand graph in Figure 8.4, such means are in the upper tail, so t_{crit} is positive.

On the other hand, say we had predicted that men score *lower* than women. Now H_a is that μ is less than 10, and H_0 is that μ is greater than or equal to 10. Because we seek a $\overline{X}$ that is *significant* and *lower* than 10, the t_{crit} is negative, so we'd have the sampling distribution on the right in Figure 8.4.

In either example, if the absolute value of t_{obt} is larger than t_{crit} and has the same sign, then the $\overline{X}$ is unlikely to be representing a μ described by H_0. Therefore, reject H_0, accept H_a, and the results are significant. If t_{obt} is not beyond t_{crit} and not in our region of rejection, then the results are not significant.

making a Type II error (incorrectly retaining H_0). And, we would apply the rules for interpreting nonsignificant results as discussed in the previous chapter, concluding that we have no evidence—one way or the other—regarding a relationship between gender and housekeeping scores.

Performing One-Tailed Tests

As usual, we perform one-tailed tests when we predict the direction of the relationship. If we had predicted that men score *higher* than women (who have a μ of 10), then H_a would be that the sample represents a popula-

Figure 8.4
H_0 Sampling Distribution of *t* for a One-Tailed Test

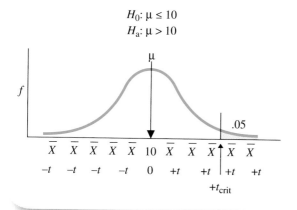
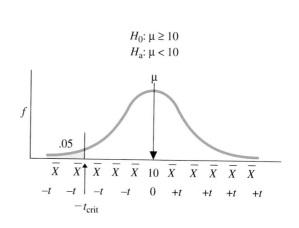

8.4 SUMMARY OF THE ONE-SAMPLE *t*-TEST

7.78

The one-sample *t*-test is used with a one-sample experiment involving normally distributed interval or ratio scores when the variability in the population is not known. Then

1. *Create either the two-tailed or one-tailed H_0 and H_a.*

2. *Compute* t_{obt}.

   a. Compute $\overline{X}$ and s_X^2.
   b. Compute $s_{\overline{X}}$.
   c. Compute t_{obt}.

3. *Envision the sampling t-distribution and use* df = N − 1 *to find* t_{crit} *in the t-table.*

4. *Compare* t_{obt} *to* t_{crit}. *If* t_{obt} *is beyond* t_{crit}, *the results are significant; describe the populations and interpret the relationship. If* t_{obt} *is not beyond* t_{crit}, *the results are not significant; make no conclusion about the relationship.*

point estimation A way to estimate a population parameter by describing a point on the variable at which the population parameter is expected to fall

interval estimation A way to estimate a population parameter by describing an interval within which the population parameter is expected to fall

confidence interval for μ A range of values of μ, within which we are confident that the actual μ is found

8.5 ESTIMATING μ BY COMPUTING A CONFIDENCE INTERVAL

As you've seen, after rejecting H_0, we estimate the population μ that the sample mean represents. There are two ways to estimate μ.

The first way is **point estimation,** in which we describe a point on the dependent variable at which the population μ is expected to fall. We base this estimate on our sample mean. Earlier, for example, we estimated that the μ of the men's population is located on the variable of housekeeping scores at the *point* equal to our men's sample mean of 7.78. However, the problem with point estimation is that it is extremely vulnerable to sampling error. Our sample of men probably does not perfectly represent the population of men, so if we actually tested the entire population, μ probably would not be exactly 7.78. This is why we have been saying only that the μ for men is probably *around* 7.78.

The other, better way to estimate a μ is to include the possibility of sampling error and perform interval estimation. With **interval estimation,** we specify a range of values within which we expect the population parameter to fall. You may encounter such intervals in real life, and they are usually phrased in terms of "plus or minus" some amount (called the *margin of error*). For example, when it is reported that a sample showed that 45% of voters support the President, the margin of error may be ±3%. This indicates the pollsters have created an interval around 45%, so that if they could ask the *entire* population, the result would be within ±3% of 45%: They believe that the actual percentage of the population that supports the President is a number between 42% and 48%.

Researchers perform interval estimation in a similar way by creating a confidence interval. The **confidence interval for** μ describes an interval within which we are confident that our population's μ falls. If we say that our sample of men represents a μ somewhere *around* 7.78, a confidence interval defines "around." To do so, we'll identify those values of μ that our sample mean is likely to represent, as shown here:

$$\mu_{low} \ldots \mu \;\; \mu \;\; \mu \;\; \mu \;\; 7.78 \;\; \mu \;\; \mu \;\; \mu \;\; \mu \ldots \mu_{high}$$

$\underbrace{\hspace{8cm}}$

values of μ, one of which is likely to be
represented by our sample mean

The μ_{low} is the lowest μ that our sample mean is likely to represent, and μ_{high} is the highest μ that the mean is likely to represent. When we compute these two values, we have the confidence interval because we are confident that the μ being represented by our sample falls between them.

When is a sample mean likely to represent a particular μ? It depends on sampling error. For example, intuitively we know that sampling error is unlikely to produce a sample mean of 7.78 if μ is, say, 500. In other words, 7.78 is significantly different from 500. But sampling error *is* likely to produce a sample mean of 7.78 if, for example, μ is 8 or 9. In other words, 7.78 is not significantly different from these μs. Thus, a sample mean is likely to represent any μ that the mean is *not* significantly different from. The logic behind a confidence interval is to compute the highest and lowest values of μ that are not significantly different from our sample mean. All μs between these two values are also not significantly different from the sample mean, so the mean is likely to represent one of them. In other words, the μ being represented by our sample mean is likely to fall within this interval.

We usually compute a confidence interval only after finding a significant t_{obt}. This is because we must be sure that our sample is not representing the μ described in H_0 before we estimate any other μ it might represent. Thus, we determined that our men represent a μ different from that of women, and so now we can go on to describe that μ. In fact, it is appropriate to compute a confidence interval in any study in which we believe a $\overline{X}$ represents a particular population μ.

Computing the Confidence Interval

The *t*-test forms the basis for the confidence interval, and it works like this. We seek the highest μ above our sample mean that is not significantly different from our mean and the lowest μ below our sample mean that is not significantly different from our mean. The most that a μ and sample mean can differ and still not be significant is when the values of μ

sample mean →

and $\overline{X}$ produce a t_{obt} that *equals* t_{crit}. We can state this using the formula for the *t*-test:

$$t_{crit} = \frac{\overline{X} - \mu}{s_{\overline{X}}}$$

To find the largest and smallest values of μ that do not differ significantly from our sample mean, we simply determine the values of μ that we can put into this formula along with our sample data. Because we are describing values above and below the sample mean, we use the two-tailed value of t_{crit}. Then we find the μ that produces a $-t_{obt}$ equal to $-t_{crit}$, and the μ that produces a $+t_{obt}$ equal to $+t_{crit}$. The μ represented by our sample is *between* these two μs, so we rearrange and combine these formulas to produce this:

THE CONFIDENCE INTERVAL FOR μ

$$(s_{\overline{X}})(-t_{crit}) + \overline{X} \leq \mu \leq (s_{\overline{X}})(+t_{crit}) + \overline{X}$$

The μ in the formula stands for our μ that we are describing. The components to the left of μ will produce μ_{low}, so we are confident our μ is greater than or equal to this value. The components to the right of μ will produce μ_{high}, so we are confident that our μ is less than or equal to this μ.

In the formula, the $\overline{X}$ and $s_{\overline{X}}$ are from your data. Find the *two-tailed* value of t_{crit} in the *t*-table at your α for $df = N - 1$, where N is the sample N.

STEP 1: *Find the two-tailed* t_{crit} *and fill in the formula*. For our housekeeping study, the two-tailed t_{crit} for $\alpha = .05$ and $df = 8$ is ± 2.306. The $\overline{X} = 7.78$ and $s_{\overline{X}} = .64$. Filling in the formula, we have $(.64)(-2.306) + 7.78 \leq \mu \leq (.64)(+2.306) + 7.78$.

REMEMBER

Use the two-tailed critical value when computing a confidence interval even if you have performed a one-tailed t-test.

STEP 2:

Multiply each t_{crit} *times* $s_{\overline{X}}$. After multiplying .64 times -2.306 and $+2.306$, we have $-1.476 + 7.78 \leq \mu \leq +1.476 + 7.78$. At this point we expect the men's mean is a μ of 7.78, plus or minus 1.476.

STEP 3:

Add the above positive and negative answers to the $\overline{X}$. After adding ± 1.476 to 7.78, we have $6.30 \leq \mu \leq 9.26$.

So, if we could measure the population of men in this situation, we expect that their μ would be a number between 6.30 and 9.26. Further, in the center of this interval is 7.78, so we are still saying that the men's μ is *around* 7.78, but now we have defined "around" as essentially plus or minus 1.476.

Note: An alternative formula for computing the interval is $\mu = \overline{X} \pm (s_{\overline{X}} \, t_{crit})$. Multiply the *positive* value of t_{crit} times $s_{\overline{X}}$ and then determine plus or minus this amount from your sample mean. You'll have the μ_{high} and μ_{low}.

Because we created our interval using the t_{crit} for an α of .05, there is a 5% chance that our μ is outside of this interval. On the other hand, there is a 95% chance, $(1 - \alpha)(100)$, that the μ is *within* the interval. Therefore, we have created what is called the *95% confidence interval:* We are 95% confident that the interval between 6.30 and 9.26 contains our μ. Usually this gives us sufficient confidence. However, had we used the t_{crit} for $\alpha = .01$, the interval would have spanned a wider range, giving us even more confidence that the interval contained the μ. Then we would have created the 99% confidence interval.

8.6 STATISTICS IN THE RESEARCH LITERATURE: REPORTING *t*

Report the results of a one- or two-tailed *t*-test in the same way that you reported the *z*-test, but also include the *df*. We would report our significant results in the original housekeeping study as

$$t(8) = -3.47, p < .05$$

The *df* is in parentheses. (Had these results not been significant, then $p > .05$.)

Note that when α is .01, t_{crit} is ± 3.355, so our t_{obt} of -3.47 also would be significant if we had used the .01 level. Therefore, instead of saying $p < .05$ above, we provide more information by reporting that $p < .01$ because we now know that the probability of a Type I error is not in the neighborhood of .04, .03, or .02. Researchers usually report the smallest values of alpha at which a result is significant.

Further, computer programs can determine the precise, minimum size of the region of rejection that our t_{obt} falls into, so researchers often report the exact probability of a Type I error. For example, you might see "$p = .04$." This indicates that t_{obt} falls into a region of rejection that is .04 of the curve, and therefore the probability of a Type I error here equals .04.

Usually, confidence intervals are reported in sentence form, but we always indicate the confidence level used. So you might see "The 95% confidence interval for μ was 45.50 to 53.72."

REMEMBER

Compute a confidence interval to estimate the μ *represented by the* $\overline{X}$ *of a condition in an experiment.*

USING SPSS

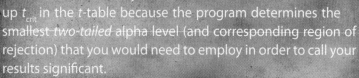

Check out your Review Cards for Chapter 8 for instructions on how to use SPSS to perform the entire one-sample *t*-test: The program computes t_{obt}, as well as automatically computing the $\overline{X}$ and s_x for the sample, the $s_{\overline{x}}$, and the 95% confidence interval. Also, you need not look up t_{crit} in the *t*-table because the program determines the smallest *two-tailed* alpha level (and corresponding region of rejection) that you would need to employ in order to call your results significant.

1. What determines whether we use the z-test or the t-test?

2. What are the requirements of the one-sample t-test?

3. a. What is the t-distribution?

    b. How is the shape of the t-distribution affected by df?

    c. What are the implications of this for selecting t_{crit}?

4. Explain when a one-tailed t-test should be used versus a two-tailed t-test.

5. a. Why do we use $N - 1$ in the t-test formula instead of N?

    b. What is $N - 1$ called?

6. a. If our results on a t-test are not significant, what should we next consider?

    b. What type of error are we in danger of committing when the results are not significant?

7. a. Describe the two ways in which we can estimate the value of μ.

    b. Which is better, and why?

    c. What information is provided by the 95% confidence interval?

8. For a research project, suppose you want to compare the mean from a sample of 21 students against a known population mean of 100. You obtain a $\overline{X} = 105$ for the 21 students. Although you do not know σ_X^2, you have calculated $s_X^2 = 400$.

    a. What are H_0 and H_a?

    b. What is your t_{obt}?

    c. What are the appropriate df?

    d. Using $\alpha = .05$ and a two-tailed test, what is t_{crit}?

    e. Comparing t_{obt} to t_{crit}, what do you conclude?

9. Justin wants to know whether a commonly prescribed prescription drug improves the attention span of students with Attention Deficit Disorder (ADD). He knows the mean attention span for students with ADD who are not taking the drug is 2.3 minutes. His sample of 12 students taking the drug yielded a $\overline{X}$ of 3.2. Justin can find no information regarding σ_X^2, so he has calculated $s_X^2 = 1.96$.

    a. What are H_0 and H_a?

    b. What is t_{obt}?

    c. What are the appropriate df?

    d. Using $\alpha = .05$, what is t_{crit}?

    e. What conclusions should Justin draw?

    f. If appropriate, compute the confidence interval for μ.

10. A researcher investigates whether daily exercise alters long-term stress levels. On a national survey, the mean stress level of those who do not exercise is 56.35. For a sample of 10 participants who exercised daily for two weeks, the researcher obtained the following stress scores:

40	59	48	36	44	43	45	45	32	37

a. What is the condition in the population? What is the condition in the sample? What is the dependent variable?

b. What are H_0 and H_a?

c. What is the value of t_{obt}?

d. With $\alpha = .05$, what is the value of t_{crit}?

e. Report the statistical results.

f. Estimate the value of μ represented by $\overline{X}$.

g. Using the preceding statistics, what conclusions should the researcher draw about the relationship between these variables?

11. A researcher investigates whether cigarette smoking by pregnant women results in babies with lower birth weights. In the population, the mean birth weight of children born to nonsmoking women is 116 ounces. In a random sample of 18 babies born to women who smoked heavily throughout their pregnancies, the mean birth weight was 113.6 ounces ($s_X^2 = 8.00$).

    a. What is the condition in the population? What is the condition in the sample? What is the dependent variable?

    b. What are H_0 and H_a?

    c. What is the value of t_{obt}?

    d. With $\alpha = .05$, what is the value of t_{crit}?

    e. What are the statistical results?

    f. If appropriate, compute the confidence interval for μ.

    g. Using the preceding statistics, what conclusions should the researcher draw about the relationship between these variables?

12. A researcher studies whether 15 minutes of meditation each day alters subjects' ability to concentrate. On a psychological test of concentration, $\mu = 17.3$ for nonmeditators. On the same test, the meditators obtained the following scores:

15	18	22	17	18	18	19	21	19	16	22	20

a. What is the condition in the population? What is the condition in the sample? What is the dependent variable?

b. Should the researcher use a one- or a two-tailed test? Why?

c. What are H_0 and H_a?

d. What is the value of t_{obt}?

e. With $\alpha = .05$, what is t_{crit}?

f. Report the statistical results.

g. If appropriate, compute the confidence interval for μ.

h. Using the preceding statistics, summarize the conclusions drawn from this study.

13. Repeat problem 12 using $\alpha = .01$.

    a. What is t_{obt}?

    b. What is t_{crit}?

c. Report the statistical results.

d. If appropriate, compute the confidence interval for μ.

e. If these results are different from what you found in problem 12, explain the difference.

14. A scientist predicts engineers are *more* creative than physicians. On a standard creativity test, the μ for physicians is 40.5. For a sample of 25 engineers, $\overline{X} = 42.1$ ($s_X^2 = 15.52$) using the same test.

   a. What is the condition in the population? What is the condition in the sample? What is the dependent variable?

   b. What are H_0 and H_a?

   c. What are the values of t_{obt} and t_{crit}?

   d. With $\alpha = .05$, what are the statistical results?

   e. What conclusions can the researcher draw from this study?

15. Repeat problem 14 using a two-tailed test.

   a. What are H_0 and H_a?

   b. What are the values of t_{obt} and t_{crit}?

   c. With $\alpha = .05$, what are the statistical results?

   d. If these results are different from what you found in problem 14, explain the difference.

16. Ms. Shoemacher has conducted a study of 10 adolescent girls. Her sample mean was $\overline{X} = 29$ with $s_X^2 = 20$. If she knows that the population mean under another condition of the independent variable is $\mu = 32$, complete the following steps to see if she has a significant difference. Use $\alpha = .05$ and a two-tailed test.

   a. What are H_0 and H_a?

   b. What is the value of t_{obt}?

   c. With $\alpha = .05$, what is t_{crit}?

   d. Report the statistical results.

17. Repeat problem 16, but change the number of participants to 31.

   a. What are H_0 and H_a?

   b. What is the value of t_{obt}?

   c. With $\alpha = .05$, what is t_{crit}?

   d. Report the statistical results.

   e. If these results are different from what you found in problem 16, explain the difference.

18. Using the following data set, conduct a two-tailed one-sample t-test for $\mu = 90$.

83	78	91	88	87	90	75	89	93	77
86	87	78	84	82	92	88	90	86	89

   a. What are H_0 and H_a?

   b. What is the value of t_{obt}?

   c. What is the value of df?

   d. With $\alpha = .01$, what is t_{crit}?

   e. What should you conclude?

" It's easy to read, it outlines important topics, and it's relevant. Thanks for the good stuff on the website, I think it will **really help with tests**.

– Thomas Scholtes, Student at University of Maryland, College Park

REVIEW HE DID

STAT puts a multitude of study aids at your fingertips. After reading the chapters, check out these resources for further help:

- **Chapter Review Cards**, found in the back of your book, include definitions, review questions, equations, topic summaries, and SPSS instructions for each chapter.

- **Printable flash cards** give you three additional ways to check your comprehension of key statistic concepts.

- **Interactive Online Problems** give you instant feedback and provide walk-thru instructions to guide you through challenging concepts.

Other great ways to study include quizzing, additional downloadable practice problems, and crossword puzzle games.

You can access it all by logging in to the CourseMate for STAT at **www.cengagebrain.com.**

Hypothesis Testing
Using the Two-Sample *t*-Test

This chapter presents the two-sample *t*-test, which is the major parametric procedure used when an experiment involves *two* samples. As the name implies, this test is similar to the one-sample *t*-test you saw in Chapter 8, except that a two-sample design requires that we use slightly different formulas. This chapter discusses (1) one version of the two-sample *t*-test called the *independent-samples* t-*test*, (2) the other version of two-sample *t*-test called the *related-samples* t-*test*, and (3) a new approach for describing a relationship that indicates *effect size*.

9.1 UNDERSTANDING THE TWO-SAMPLE EXPERIMENT

The one-sample experiment discussed in previous chapters requires that we know the value of μ for a population under one condition of the independent variable. However, because we explore new behaviors and variables, we usually do not know μ ahead of time. Instead, the simplest alternative is to conduct a two-sample experiment, measuring participants' scores under two conditions of the independent variable. Condition 1 produces one sample mean—call it $\overline{X}_1$—that represents μ_1, the μ we would find if we tested everyone in the population under Condition 1. Condition 2 produces another sample mean—call it $\overline{X}_2$—that

Looking Back

- From Chapter 1, know what a condition, independent variable, and dependent variable are.
- From Chapter 8, know how to perform the one-sample *t*-test using the *t*-distribution and *df*.

© Nicholas Rigg/Photographer's Choice RF/Getty Images

represents μ_2, the μ we would find if we tested everyone in the population under Condition 2. A possible outcome from such an experiment is shown in Figure 9.1 on the next page. If each condition represents a different population, then the experiment has demonstrated a relationship in nature.

However, there's the usual problem of sampling error. Even though we may have different sample means, the independent variable may not really change scores. Instead, we might find the *same* population of scores under both conditions, but the scores in one or the other of our conditions poorly represent this population. Thus, in Figure 9.1 we might find only the lower or upper distribution, or we might find one in between. Therefore, before we make any conclusions about the experiment, we must determine whether the difference between the sample means reflects sampling error.

The parametric statistical procedure for determining whether the results of a two-sample experiment are significant is the two-sample *t*-test. However, we have two different ways to create

independent-samples *t*-test The parametric procedure used for significance testing of sample means from two independent samples

independent samples Samples created by selecting each participant for one condition, without regard to the participants selected for any other condition

Figure 9.1

Relationship in the Population in a Two-Sample Experiment
As the conditions change, the population tends to change in a consistent fashion.

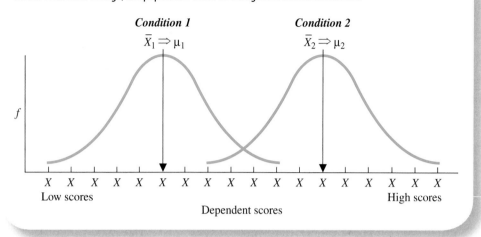

the samples, so we have two different versions of the *t*-test: One is called the *independent-samples* t-*test*, and the other is the *related-samples* t-*test*.

REMEMBER: *The two ways to calculate the two-sample* t-*test are the* independent-samples *t*-test *and the* related-samples *t*-test.

9.2 THE INDEPENDENT-SAMPLES *t*-TEST

The **independent-samples *t*-test** is the parametric procedure for testing two sample means from **independent samples.** Two samples are independent when we randomly select participants for a condition, without regard to who else has been selected for either condition. Then the scores in one sample are not influenced by—are "independent" of—the scores in the other sample. You can recognize independent samples by the absence of things such as matching the participants in one condition with those

in the other or repeatedly testing the same participants in both conditions.

Here is a study that calls for the independent-samples *t*-test. We propose that people may recall an event differently when they are hypnotized. To test this, we'll have two groups watch a videotape of a supposed robbery. Later, one group will be hypnotized and then answer 30 questions about the event. The other group will answer the questions without being hypnotized. Thus, the conditions of the independent variable are the presence or absence of hypnosis, and the dependent variable is the amount of information correctly recalled. This design is shown in Table 9.1. We will compute the mean of each condition (in each column). If the means

Table 9.1

Diagram of Hypnosis Study Using an Independent-Samples Design
The independent variable is amount of hypnosis, and the dependent variable is recall.

	No Hypnosis	Hypnosis
Recall Scores →	X	X
	X	X
	X	X
	X	X
	X	X
	$\bar{X}$	$\bar{X}$

differ, we'll have evidence of a relationship where, as amount of hypnosis changes, recall scores also change.

First we check that our study meets the assumptions of the statistical test. In addition to requiring independent samples, this *t*-test has two other requirements: (1) the dependent scores are normally distributed, interval or ratio scores, and (2) the populations have homogeneous variance. **Homogeneity of variance** means that the variances (σ_X^2) of the populations being represented are equal.

Determine if your data meet these assumptions by seeing how your variables are treated in the research literature.

Note: You are not required to have the same number of participants in each condition.

homogeneity of variance A characteristic of data describing populations represented by samples in a study that have the same variance

Statistical Hypotheses for the Independent-Samples *t*-Test

Depending on our experimental hypotheses, we may perform either a one- or two-tailed test. Let's begin with a two-tailed test: We predict that hypnosis will simply produce different recall scores compared to those in the no-hypnosis condition, so our samples represent different populations that have different μs.

First, the alternative hypothesis: The predicted relationship exists if one population mean (μ_1) is larger or smaller than the other (μ_2); that is, μ_1 should not equal μ_2. We could state this as H_a: $\mu_1 \neq \mu_2$, but there is a better way. If the two μs are not equal, then their *difference* does not equal zero. Thus, the two-tailed alternative hypothesis is

$$H_a: \mu_1 - \mu_2 \neq 0$$

H_a implies that the means from our conditions each represent a different population of recall scores, so a relationship is present.

Now, the null hypothesis: If no relationship exists, then if we tested everyone under the two conditions, we would find the same population and μ. In other words, μ_1 *equals* μ_2. We could state this as H_0: $\mu_1 = \mu_2$, but, again, there is a better way. If the two μs are equal, then their *difference* is zero. Thus, the two-tailed null hypothesis is

$$H_0: \mu_1 - \mu_2 = 0$$

H_0 implies that both sample means represent the same population of recall scores, which have the same μ, so no relationship is present. If our sample means differ, H_0 maintains that this is due to sampling error in representing that one μ.

Notice that the above hypotheses do not contain a specific value of μ. Therefore, they are the two-tailed hypotheses for *any* independent variable. However, this is true only when you test whether the data might actually represent *zero difference* between the populations. This is the most common approach and the one we will use. (You can also test for nonzero differences: You might know of an existing difference between two populations and test if the independent variable alters that difference. Consult an advanced statistics book for the details of this test.)

As usual, we test the null hypothesis, and to do that we examine the sampling distribution.

The Sampling Distribution for the Independent-Samples *t*-Test

To understand the sampling distribution, let's say that we find a mean recall score of 20 in the no-hypnosis condition and a mean of 23 in the hypnosis condition. We summarize these results using the *difference* between our means. Here, changing from no hypnosis to hypnosis results in a difference in mean recall of 3 points. We always test H_0 by finding the probability of obtaining our results when no relationship is present, so here we will determine the probability of obtaining a difference of 3 between our $\overline{X}$s when they actually represent the same μ.

Think of the sampling distribution as being created in the following way. We select *two* random samples from *one* raw score population, compute the means, and arbitrarily subtract one from the other. The result is the *difference between the means*, symbolized by $\overline{X}_1 - \overline{X}_2$. We do this an infinite number of times and

REMEMBER

The independent-samples t-test determines the probability of obtaining our difference between $\bar{X}$s when H_0 is true.

plot a frequency distribution of these differences. We then have the **sampling distribution of differences between means.** This is the distribution of all possible differences between two means when both samples come from the raw score population described by H_0. You can envision this sampling distribution as in Figure 9.2.

The mean of the sampling distribution is zero because, most often, both sample means will equal the μ of the raw score population, so their difference will be zero. Sometimes, however, both sample means will not equal μ or each other. Depending on whether $\bar{X}_1$ or $\bar{X}_2$ is larger, the difference will be positive or negative. Small negative or positive differences occur frequently when H_0 is true, but larger ones do not.

To test H_0, we compute a new version of t_{obt} to determine where on this sampling distribution the difference between our sample means lies. As in Figure 9.2, the larger the value of $\pm t_{obt}$, the farther into the tail of the distribution our difference lies.

Computing the Independent-Samples t-Test

In the previous chapter you computed t_{obt} by performing three steps: (1) estimating the variability of the raw score population, (2) computing the estimated standard error of the sampling distribution, and (3) computing t_{obt}. You perform similar steps in the two-sample t-test.

STEP 1: *Compute the mean and estimated population variance in each condition.* First decide which condition will be Condition 1 and which will be Condition 2. Then using the scores in Condition 1, compute $\bar{X}_1$; using the scores in Condition 2, compute $\bar{X}_2$.

Next compute an estimated population *variance* for each condition. We will use the symbol s_1^2 for the estimated variance in Condition 1 and s_2^2 for the estimated variance in Condition 2. Also, we must change our use of the symbol N. Previously, N was the number of scores in our sample, but actually N is the number of scores in the *study*. With only one sample, N was also the number of scores in the sample. However, now we have two samples, so a lowercase n with a subscript stands for the number of scores in each sample. Thus, n_1 is the number of scores in Condition 1, and n_2 is the number of scores in Condition 2. N is the total number of scores in the experiment, so adding the ns together equals N.

sampling distribution of differences between means A frequency distribution showing all possible differences between two means that occur when two independent samples of a particular size are drawn from the population of scores described by the null hypothesis

Figure 9.2

Sampling Distribution of Differences between Means When H_0: $\mu_1 - \mu_2 = 0$

The X axis has two labels: Each $\bar{X}_1 - \bar{X}_2$ symbolizes the difference between two sample means; when labeled as t, a larger $\pm t$ indicates a larger difference between means that is less likely when H_0 is true.

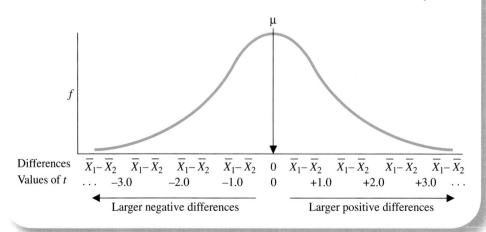

REMEMBER

N stands for the total number of scores in an experiment; n stands for the number of scores in a condition.

Using n, the generic formula for calculating the variance in a condition is

$$s_X^2 = \frac{\sum X^2 - \frac{(\sum X)^2}{n}}{n - 1}$$

Use this formula to compute s_1^2 and s_2^2. Each time, use the Xs from only one condition, and n is the number of scores in that condition.

STEP 2:

Compute the pooled variance. Both s_1^2 and s_2^2 estimate the population variance, but each may contain sampling error. Therefore, to obtain the best estimate, we compute an average of the two. Each variance is "weighted" based on the size of its sample. This weighted average is called the **pooled variance,** and its symbol is s_{pool}^2.

THE POOLED VARIANCE

$$s_{pool}^2 = \frac{(n_1 - 1)s_1^2 + (n_2 - 1)s_2^2}{(n_1 - 1) + (n_2 - 1)}$$

This says to determine $n_1 - 1$ and multiply this by the s_1^2 that you computed above. Likewise, find $n_2 - 1$ and multiply it by s_2^2. Add the results together and divide by the sum of $(n_1 - 1) + (n_2 - 1)$.

For example, say that the hypnosis study produced the results shown in Table 9.2. The hypnosis condition produces $\overline{X}_1$, s_1^2, and n_1. The no-hypnosis condition produces $\overline{X}_2$,

pooled variance (s_{pool}^2) The weighted average of the sample variances in a two-sample experiment

standard error of the difference ($s_{\overline{X}_1 - \overline{X}_2}$) The estimated standard deviation of the sampling distribution of differences between the means of independent samples in a two-sample experiment

s_2^2, and n_2. Filling in the formula, we have

$$s_{pool}^2 = \frac{(17 - 1)9.0 + (15 - 1)7.5}{(17 - 1) + (15 - 1)}$$

After subtracting, we have

$$s_{pool}^2 = \frac{(16)9.0 + (14)7.5}{16 + 14}$$

In the numerator, 16 times 9 is 144, and 14 times 7.5 is 105. In the denominator, 16 plus 14 is 30, so

$$s_{pool}^2 = \frac{144 + 105}{30} = \frac{249}{30} = 8.3$$

The s_{pool}^2 is our estimate of the variability in the raw score population that H_0 says we are representing. As in previous procedures, once we know how spread out the underlying raw score population is, we can determine how spread out the sampling distribution is by computing the standard error.

STEP 3:

Compute the standard error of the difference. The **standard error of the difference** is the "standard deviation" of the sampling distribution of differences between means (of the distribution back in Figure 9.2). The symbol for the standard error of the difference is $s_{\overline{X}_1 - \overline{X}_2}$.

Table 9.2
Data from the Hypnosis Study

	Condition 1: Hypnosis	Condition 2: No Hypnosis
Mean recall score	$\overline{X}_1 = 23$	$\overline{X}_2 = 20$
Number of participants	$n_1 = 17$	$n_2 = 15$
Estimated variance	$s_1^2 = 9.0$	$s_2^2 = 7.5$

In previous chapters we computed the standard error by dividing the variance by N and then taking the square root. However, instead of dividing by N we can multiply by $1/N$. Then for the two-sample t-test, we substitute the pooled variance and our two ns, producing this formula:

THE STANDARD ERROR
OF THE DIFFERENCE

$$s_{\overline{X}_1 - \overline{X}_2} = \sqrt{(s_{pool}^2)\left(\frac{1}{n_1} + \frac{1}{n_2}\right)}$$

To compute $s_{\overline{X}_1 - \overline{X}_2}$, first reduce the fractions $1/n_1$ and $1/n_2$ to decimals. Then add them together. Then multiply the sum times s_{pool}^2, which you computed in Step 2. Then find the square root.

For the hypnosis study, s_{pool}^2 is 8.3, n_1 is 17, and n_2 is 15. Filling in the formula gives

$$s_{\overline{X}_1 - \overline{X}_2} = \sqrt{(8.3)\left(\frac{1}{17} + \frac{1}{15}\right)}$$

First, 1/17 is .059 and 1/15 is .067, so

$$s_{\overline{X}_1 - \overline{X}_2} = \sqrt{8.3(.059 + .067)}$$

After adding,

$$s_{\overline{X}_1 - \overline{X}_2} = \sqrt{8.3(.126)} = \sqrt{1.046} = 1.023$$

STEP 4: *Compute* t_{obt}. In previous chapters we've calculated how far the result of our study ($\overline{X}$) was from the mean of the H_0 sampling distribution (μ), measured in standard error units. Now the "result of our study" is the *difference* between our two sample means, which we symbolize as ($\overline{X}_1 - \overline{X}_2$). The mean of the H_0 sampling distribution is the *difference* between the μs described by H_0 symbolized by ($\mu_1 - \mu_2$). Finally, our standard error is $s_{\overline{X}_1 - \overline{X}_2}$. All together, we have

THE INDEPENDENT-SAMPLES t-TEST

$$t_{obt} = \frac{(\overline{X}_1 - \overline{X}_2) - (\mu_1 - \mu_2)}{s_{\overline{X}_1 - \overline{X}_2}}$$

Here, $\overline{X}_1$ and $\overline{X}_2$ are the sample means, $s_{\overline{X}_1 - \overline{X}_2}$ is computed as in Step 3, and the value of $\mu_1 - \mu_2$ is specified by the null hypothesis. We write H_0 as $\mu_1 - \mu_2 = 0$ so that it indicates the value of $\mu_1 - \mu_2$ to put into this formula. Then the formula measures how far our difference between $\overline{X}$s is from the zero difference between the μs that H_0 says we are representing, measured in standard error units.

For the hypnosis study, our sample means were 23 and 20, the difference between μ_1 and μ_2 is 0, and $s_{\overline{X}_1 - \overline{X}_2}$ is 1.023. Putting these values into the formula gives

$$t_{obt} = \frac{(23 - 20) - 0}{1.023}$$

After subtracting the means

$$t_{obt} = \frac{(+3) - 0}{1.023} = \frac{+3}{1.023} = +2.93$$

Our t_{obt} is +2.93. Thus, the difference between our sample means is located at something like a z-score of +2.93 on the sampling distribution of differences when H_0 is true and both samples represent the same population.

Interpreting the Independent-Samples *t*-Test

To determine if t_{obt} is significant we compare it to t_{crit}, which is found in the t-table (Table 2 in Appendix B). We obtain t_{crit} using degrees of freedom, but with two samples, the df are computed differently.

DEGREES OF FREEDOM IN THE
INDEPENDENT-SAMPLES t-TEST

$df = (n_1 - 1) + (n_2 - 1)$, where each n is the number of scores in a condition.

For the hypnosis study, n_1 = 17 and n_2 = 15, so df equals $(17 - 1) + (15 - 1)$, which is 30. With alpha at .05, the two-tailed t_{crit} is ±2.042.

The complete sampling distribution is in Figure 9.3. It shows all differences between sample means that occur through sampling error when the samples represent no difference in the population (when hypnosis does not influence recall). Our H_0 says that the difference of $+3$ between our sample means is merely a poor representation of no difference. But, the sampling distribution shows that a difference between means of $+3$ hardly ever occurs when the samples represent no difference. Therefore, it is difficult to believe that *our* difference of $+3$ represents no difference. In fact, our t_{obt} lies beyond t_{crit}, so the results are significant: Our difference of $+3$ is so unlikely to occur if our samples were representing no difference in the population that we reject that this is what they represent.

Thus, we reject H_0 and accept H_a that our data represent a difference between μs that is not zero. To be precise, from the sample means we expect that if we tested the entire population in this experiment, we would find the μ for no hypnosis to be around 20 and the μ for hypnosis to be around 23. Therefore, we conclude that we have demonstrated a relationship such that increasing the amount of hypnosis leads to significantly higher recall scores. (As usual, with $\alpha = .05$, the probability of a Type I error is $p < .05$.)

To more precisely describe these μs, we could compute a *confidence interval* for each μ, using the formula in the previous chapter and the data from one condition at a time. Alternatively, we can compute the *confidence interval for the difference between two μs*. It will describe a range of possible differences between μs, within which we are confident that the actual difference between our μs falls. The computations for this confidence interval are presented in Appendix A.2.

Figure 9.3

H_0 Sampling Distribution of Differences between Means When $\mu_1 - \mu_2 = 0$
The t_{obt} shows the location of a difference of $+3.0$.

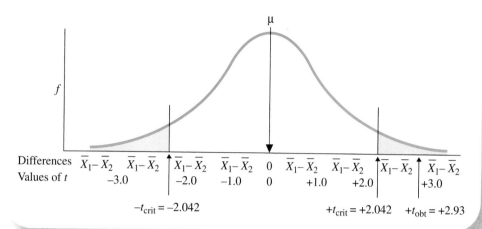

| Differences | $\overline{X}_1 - \overline{X}_2$ | $\overline{X}_1 - \overline{X}_2$ | $\overline{X}_1 - \overline{X}_2$ | $\overline{X}_1 - \overline{X}_2$ | 0 | $\overline{X}_1 - \overline{X}_2$ | $\overline{X}_1 - \overline{X}_2$ | $\overline{X}_1 - \overline{X}_2$ | $\overline{X}_1 - \overline{X}_2$ |
| Values of t | -3.0 | | -2.0 | -1.0 | 0 | $+1.0$ | $+2.0$ | | $+3.0$ |

$-t_{crit} = -2.042$ $\qquad$ $+t_{crit} = +2.042$ $\quad$ $+t_{obt} = +2.93$

If t_{obt} was not beyond t_{crit} we would not reject H_0 and we would make no conclusions about the relationship between hypnosis and recall. Then we'd be concerned with whether we had sufficient *power* so that we were unlikely to make a Type II error.

Performing One-Tailed Tests on Independent Samples

We could have conducted the hypnosis study using a one-tailed test if, for example, we predicted a relationship such that hypnosis results in *higher* recall scores than no hypnosis. Everything discussed above applies here, but to prevent confusion, use more meaningful subscripts than 1 and 2. For example, use the subscript h for hypnosis and n for no-hypnosis. Then follow these steps:

1. Decide which $\overline{X}$ and corresponding μ is expected to be larger. (We think the μ for hypnosis is larger.)

2. Arbitrarily decide which condition to subtract from the other. (We'll subtract no-hypnosis *from* hypnosis.)

3. Decide whether the difference will be positive or negative. (Subtracting the smaller μ_n from the larger μ_h should produce a positive difference, *greater* than zero.)

4. Create H_a and H_0 to match this prediction. (Our H_a is that $\mu_h - \mu_n > 0$; H_0 is that $\mu_h - \mu_n \leq 0$.)

5. Locate the region of rejection based on your predictions and subtraction. (We expect a positive difference that is in the right-hand tail of the sampling distribution, so t_{crit} is positive.)

6. Complete the *t*-test as above. Subtract the $\overline{X}$s in the same way the μs are subtracted! (We used $\mu_h - \mu_n$, so we'd compute $\overline{X}_h - \overline{X}_n$.)

Confusion arises because, while *still predicting a larger* μ_h, we could have reversed H_a, saying $\mu_n - \mu_h < 0$. Subtracting the larger μ_h from the smaller μ_n should produce a difference less than zero, so now the region of rejection is in the left-hand tail, and t_{crit} is negative.

9.3 SUMMARY OF THE INDEPENDENT-SAMPLES *t*-TEST

A fter checking that the study meets the assumptions, the independent-samples *t*-test involves the following.

1. *Create either the two-tailed or one-tailed* H_0 *and* H_a.

2. *Compute* t_{obt} *by following these four steps.*

   a. Compute $\overline{X}_1$, s_1^2, and n_1; $\overline{X}_2$, s_2^2, and n_2.

   b. Compute the pooled variance (s_{pool}^2).

   c. Compute the standard error of the difference ($s_{\overline{X}_1 - \overline{X}_2}$).

   d. Compute t_{obt}.

3. *Using* df $= (n_1 - 1) + (n_2 - 1)$, *find* t_{crit} *in the t-table.*

4. *Compare* t_{obt} *to* t_{crit}. If t_{obt} is beyond t_{crit}, the results are significant; describe the relationship. If t_{obt} is not beyond t_{crit}, the results are not significant; make no conclusion about the relationship.

9.4 THE RELATED-SAMPLES *t*-TEST

N ow we will discuss the other version of the two-sample *t*-test. The **related-samples *t*-test** is the parametric procedure used with two related samples. **Related samples** occur when we pair each score in one sample with a particular score in the other sample. Researchers create related samples to have more equivalent and thus comparable samples. The two types of research designs that produce related samples are *matched-samples designs* and *repeated-measures designs*.

In a **matched-samples design,** the researcher matches each participant in one condition with a particular participant in the other condition. For example, say we wish to match participants on the variable of their height. We would select pairs of people who are the same height and assign a member of the pair to each condition. Thus, if two people are six feet tall, one is assigned to one condition and one to the other condition. Likewise, a four-foot person in one condition is matched with a four-footer in the other condition, and so on. Then, overall the conditions are comparable in height so we'd proceed with the experiment. In the same way, we might match participants using their age or physical ability, or we might use naturally occurring pairs, such as roommates or identical twins.

The other way to create related samples is with a **repeated-measures design,** in which each participant is tested under all conditions of the independent variable. That is, first participants are tested under Condition 1 and then the *same* participants are tested under Condition 2. Although we have one sample of participants, we have two samples of scores.

Matched-groups and repeated-measures designs are analyzed in the same way. (Such related samples are also called *dependent samples.*) Except for requiring related samples, the assumptions of the related-samples *t*-test are the same as those for the independent-samples *t*-test: (1) The dependent variable involves normally distributed, interval or ratio scores and (2) the populations being represented have homogeneous variance. Because related samples form pairs of scores, the *n*s in the two samples must be equal.

The Logic of the Related-Samples *t*-Test

Let's say that we have a new therapy to test on arachnophobes—people who are overly frightened by spiders. From the local phobia club we randomly select the unpowerful *N* of five arachnophobes and test our therapy using repeated measures of two conditions: before therapy and after therapy. Before therapy we measure each person's fear response to a picture of a spider, measuring heart rate, perspiration, etc., and compute a "fear" score between 0 (no fear) and 20 (holy terror!). After giving each participant the therapy, we again measure the person's fear response to the picture. (A before-and-after, or *pretest/posttest*, design such as this always uses the related-samples *t*-test.)

In Table 9.3, the two left-hand columns show the fear scores from the two conditions. First, we compute the mean of each condition. Before therapy the mean fear score is 14.80, and after therapy the mean is 11.20. It looks as if therapy reduces fear scores by an average of $14.80 - 11.20 = 3.6$ points. But, conversely, maybe therapy does nothing; maybe this difference reflects sampling

Table 9.3
Finding the Difference Scores in the Phobia Study
Each D = Before − After

	Before Therapy	−	After Therapy	=	D	D^2
1 (Maude)	11	−	8	=	+3	9
2 (Alonzo)	16	−	11	=	+5	25
3 (Millie)	20	−	15	=	+5	25
4 (Althea)	17	−	11	=	+6	36
5 (Ezra)	10	−	11	=	−1	1
	$\bar{X} = 14.80$		$\bar{X} = 11.20$		$\Sigma D = +18$	$\Sigma D^2 = 96$
$N = 5$					$\bar{D} = +3.6$	

error from the one population of fear scores we'd have with or without therapy. To test these hypotheses, we first transform the data. Then we perform the *t*-test using the transformed scores.

As in the right side of Table 9.3, we transform the data by first finding the difference between the two fear scores for each participant. Thus, we subtract Maude's after score (8) from her before score (11) for a difference of +3; we subtract Alonzo's after score (11) from his before score (16) for a difference of +5; and so on. Notice that the symbol for *difference score* is symbolized by *D*. Here we arbitrarily subtracted after-therapy from before-therapy. You could subtract in the opposite way, but subtract all scores in the same way. If this were a matched-samples design, we'd subtract the scores in each pair of matched participants.

Next, compute the *mean difference,* symbolized as $\bar{D}$. Add the positive and negative differences to find the sum of the differences, symbolized by ΣD. Then divide by *N*, the number of difference scores. In Table 9.3, $\bar{D} = 18/5 = +3.6$: The before scores were, on average, 3.6 points higher than the after scores. (As in the far right-hand column of Table 9.3, later we'll need to square each difference and then find the sum, finding ΣD^2.)

Finally, here's a surprise: Because now we have one sample mean from one sample of scores, we perform the one-sample *t*-test! The fact that we have difference scores is irrelevant, so we create the statistical hypotheses and test them in virtually the same way that we did with the one-sample *t*-test in the previous chapter.

cal fluctuations, so their Ds will not equal 0. For example, H_0 says that the therapy did not reduce Maude's fear, but because of random factors maybe she happened to be less fearful at our second measurement, so it *looked* as if the therapy reduced her fear. Essentially H_0 argues that we have sampling error in measuring her true fear score each time, so depending on the luck of the draw, we obtained one higher and one lower score.

Thus, in the same way that we all have good days and bad days, H_0 says that everyone in the population may, by chance, have had a higher or lower fear score on one or the other measurement. This will produce Ds that are sometimes positive numbers and sometimes negative numbers, so that we have a population of different Ds, as shown on the top of Figure 9.4.

Because chance produces the positive and negative Ds, over the long run in the population they will cancel

REMEMBER: *The* related-samples *t*-test *is performed by applying the one-sample* t-*test to the difference scores.*

Statistical Hypotheses for the Related-Samples *t*-Test

Our sample of difference scores represents the population of difference scores that would result if we measured everyone in the population before therapy and again after therapy, and then computed their difference scores. This population of difference scores has a μ that we identify as μ_D. To create the statistical hypotheses, we determine the predicted values of μ_D in H_0 and H_a.

Let's first perform a two-tailed test (although realistically the therapy should only lower scores). The H_0 always says that our independent variable does not work as predicted, so here it is as if we had provided no therapy and simply measured everyone's fear score twice. Ideally, everyone should have the same score on both occasions so, after subtracting, everyone in the population should have a D of zero. However, this will not happen. Participants will not exhibit identical fear scores on the two occasions because of random physiological and psychologi-

Figure 9.4

Population of Difference Scores Described by H_0 and the Resulting Sampling Distribution of Mean Differences

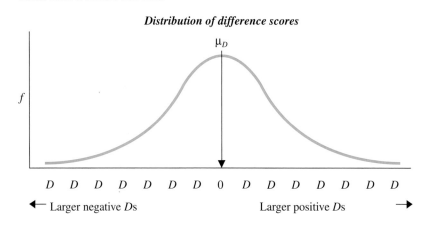

Distribution of difference scores

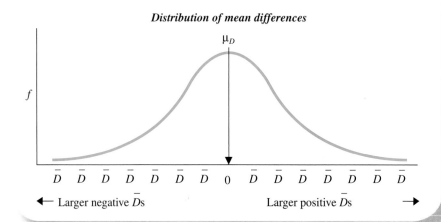

Distribution of mean differences

out to produce a $\mu_D = 0$. Therefore, $H_0: \mu_D = 0$. This says that our sample with a $\overline{D}$ of $+3.6$ is, because of sampling error, poorly representing the population of Ds where μ_D is 0. (Likewise, in a matched-pairs design, each pair of individuals would not perform identically without the independent variable, so H_0 would still say we have a population of Ds with a $\mu_D = 0$.)

For the alternative hypothesis, if the therapy alters fear scores in the population, then either the before scores or the after scores will be consistently higher. Then, after subtracting them, the population of Ds will tend to contain only positive or only negative scores. Therefore, μ_D will be a positive or negative number and not zero. So, $H_a: \mu_D \neq 0$.

We test H_0 by examining the *sampling distribution of mean differences*. Shown on the bottom of Figure 9.4, it is as if we infinitely sampled the population of Ds on the top of the figure that H_0 says our sample represents. The sampling distribution shows the different values of $\overline{D}$ that occur through sampling error when H_0 is true. For the phobia study, it essentially shows all values of $\overline{D}$ we might get by chance when the therapy does not work. The $\overline{D}$s that are farther into the tails of the distribution are less likely to occur if H_0 was true and the therapy did not work. We test H_0 by locating our $\overline{D}$ on this sampling distribution by computing t_{obt}.

Notice that the hypotheses $H_0: \mu_D = 0$ and $H_a: \mu_D \neq 0$ and the above sampling distribution are for the two-tailed test for *any* dependent variable when you test whether there is *zero difference* between your conditions. This is the most common approach and the one we'll discuss. (You can also test whether you've altered a nonzero difference. Consult an advanced statistics book for the details.)

Computing the Related-Samples *t*-Test

Computing t_{obt} here is identical to computing the one-sample *t*-test discussed in Chapter 8—only the symbols have been changed from X to D. There, we first computed the estimated population variance, then the standard error of the mean, and then t_{obt}. We perform the same three steps here.

STEP **1:** Find s_D^2, which is the estimated variance of the population of difference scores shown on the top in Figure 9.4. Replacing the Xs in our old formula for variance with Ds gives

$$s_D^2 = \frac{\Sigma D^2 - \dfrac{(\Sigma D)^2}{N}}{N-1}$$

Note: For all computations in this *t*-test, N equals the number of *difference* scores.

Using the phobia data from Table 9.3, we have

$$s_D^2 = \frac{\Sigma D^2 - \dfrac{(\Sigma D)^2}{N}}{N-1}$$

$$= \frac{96 - \dfrac{(18)^2}{5}}{4} = 7.8$$

> **standard error of the mean difference ($s_{\overline{D}}$)** The standard deviation of the sampling distribution of mean differences between related samples in a two-sample experiment

STEP **2:** Compute the **standard error of the mean difference.** This is the "standard deviation" of the sampling distribution of $\overline{D}$ (shown on the bottom of Figure 9.4). Its symbol is $s_{\overline{D}}$.

THE STANDARD ERROR OF THE MEAN DIFFERENCE

$$s_{\overline{D}} = \sqrt{\frac{s_D^2}{N}}$$

For the phobia study, $s_D^2 = 7.8$, and $N = 5$, so

$$s_{\overline{D}} = \sqrt{\frac{s_D^2}{N}} = \sqrt{\frac{7.8}{5}} = \sqrt{1.56} = 1.249$$

STEP **3:** Find t_{obt}.

THE RELATED-SAMPLES *t*-TEST

$$t_{obt} = \frac{\overline{D} - \mu_D}{s_{\overline{D}}}$$

In the formula, $\overline{D}$ is the mean of your difference scores, $s_{\overline{D}}$ is computed as above, and μ is the value given in H_0. (It is always 0 unless you are testing for a nonzero difference.) Then t_{obt} is again like a *z*-score, indicating how far our $\overline{D}$ is from the μ_D of the sampling distribution when measured in standard error units.

For the phobia study, $\overline{D}$ is $+3.6$, $s_{\overline{D}}$ is 1.249, and μ_D equals 0, so

$$t_{obt} = \frac{\overline{D} - \mu_D}{s_{\overline{D}}} = \frac{+3.6 - 0}{1.249} = +2.88$$

So, $t_{obt} = +2.88$.

Figure 9.5
Two-Tailed Sampling Distribution of $\overline{D}$s When $\mu_D = 0$

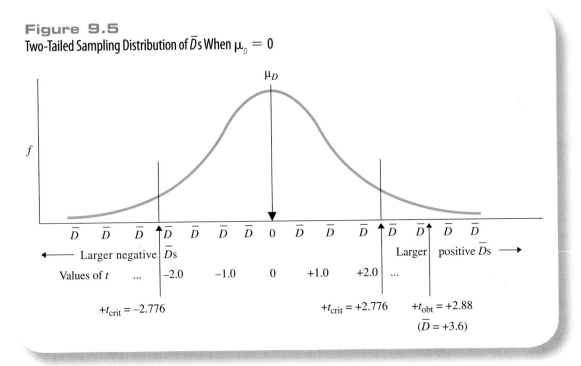

μ_D

f

$\overline{D}$ $\overline{D}$ $\overline{D}$ $\overline{D}$ $\overline{D}$ $\overline{D}$ $\overline{D}$ 0 $\overline{D}$ $\overline{D}$ $\overline{D}$ $\overline{D}$ $\overline{D}$ $\overline{D}$ $\overline{D}$

← Larger negative $\overline{D}$s   Larger positive $\overline{D}$s →

Values of t ...   −2.0   −1.0   0   +1.0   +2.0   ...

$+t_{crit} = -2.776$ $+t_{crit} = +2.776$ $+t_{obt} = +2.88$
                                                       $(\overline{D} = +3.6)$

Interpreting the Related-Samples *t*-Test

Interpret t_{obt} by comparing it to t_{crit} from the *t*-table in Appendix B.

> THE DEGREES OF FREEDOM IN THE RELATED-SAMPLES *t*-TEST
>
> $df = N - 1$, where N is the number of difference scores.

For the phobia study, with $\alpha = .05$ and $df = 4$, the t_{crit} is ±2.776. The complete sampling distribution is shown in Figure 9.5. It shows the distribution of all $\overline{D}$s that occur when we are representing a population of Ds where μ_D is 0 (when our therapy does not influence fear).

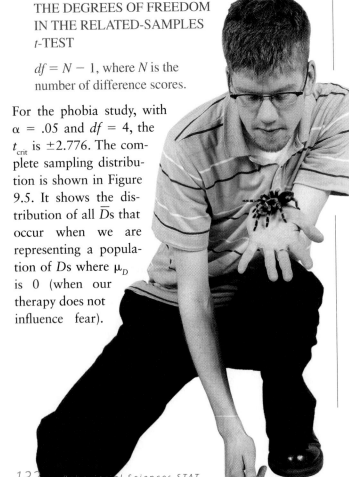

But, we see that a $\overline{D}$ like ours hardly ever occurs when the sample represents this population. In fact, our t_{obt} lies beyond t_{crit}, so we conclude that our $\overline{D}$ of +3.6 is unlikely to represent the population of Ds where $\mu_D = 0$. Therefore, the results are significant: We reject H_0 and accept H_a, concluding that the sample represents a μ_D around +3.6.

Now we work backwards to our original raw fear scores. Whenever a sample of Ds are significant, then the two samples of raw scores that produced the Ds also differ significantly. As usual, this means that if we could test everyone in the population in our experiment, we'd expect to find one population of raw scores for one condition and a *different* population of scores for the other condition. Thus, we conclude that our before scores differ significantly from our after scores. Based on the mean fear scores in our samples, we would expect a population of before-therapy scores at a μ around 14.80 and a population of after-therapy scores at a μ around 11.20. We conclude that the sample data represent a relationship in the population such that fear scores go from around 14.80 before therapy to around 11.20 after therapy. And, if we subtracted everyone's after score from their before score, the average difference score in the population (the μ_D) would be around +3.6. (As usual, with $\alpha = .05$, the probability of a Type I error is $p < .05$.)

Note: It is incorrect to compute a confidence interval for the μ of each raw score population we've estimated above. That procedure is appropriate only for *independent* samples. Instead, we can compute the *confidence interval for* μ_D. It describes a range of possible values of μ_D, within which we are confident the μ_D represented by our $\overline{D}$ falls. The computations for this confidence interval are presented in Appendix A.2.

If t_{obt} had not been beyond t_{crit} the results would not be significant and we would make no conclusions about whether our therapy influences fear scores (and we'd again consider our power).

One-Tailed Hypotheses with the Related-Samples *t*-Test

The phobia study would have involved a one-tailed test if, for example, we had predicted that fear scores would be lower after therapy. For H_0 and H_a, first decide which condition to subtract from which. We subtracted *after* from *before*, so lower after-therapy scores would produce positive differences. Then $\overline{D}$ should be positive, representing a positive, μ_D. Therefore, $H_a: \mu_D > 0$. Then $H_0: \mu_D \leq 0$.

We again examine the sampling distribution that occurs when $\mu_D = 0$. Obtain the one-tailed t_{crit} from the *t*-table. We predict a positive $\overline{D}$, so the region of rejection is in the upper tail and t_{crit} is positive.

Had we predicted *higher* after scores, then by subtracting before from after, the Ds and their $\overline{D}$ should be negative, representing a negative μ_D. Thus, $H_a: \mu_D < 0$, and $H_0: \mu_D \geq 0$. Now the region of rejection is in the lower tail and t_{crit} is negative.

For either test, compute t_{obt} using the previous formula. Be sure you subtract to get your Ds in the same way as when you created your hypotheses.

9.5 SUMMARY OF THE RELATED-SAMPLES *t*-TEST

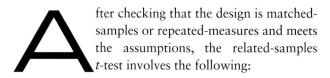

fter checking that the design is matched-samples or repeated-measures and meets the assumptions, the related-samples *t*-test involves the following:

1. *Create either the two-tailed or one-tailed* H_0 *and* H_a.

2. *Compute* t_{obt}.

    a. Compute the *difference score* for each pair of scores.

    b. Compute $\overline{D}$ and s_D^2.

    c. Compute $s_{\overline{D}}$.

    d. Compute t_{obt}.

3. *Create the sampling distribution and, using* df = N − 1, *find* t_{crit} *in the* t-table.

4. *Compare* t_{obt} *to* t_{crit}. *If* t_{obt} *is beyond* t_{crit}, the results are significant; describe the populations of raw scores and interpret the relationship. If t_{obt} is not beyond t_{crit}, the results are not significant; make no conclusion about the relationship.

9.6 STATISTICS IN THE RESEARCH LITERATURE: REPORTING A TWO-SAMPLE STUDY

eport the results of an independent- or related-samples *t*-test using the same format used previously. For example, in our hypnosis study, the t_{obt} of +2.93 was significant with 30 *df*, so we report $t(30) = +2.93$, $p < .05$. As usual, *df* is in parentheses, and because $\alpha = .05$, the probability is less than .05 that we've made a Type I error.

In addition, as in Chapter 3, we report the mean and standard deviation from each condition. Also, with two or more conditions, researchers often include graphs of their results. Recall that we graph the results of an experiment by plotting the mean of each condition on the *Y* axis and the conditions of the independent variable on the *X* axis.

Note: In a related-samples study, report the means and standard deviations of the original raw scores—not the Ds—and graph these means.

An important statistic for describing a significant relationship is called a *measure of effect size*. The "effect" is the influence that the independent variable had on dependent scores. **Effect size** indicates the amount of influence that changing the conditions of the independent variable had on dependent scores. Thus, for example, the extent to which changing the amount of hypnosis produced differences in recall scores is the effect size of hypnosis.

We want to identify those variables that most influence a behavior, so **the larger the effect size, the more scientifically important the independent variable is.** Remember that *significant* does not mean important, but only that the sample relationship is unlikely to reflect sampling error. Although a relationship must be significant to be potentially important, it can be significant and still be unimportant. Therefore, you should always compute a measure of effect size for any significant result. In fact, the American Psychological Association requires published research to report effect size.

We will discuss two methods for measuring effect size. The first is to compute Cohen's *d*.

Effect Size Using Cohen's *d*

effect size An indicator of the amount of influence that changing the conditions of the independent variable had on dependent scores

Cohen's *d* A measure of effect size in a two-sample experiment that reflects the magnitude of the differences between the means of the conditions, relative to the variability of the scores

One approach for describing effect size is in terms of how *big* the difference is between the means of the conditions. For example, the presence/absence of hypnosis produced a difference between the means of 3. To interpret this, however, we need a frame of reference, and so we also consider the estimated population standard deviation. Recall that a standard deviation reflects the "average" amount that scores differ. Thus, if individual scores differ by an "average" of 30, then large differences between scores frequently occur, so a difference of 3 between their means is not all that impressive. However, if scores differ by an "average" of only 5, then a difference between their means of 3 is more impressive.

Cohen's *d* measures effect size by describing the magnitude of the difference between the means relative to the population standard deviation. We have two versions of how it is computed, depending on which two-sample *t*-test we have performed.

The formulas for Cohen's *d* are:

INDEPENDENT-SAMPLES *t*-TEST	RELATED-SAMPLES *t*-TEST
$d = \dfrac{\overline{X}_1 - \overline{X}_2}{\sqrt{s^2_{pool}}}$	$d = \dfrac{\overline{D}}{\sqrt{s^2_D}}$

For the independent samples *t*-test, the difference between the conditions is measured as $\overline{X}_1 - \overline{X}_2$, and the standard deviation comes from the square root of the pooled variance. For our hypnosis study, the means were 23 and 20, and s^2_{pool} was 8.3, so

$$d = \frac{\overline{X}_1 - \overline{X}_2}{\sqrt{s^2_{pool}}} = \frac{23 - 20}{\sqrt{8.3}} = \frac{+3}{2.88} = 1.04$$

This tells us that the effect of changing our conditions was to change scores by an amount that is slightly larger than one standard deviation.

For the related samples *t*-test, the difference between the conditions is measured by $\overline{D}$ and the standard deviation comes from finding the square root of the estimated variance (s^2_D.) In our phobia study, $\overline{D} = +3.6$ and $s^2_D = 7.8$, so

$$d = \frac{\overline{D}}{\sqrt{s^2_D}} = \frac{+3.6}{\sqrt{7.8}} = \frac{+3.6}{2.79} = 1.29$$

Thus, the effect size of the therapy was 1.29.

REMEMBER

The larger the effect size, the greater the influence that an independent variable has on dependent scores and thus the more important the variable is.

We can interpret the above ds in two ways. First, the larger the *absolute* size of d, the larger the impact of the independent variable. In fact, Cohen[1] proposed the following guidelines.

Values of d	Interpretation of Effect Size
$d = .2$	small effect
$d = .5$	medium effect
$d = .8$	large effect

Thus, in our previous examples we found two *very* large effects.

Second, we can compare the relative size of different ds to determine the relative impact of different independent variables. In our previous examples, the d for hypnosis was 1.04, but for therapy it was 1.29. Therefore, in the respective studies, the therapy manipulation had a slightly larger impact on the dependent variable.

Another way to measure effect size is by computing the *proportion of variance accounted for*.

Effect Size Using Proportion of Variance Accounted for

The other approach for measuring effect size is not in terms of the *size* of the changes in scores, but in terms of how *consistently* the scores change. Here, a variable has a greater impact the more that everyone behaves in the same way, producing virtually the same score in a condition. A variable is more minor, however, when it exhibits less control of a behavior and scores.

For example, here are some possible fear scores from our phobia study.

Before Therapy	After Therapy
10	5
11	6
12	7

The after scores are each reduced by 5 points. These are differences attributable to changing the conditions of our independent variable. However, we also see differences *within* each condition: In the before scores, for example, one participant had a 10 while someone else had an 11. These differences are not attributable to our

[1]Cohen, J. (1988) *Statistical power analysis for the behavioral sciences*. Hillsdale, NJ: Lawrence Erlbaum Associates.

independent variable. Thus, out of all the differences among these 6 scores, some differences seem to be produced by changing our independent variable while others are not. In an experiment, the **proportion of variance accounted for** is the proportion of all differences in dependent scores that can be attributed to changing our conditions. The larger the proportion, the more that differences in scores seem to be consistently caused by changing the independent variable, so the larger is the variable's *effect* in determining scores.

However, adjust your expectations about what is "large." Any behavior is influenced by many variables, so one variable by itself will have a modest effect. Therefore, here is a rough guide of what to expect in real research: An effect size less than .09 is considered small; between about .10 and .25 is considered moderate and is relatively common, and above .25 is considered large and is rare.

In the next chapter, we will see that the proportion of variance accounted for depends on the consistency of the relationship. There we will also discuss the statistic for summarizing a relationship called the *correlation coefficient*. We compute the proportion of variance accounted for by squaring the appropriate correlation coefficient. In the two-sample experiment we compute the **squared point-biserial correlation coefficient**, and its symbol is r_{pb}^2.

COMPUTING EFFECT SIZE IN A TWO-SAMPLE EXPERIMENT

$$r_{pb}^2 = \frac{(t_{obt})^2}{(t_{obt})^2 + df}$$

This can be used with either the independent-samples or related-samples t-test. In the numerator, square t_{obt}. In the denominator, add $(t_{obt})^2$ to the df from the study. For independent samples $df = (n_1 - 1) + (n_2 - 1)$, but for related samples $df = N - 1$.

In our hypnosis study, $t_{obt} = +2.93$ with $df = 30$. So

$$r_{pb}^2 = \frac{(t_{obt})^2}{(t_{obt})^2 + df} = \frac{(2.93)^2}{(2.93)^2 + 30} = \frac{8.585}{38.585} = .22$$

proportion of variance accounted for The proportion of the differences in scores that is associated with changes in the X variable

squared point-biserial correlation coefficient (r_{pb}^2) A statistic for describing a relationship that when squared, indicates the proportion of variance in dependent scores that is accounted for by the independent variable in a two-sample experiment

Thus, .22 or 22% of the differences in our recall scores are accounted for by changing our hypnosis conditions. You can understand why this is a moderate effect by considering that if 22% of the differences in scores are attributable to changing our conditions, then 78% of the differences are due to something else (perhaps motivation or the participant's IQ played a role). Therefore, hypnosis is only one of a number of variables that influence memory here, and thus, it is only somewhat important in determining scores.

On the other hand, in the phobia study $t_{obt} = +2.88$ and $df = 4$, so

$$r^2_{pb} = \frac{(t_{obt})^2}{(t_{obt})^2 + df} = \frac{(2.88)^2}{(2.88)^2 + 4} = .68$$

This indicates that 68% of all differences in our fear scores are associated with before- or after-therapy. Therefore, our therapy had an extremely large effect size and is an important variable in determining fear scores.

Further, we use effect size to compare the importance of different independent variables: The therapy accounts for 68% of the variance in fear scores, but hypnosis accounts for only 22% of the variance in recall scores. Therefore, in their respective relationships, the therapy variable had a much larger effect and is more important.

REMEMBER

Effect size *as measured by the* proportion of variance accounted for *indicates how important a role an independent variable plays in determining dependent scores.*

using what you know

1. a. What are the two types of two-sample t-tests?
   b. What question are both types used to answer?
2. What is the difference between independent samples and related samples?
3. What are the assumptions of the two-sample t-test?
4. What is meant by *homogeneity of variance*?
5. a. What information does the standard error of the difference provide?
   b. What information does the standard error of the mean difference provide?
6. a. What is effect size?
   b. How is effect size measured for the independent samples t-test? For the related-measures t-test?

7. For each of the following, which type of t-test is required?
   a. Studying whether males or females are more prone to simple mathematical errors on a statistics exam.
   b. The study described in part a, but for each male, there is a female with the same reading level.
   c. An investigation of spending habits of teenagers, comparing the amount of money each spends in a video store and in a clothing store.
   d. An investigation of the effects of a new anti-anxiety drug, measuring subjects' anxiety before and again after administration of the drug.
   e. Testing whether males in the U.S. Army are more aggressive than males in the U.S. Marine Corps.

8. A recent study examined the drinking behaviors of underage college male and female freshmen. Each participant was asked how many alcoholic beverages they consumed during the past 7 days. The researchers wish to see if there is a difference in the drinking habits of males and females.

   Sample 1—females: $\overline{X}_1 = 12.9$, $s_1^2 = 3.30$, $n_1 = 13$
   Sample 2—males: $\overline{X}_2 = 11.2$, $s_2^2 = 2.21$, $n_2 = 16$

   a. What are H_0 and H_a?
   b. Compute the appropriate t-test.
   c. With $\alpha = .05$, report the statistical results.
   d. What should the researchers conclude?

9. On a standard test of attention span in children, a researcher wants to know whether there is a difference between children who play video games extensively and those who do not. She selects a sample of children who regularly play video games and those who do not and obtains the following attention-span data:

   Sample 1—play video games:
   $\overline{X}_1 = 12.1$, $s_1^2 = 2.94$, $n_1 = 23$

   Sample 2—do not play video games:
   $\overline{X}_2 = 9.6$, $s_2^2 = 1.82$, $n_2 = 19$

   a. What are H_0 and H_a?
   b. Compute the appropriate t-test.
   c. With $\alpha = .05$, report the statistical results.
   d. What should the researcher conclude about these results?

10. Martha believes a relaxation technique involving visualization will help people with mild insomnia fall asleep faster. She randomly selects a sample of 20 participants from a group of mild insomnia patients and randomly assigns 10 to receive visualization therapy. The other 10 participants receive no treatment. Each participant is then measured to see how long (in minutes) it takes him or her to fall asleep. Her data are below.

No Treatment (X_1)	Treatment (X_2)
22	19
18	17
27	24
20	21
23	27
26	21
27	23
22	18
24	19
22	22

a. Should she use an independent-samples or a related-samples t-test? Explain your answer.
b. What are the independent and dependent variables?
c. Using the fact that Martha believes the treatment will reduce the amount of time to fall asleep, state the null and alternative hypotheses.
d. Use $\alpha = .05$. What is t_{crit}?
e. Calculate t_{obt}.
f. Report the statistical results using the correct format.
g. What should Martha conclude?
h. If appropriate, calculate the effect size using the Cohen's d.

11. Martha believes a relaxation technique involving visualization will help people with mild insomnia fall asleep faster. She randomly selects 10 patients from a group of mild insomnia patients and measures how many minutes it takes each one to fall asleep. Each participant is then taught the visualization technique and measured again to see how long it takes him or her to fall asleep. Her data are below.

Before Treatment (X_1)	After Treatment (X_2)
22	19
18	17
27	24
20	21
23	27
26	21
27	23
22	18
24	19
22	22

a. Should she use an independent-samples t-test or a related-samples t-test? Explain.
b. What are the independent and dependent variables?
c. Using the fact that Martha believes the treatment will reduce the amount of time to fall asleep, state the null and alternative hypotheses.
d. Use $\alpha = .05$. What is t_{crit}?
e. Calculate t_{obt}.
f. Report the statistical results using the correct format.
g. What should Martha conclude?
h. If appropriate, calculate the effect size using the Cohen's d.

Describing Relationships Using Correlation and Regression

Recall that in a relationship, as the scores on one variable change, we see a consistent pattern of change in the scores on the other variable. In research, in addition to demonstrating a relationship, we also want to describe and summarize the relationship. This chapter presents a new descriptive statistic for summarizing a relationship called the *correlation coefficient*. In the following sections we will discuss (1) what a correlation coefficient is and how to interpret it, (2) how to compute the most common coefficient, (3) how to perform inferential hypothesis testing of it, and (4) how we use a relationship to predict unknown scores.

SECTIONS

10.1 Understanding Correlations

10.2 The Pearson Correlation Coefficient

10.3 Statistics in the Research Literature: Reporting *r*

10.4 Understanding Linear Regression

10.5 The Proportion of Variance Accounted For: r^2

10.1 UNDERSTANDING CORRELATIONS

Whenever we find a relationship, we then want to know its characteristics: What pattern is formed, how consistently do the scores change together, what direction do the scores change, and so on. The best—and easiest—way to answer these questions is by computing a correlation coefficient. The term *correlation* means relationship, and a correlation coefficient is a number that describes a relationship. The **correlation coefficient**

correlation coefficient A number that describes the type and the strength of the relationship present in a set of data

Looking Back

- From Chapter 1, understand the difference between an experiment and a correlational study, and how to recognize a relationship between the scores of two variables.

- From Chapter 4, understand that greater variability indicates that scores are not close to each other.

- From the previous chapters, know the basics of significance testing.

© Martin Holtkamp/Taxi Japan/Getty Images

is the statistic that describes the important characteristics of a relationship. The correlation coefficient is important because it simplifies a complex pattern involving many scores into one easily interpreted statistic.

Correlation coefficients are most commonly used to summarize the relationship found in a *correlational design,* but computing a correlation coefficient does not create this type of study. Recall from Chapter 1 that, in a correlational design, we simply measure participants' scores on two variables. For example, as people drink more coffee, they typically become more nervous. To study this in a correlational study, we might simply ask participants the amount of coffee they had consumed that day and also measure how nervous they were. In an experiment, however, we would *manipulate* coffee consumption by assigning some people to a one-cup condition, others to a two-cup condition, and so on, and then measure their nervousness. In either case a correlation coefficient would, in one number, summarize the important aspects of the relationship.

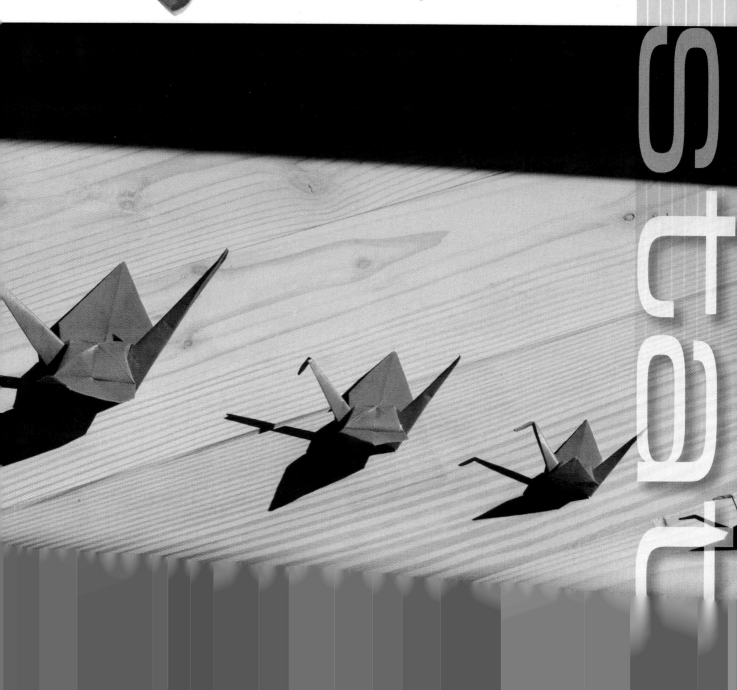

scatterplot A graph of the individual data points from a set of *X-Y* pairs

Distinguishing Characteristics of Correlational Analysis

There are four major differences between how we handle data in a correlational analysis versus in an experiment. First, in our coffee experiment we would examine the mean nervousness score (the *Y* scores) for each condition of the amount of coffee consumed (each *X*). Then we examine the relationship they show. However, with a correlational study we typically have a rather large range of different *X* scores: People will probably report many different amounts of coffee consumed in a day. Comparing the mean nervousness scores for so many amounts would be very difficult. Therefore, we compute the correlation coefficient, which summarizes the *entire* relationship at once that is formed by all pairs of *X-Y* scores in the data.

A second difference is that, because we examine all pairs of *X-Y* scores, correlational procedures involve *one* sample:

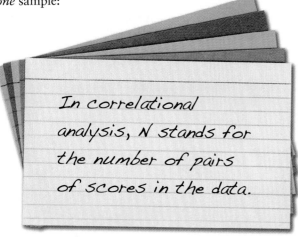

In correlational analysis, N stands for the number of pairs of scores in the data.

Third, in a correlational study, either variable may be labeled as the *X* or *Y* variable. How do we decide which is our *X* or *Y*? Any relationship may be seen as asking "For a *given X*, what are the *Y* scores?" So, simply identify your "given" variable and it is *X*. Thus, if we ask, "For a given amount of coffee, what are the nervousness scores?" then amount of coffee is the *X* variable and nervousness is the *Y*. Conversely, if we ask, "For a given nervousness score, what is the amount of coffee consumed?" then nervousness is *X* and amount of coffee is *Y*. (In fact, some researchers may then call the *X* variable the independent variable, even though they don't manipulate it.)

Finally, the data are graphed differently in correlational research. We use the individual pairs of scores to create a *scatterplot*. A **scatterplot** is a graph that shows the location of each data point formed by a pair of *X-Y* scores. The scatterplot in Figure 10.1 shows data from studying nervousness and coffee consumption. (Real research typically involves a larger *N*, and the data points will not form such a pretty pattern.) To create this scatterplot, the first person in the table had 1 cup of coffee and a nervousness score of 1, so we place a data point above an *X* of 1 cup at the height of a score of 1 on the *Y* variable of nervousness. The second participant had the same *X* and *Y* scores, so that data point is on top of the previous one. The third participant scored an *X* of 1 and *Y* of 2, and so on. Then we examine the overall pattern formed by the data points. We read a graph from left to right and see that the *X*s become larger. Therefore, get in the habit of describing a relationship by asking, "As the

Figure 10.1

Scatterplot Showing Nervousness as a Function of Coffee Consumption

Each data point is created using a participant's coffee consumption as the *X* score and nervousness as the *Y* score.

Cups of Coffee: *X*	Nervousness Scores: *Y*
1	1
1	1
1	2
2	2
2	3
3	4
3	5
4	5
4	6
5	8
5	9
6	9
6	10

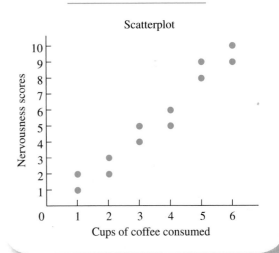

Scatterplot

X scores become larger, what happens to the *Y*s? In Figure 10.1, as the *X* scores increase, the data points move *higher* on the graph, indicating that the corresponding *Y* scores are higher. Thus, the scatterplot reflects a relationship. Recall that in any relationship as the *X* scores increase, the *Y* scores change such that a different value of *Y* tends to be paired with a different value of *X*.

Drawing the scatterplot allows you to see your particular relationship and to map out the best way to summarize it. This is because the shape and orientation of a scatterplot reflect the characteristics of the relationship formed by the data, and it is these characteristics that are summarized by the correlation coefficient. There are two important characteristics of any relationship that we wish to know about: the *type* and the *strength* of the relationship. The following sections discuss these characteristics.

Types of Relationships

The *type* of relationship that is present in a set of data is the overall manner in which the *Y* scores change as the *X* scores change. There are two general types of relationships: *linear* and *nonlinear* relationships.

Linear Relationships The term *linear* means "straight line," and a linear relationship forms a pattern that follows one straight line. This is because in a **linear relationship,** as the *X* scores increase, the *Y* scores tend to change in only *one* direction. Figure 10.2 shows two linear relationships: between the amount of time that students study and their test performance, and between the number of hours that students watch television and the amount of time they sleep. These are linear because as students study longer, their grades tend only to increase, and as they watch more television, their sleep time tends only to decrease.

Sleep time decreases as the amount of TV watched increases.

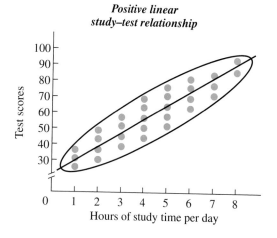
Positive linear study–test relationship

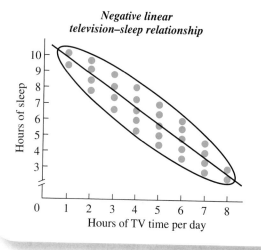
Negative linear television–sleep relationship

To better see the overall pattern in a scatterplot, visually summarize it by drawing a line around its outer edges. As in Figure 10.2, a scatterplot that forms a slanted ellipse that follows a straight line indicates a linear relationship: By slanting, it indicates that the *Y* scores are changing as the *X* scores increase; the straight line indicates it is a linear relationship.

Further, we also summarize the relationship by drawing a line through the scatterplot. This line is called the *regression line.* While the correlation coefficient is the *statistic* that

linear relationship
A relationship in which the *Y* scores tend to change in only one direction as the *X* scores increase, forming a slanted straight regression line on a scatter plot

summarizes a relationship, the regression line is the *line* that summarizes the relationship. The **linear regression line** is the straight line that summarizes a relationship by passing through the center of the scatterplot. That is, although not all data points are *on* the line, the distance that some are above the line equals the distance that others are below it, so that the regression line passes through the center of the scatterplot. Therefore, think of the regression line as reflecting the pattern that all data points more or less follow, so it shows the linear—straight-line—relationship hidden in the data.

The different ways that the scatterplots slant in Figure 10.2 illustrate that one other very important piece of information to learn about a relationship is its *direction*. The direction of the relationship is determined by the direction in which the *Y* scores change. The study–test relationship is a positive relationship. In a **positive linear relationship,** as the *X* scores increase, the *Y* scores also tend to increase. Any relationship that fits the pattern "the more *X*, the more *Y*" is a positive linear relationship. On the other hand, the television–sleep relationship is a negative relationship. In a **negative linear relationship,** as the *X* scores increase, the *Y* scores tend to decrease. Any relationship that fits the pattern "the more *X*, the less *Y*" is a negative linear relationship. (*Note:* the term *negative* does not mean there is something wrong with the relationship: It merely indi-cates the direction in which the *Y* scores change as the *X* scores increase.)

Nonlinear Relationships If a relationship is not linear, then it is *nonlinear*, meaning that the data cannot be summarized by *one straight* line. Thus, another name for a nonlinear relationship is a curvilinear relationship. In a **nonlinear,** or **curvilinear, relationship,** as the *X* scores change, the *Y* scores do not *only* increase or *only* decrease: At some point, the *Y* scores alter their direction of change.

Nonlinear relationships come in many different shapes, but Figure 10.3 shows two common ones. On the left is the relationship between a person's age and the amount of time required to move from one place to another. At first, as age increases, movement time decreases; but, beyond a certain age, the time scores change direction and begin to increase. (This is called a *U-shaped* pattern.) The scatterplot on the right shows the relationship between the number of alcoholic drinks consumed and feeling well. At first, people tend to feel better as they drink, but beyond a certain point, drinking more makes them feel progressively worse. (This pattern reflects an *inverted U-shaped relationship*.) Curvilinear relationships may be more complex than those above, producing a wavy pattern that repeatedly changes direction. Also, the scatterplot does not need to be curved to be non-linear. Scatterplots similar to those in Figure 10.3 might be best summarized by two straight regression lines that form a V and inverted V, respectively. Or we might see regression lines that form angles like $\searrow$ or $\diagup$. All are still nonlinear relationships, because they cannot be summarized by *one* straight line.

Note that the preceding terminology is also used to describe the type of relationship found in experiments. If, as the amount of the independent variable (*X*) increases the dependent scores (*Y*) also increase, then you have a positive relationship. If the dependent scores decrease as the independent variable (*X*) increases, then you have a

linear regression line The straight line that summarizes the scatter plot of a linear relationship by passing through the center of the scatterplot

positive linear relationship A linear relationship in which the *Y* scores tend to increase as the *X* scores increase

negative linear relationship A linear relationship in which the *Y* scores tend to decrease as the *X* scores increase

nonlinear (curvilinear) relationship A relationship in which the *Y* scores change their direction of change as the *X* scores change

© iStockphoto.com/Thomas Flügge

REMEMBER

A linear relationship follows one straight line and may be positive (with increasing Y scores) or negative (with decreasing Y scores).

Figure 10.3

Scatterplots Showing Nonlinear Relationships

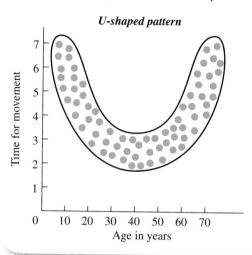

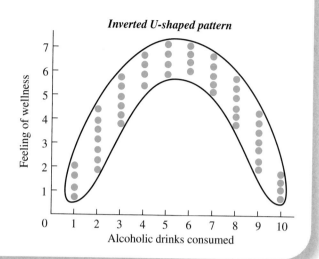

negative relationship. And if, as the independent variable (X) increases, the dependent scores alter their direction of change, then you have a nonlinear relationship.

How the Coefficient Describes the Type of Relationship Behavioral research focuses primarily on linear relationships, so we'll discuss only linear correlation coefficients. How do you know whether data form a linear relationship? If the scatterplot generally follows a straight line, then linear correlation is appropriate. Do not summarize a nonlinear relationship by computing a linear correlation coefficient. The relationship won't fit the straight line very well, so the coefficient won't accurately describe the relationship.

A linear correlation coefficient communicates two things about the relationship. First, a linear correlation coefficient communicates that we are describing a linear relationship. Second, the coefficient itself indicates whether the relationship is positive or negative. The coefficient—the number we compute—either will have a minus sign ($-$) in front of it, or it will have no sign because it's positive, so we add a plus sign ($+$). A positive correlation coefficient indicates a positive linear relationship, and a negative correlation coefficient indicates a negative linear relationship. The other characteristic of a relationship communicated by the correlation coefficient is the *strength* of the relationship.

Strength of the Relationship

Recall that a relationship can exhibit varying degrees of consistency. The **strength of a relationship** is the

REMEMBER

A linear correlation coefficient *has two components: the sign, indicating a positive or a negative relationship, and the absolute value, indicating the strength of the relationship.*

extent to which one value of Y is consistently paired with one and only one value of X. (This is also referred to as the *degree of association*.) The strength of a relationship is indicated by the absolute value of its correlation coefficient (ignoring the sign). The larger the coefficient, the stronger, more consistent, the relationship is. The largest possible value of a correlation coefficient is 1, and the smallest value is 0. When you include the sign, the correlation coefficient can be any value between -1 and $+1$. Thus, the closer the coefficient is to ± 1, the more consistently one value of Y is paired with one and only one value of X.

strength of a relationship The extent to which one value of Y within a relationship is consistently associated with one and only one value of X; also called the *degree of association*

Recognize that correlation coefficients do not directly measure units of "consistency." Thus, if one correlation coefficient is +.40 and another is +.80, you *cannot* conclude that the first is only half as consistent as the second. Instead, evaluate any correlation coefficient by comparing it to the extreme values of 0 and ±1. The starting point is a perfect relationship.

The Perfect Correlation A correlation coefficient of +1 or −1 describes a perfectly consistent linear relationship. Figure 10.4 shows an example of each. In this and the following figures, first look at the scores to see how they pair up; then look at the scatterplot. Other data having the same correlation coefficient will produce similar patterns having similar scatterplots. Therefore, interpreting the correlation coefficient involves envisioning the scatterplot that is present. Here are four related ways to think about what a coefficient tells you about the relationship.

First, the coefficient indicates *the relative degree of consistency* with which Ys are paired with Xs. A coefficient of ±1 indicates that *everyone* who obtains a particular X obtains the same, one Y. Then, every time X changes, the Y scores all change to one new value.

Second, the opposite of consistency is variability, so *the coefficient communicates the variability in the* Y *scores paired with each* X. When the coefficient is ±1, only one Y is paired

Figure 10.4

Data and Scatterplots Reflecting Perfect Positive and Negative Correlations

Perfect positive coefficient = +1

X	Y
1	2
1	2
1	2
3	5
3	5
3	5
5	8
5	8
5	8

Perfect negative coefficient = −1

X	Y
1	8
1	8
1	8
3	5
3	5
3	5
5	2
5	2
5	2

with an X, so there is no variability among the Y scores paired with each X.

Third, *the coefficient communicates the relative accuracy of our predictions*. That is, a goal of behavioral science is to predict the specific behaviors—and the scores that reflect them—that occur in a particular situation. We do this using relationships because a particular Y score is naturally paired with a particular X score. Therefore, once we've identified a relationship, if we know someone's X score, we use the relationship to predict that individual's Y score. A coefficient of ±1 indicates perfect accuracy in our predictions: Because in a perfect relationship only one value of Y occurs with an X, by knowing someone's X, we can predict exactly what his or her Y will be. For example, in both graphs an X of 3 is always paired with a Y of 5, so we predict that *anyone* scoring an X of 3 will have a Y score of 5.

Fourth, *the coefficient indicates how closely the scatterplot fits the regression line*. Because a coefficient equal to ±1 indicates zero variability or *spread* among

REMEMBER

The correlation coefficient communicates how consistently the Ys are paired with X, the variability in Ys at each X, the accuracy in our predictions of Y, and how closely the scatterplot fits the regression line.

the *Y* scores at each *X*, we can envision that the data points form a perfect straight-line relationship so that they all lie *on* the regression line.

Intermediate Strength A correlation coefficient that is not ±1 indicates that the data form a linear relationship to only some degree. The key to understanding the strength of any relationship is this:

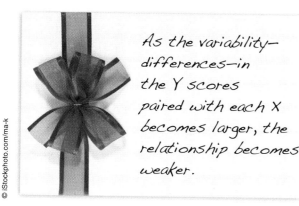

As the variability— differences—in the Y scores paired with each X becomes larger, the relationship becomes weaker.

The correlation coefficient communicates this because, as the variability in the *Y*s at each *X* becomes larger, the correlation coefficient becomes smaller.

For example, Figure 10.5 shows data that produce a correlation coefficient of +.98. Again, interpret the coefficient in four ways:

 Consistency: An absolute value less than ±1 indicates that not every participant obtaining a particular *X* obtained the same *Y*. However, a coefficient of +.98 is close to +1, so here we have close to perfect association between the *X* and *Y* scores.

 Variability: A coefficient less than ±1 indicates that there *is* variability among the *Y* scores at each *X*. In other words, a *group* of different *Y* scores are now paired with an *X*. However, +.98 is close to +1, indicating this variability is relatively small

so that the different *Y*s at an *X* are relatively close to each other.

 Predictions: When the coefficient is not ±1, there is not one *Y* score for a particular *X*, so we can predict only *around* what someone's *Y* score will be. In Figure 10.5, at an *X* of 1 are *Y* scores of 1 and 2. Split the difference and for each person here we'd predict a *Y* of around 1.5. But no one scored exactly 1.5, so we'll have some error in our predictions. However, +.98 is close to +1, indicating that, overall, our error will be relatively small.

 The scatterplot: A coefficient less than +1 indicates variability in *Y* at each *X* so the data points at an *X* are vertically spread out above and below the regression line. However, a coefficient of +.98 is close to +1, so we know the *Y* scores are close to the regression line, resulting in a scatterplot that is a narrow ellipse.

On the other hand, Figure 10.6 shows data that produce a coefficient of −.28. Because −.28 is not very close

Figure 10.5
Data and Scatterplot Reflecting a Correlation Coefficient of +.98

X	Y
1	1
1	2
1	2
3	4
3	5
3	5
5	7
5	8
5	8

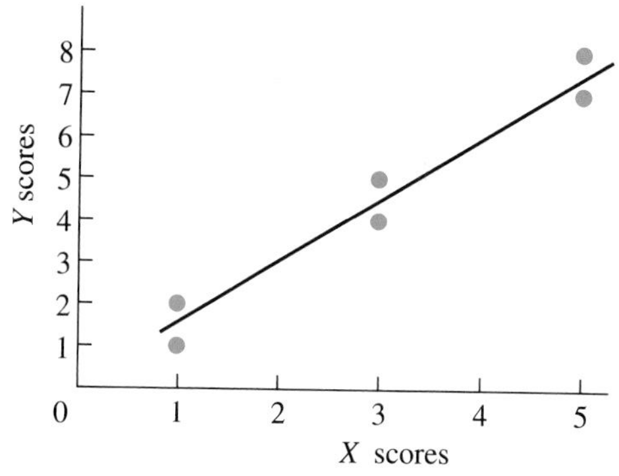

Figure 10.6
Data and Scatterplot Reflecting a Correlation Coefficient of −.28

X	Y
1	9
1	6
1	3
3	8
3	6
3	3
5	7
5	5
5	1

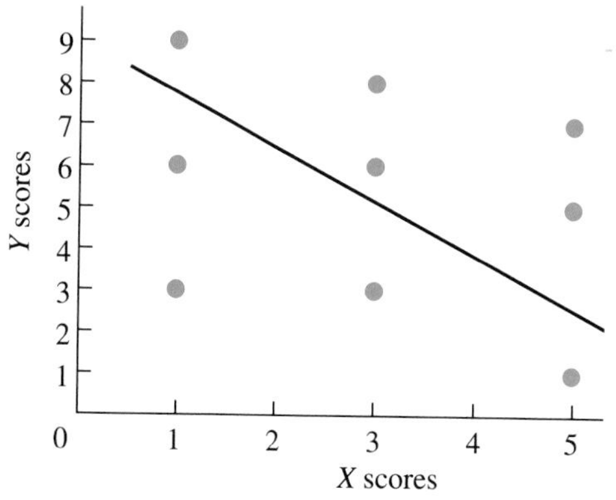

to -1, this tells us: (1) this relationship is not close to showing perfectly consistent association; (2) the variability in the Y scores paired with each X is relatively large; (3) because each X is paired with many different Ys, knowing a participant's X will not get us close to his or her Y, so our predictions will contain large amounts of error; and (4) the large variability produces data points on the scatterplot at each X that are vertically very spread out above and below the regression, forming a relatively wide scatterplot.

Zero Correlation The lowest possible value of the correlation coefficient is 0, indicating that no relationship is present. Figure 10.7 shows data that produce such a coefficient. A correlation coefficient of 0 is as far as possible from ± 1, telling us the scatterplot is as far as possible from forming a slanted straight line. Therefore, we know the following: (1) No Y score tends to be consistently associated with only one X, and instead, virtually the same batch of Y scores is paired with every X. (2) The spread in Y at any X is at maximum and equals the overall spread of Y in the data. (3) Because each X is paired with virtually all Y scores, knowing someone's X score is no help in predicting his or her Y score. (4) The scatterplot is horizontal and elliptical, or circular, and it in no way hugs the regression line.

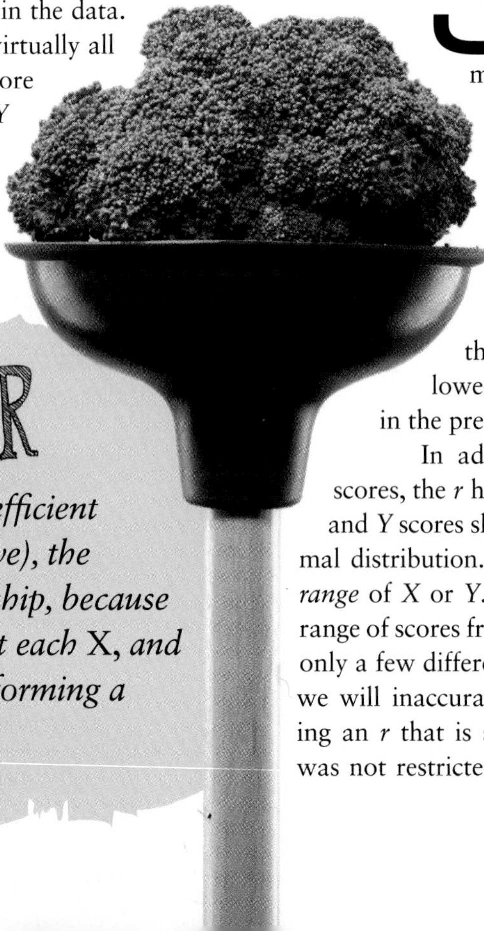

REMEMBER

The larger a correlation coefficient (whether positive or negative), the stronger the linear relationship, because the Ys are spread out less at each X, and so the data come closer to forming a straight line.

Figure 10.7
Data and Scatterplot Reflecting a Correlation Coefficient of 0

X	Y
1	3
1	5
1	7
3	3
3	5
3	7
5	3
5	5
5	7

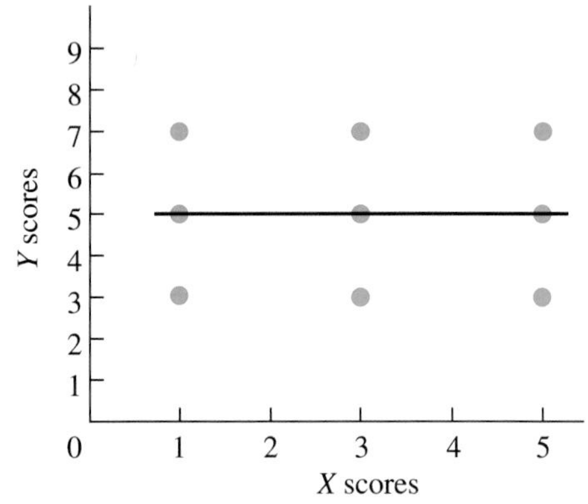

10.2 THE PEARSON CORRELATION COEFFICIENT

Statisticians have developed several different correlation coefficients that are used with different scales of measurement and in different kinds of studies. However, the most common correlation in behavioral research is the **Pearson correlation coefficient**, which describes the linear relationship between two interval variables, two ratio variables, or one interval and one ratio variable. The symbol for the Pearson correlation coefficient is the lowercase **r**. All of the example coefficients in the previous section were rs.

In addition to requiring interval or ratio scores, the r has two other requirements. First, the X and Y scores should each form an approximately normal distribution. Second, we should avoid a *restricted range* of X or Y. A **restricted range** occurs when the range of scores from a variable is limited so that we have only a few different scores that are close together. Then we will inaccurately describe the relationship, obtaining an r that is smaller than it would be if the range was not restricted. Generally, a restricted range occurs

© Chris Stein/Stone/Getty Images

GRADE A STUDENT

when researchers are too selective when obtaining participants. Thus, if we are interested in the relationship between participants' high school grades and their subsequent salaries, we should not restrict the range of grades by studying only honor students. Instead, we should include all students to get the widest range of grades possible.

In the following sections we discuss how to compute the Pearson r and how to perform significance testing of it.

Computing the Pearson r

Computing r involves several new symbols: We have ΣXY, called the *sum of the cross products*. This says to first multiply each X score in a pair times its corresponding Y score. Then sum all of the resulting products. We also have $(\Sigma X)(\Sigma Y)$. This says to find the sum of the Xs and the sum of the Ys. Then multiply the two sums together. And we have the *sum of the squared Y*s, symbolized by ΣY^2: square each Y score and then sum the squared Ys. Finally, we have the *squared sum of Y*s, symbolized by $(\Sigma Y)^2$: sum all Y scores and then square that sum.

The basis for r is that it compares how consistently each value of Y is paired with each value of X. In Chapter 5 you saw that we compare scores from different variables by transforming them into z-scores. Essentially, calculating r involves transforming each Y score and each X score into a z-score, and then determining the "average" amount

of correspondence between all pairs of z-scores. The following formula simultaneously accomplishes all of that.

THE PEARSON CORRELATION COEFFICIENT

$$r = \frac{N(\Sigma XY) - (\Sigma X)(\Sigma Y)}{\sqrt{[N(\Sigma X^2) - (\Sigma X)^2][N(\Sigma Y^2) - (\Sigma Y)^2]}}$$

In the numerator, the N (the number of pairs of scores) is multiplied by ΣXY. Then subtract the quantity $(\Sigma X)(\Sigma Y)$. In the denominator, in the left brackets multiply N by ΣX^2, and from that subtract $(\Sigma X)^2$. In the right bracket, multiply N by ΣY^2, and from that subtract $(\Sigma Y)^2$. Multiply the answers from the two brackets together, and then find the square root. Then divide the denominator into the numerator and, voilà, the answer is the Pearson r.

For example, say that we ask 10 people the number of times they visited a doctor in the last year and the number of glasses of orange juice they drink daily. To describe the linear relationship between juice drinking and doctor visits, we compute r. Table 10.1 shows a good way to set up the data.

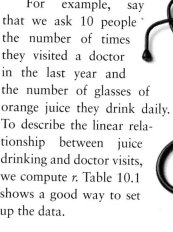

Table 10.1

Sample Data for Computing the r between Orange Juice Consumed (the X variable) and Doctor Visits (the Y variable)

Participant	Glasses of Juice per Day		Doctor Visits per Year		
	X	X^2	Y	Y^2	XY
1	0	0	8	64	0
2	0	0	7	49	0
3	1	1	7	49	7
4	1	1	6	36	6
5	1	1	5	25	5
6	2	4	4	16	8
7	2	4	4	16	8
8	3	9	4	16	12
9	3	9	2	4	6
10	4	16	0	0	0
$N = 10$	$\Sigma X = 17$	$\Sigma X^2 = 45$	$\Sigma Y = 47$	$\Sigma Y^2 = 275$	$\Sigma XY = 52$
	$(\Sigma X)^2 = 289$		$(\Sigma Y)^2 = 2209$		

STEP 1:

Compute ΣX, (ΣX)², ΣX², ΣY, (ΣY)², ΣY², ΣXY, and N. As in Table 10.1, in addition to the columns for X and Y, make columns for X² and Y². Also, make a column for XY and multiply each X times its paired Y. Then sum all of the columns. Then square ΣX and ΣY.

Filling in the formula for *r* we get

$$r = \frac{N(\Sigma XY) - (\Sigma X)(\Sigma Y)}{\sqrt{[N(\Sigma X^2) - (\Sigma X)^2][N(\Sigma Y^2) - (\Sigma Y)^2]}}$$

$$= \frac{10(52) - (17)(47)}{\sqrt{[10(45) - 289][10(275) - 2209]}}$$

STEP 2:

Compute the numerator. 10 times 52 is 520, and 17 times 47 is 799. Now, we have

$$r = \frac{520 - 799}{\sqrt{[10(45) - 289][10(275) - 2209]}}$$

Complete the numerator: 799 *from* 520 is −279. (Note the negative sign.)

STEP 3:

Compute the denominator and then divide. First perform the operations within each bracket. In the left bracket above, 10 times 45 is 450. In the right bracket, 10 times 275 is 2750.

This gives

$$r = \frac{-279}{\sqrt{(450 - 289)(2750 - 2209)}}$$

On the left, subtracting 450 − 289 gives 161. On the right, subtracting 2750 − 2209 gives 541. So

$$r = \frac{-279}{\sqrt{(161)(541)}}$$

Now multiply the quantities in the parentheses together: 161 times 541 equals 87,101. After taking the square root we have

$$r = \frac{-279}{295.129} = -.95$$

Our correlation coefficient between orange juice drinks and doctor visits is −.95. (*Note:* We usually round off a correlation coefficient to two decimals.) On a scale of 0 to ±1, our −.95 indi-

cates an extremely strong negative linear relationship: Each amount of orange juice is associated with a very small range of doctor visits, and as juice scores increase, doctor visits consistently decrease. Therefore we envision a very narrow scatterplot that slants downward. Further, based on participants' juice scores, we can very accurately predict their frequency of doctor visits.

Significance Testing of the Pearson *r*

The Pearson *r* describes a *sample*. Ultimately, however, we wish to describe the relationship that occurs in nature—in the population. Therefore, we use the sample's correlation coefficient to estimate or infer the coefficient we'd find if we could study everyone in the population. But before we can believe that the sample correlation represents the relationship found in the population, we must first perform statistical hypothesis testing and conclude that *r* is significant.

Here's a new example. We are interested in the relationship between a man's age and his physical agility. We select 25 men, measure their age and their agility, and using the previous formula, compute that $r = -.45$. This suggests that the correlation in the population would also be −.45.

The symbol for the Pearson population correlation coefficient is the Greek letter **ρ**, called "rho." A ρ is interpreted in the same way as *r*: It is a number between 0 and ±1, indicating either a positive or a negative linear relationship in the population. The larger the absolute value of ρ, the stronger the relationship and the more closely the population's scatterplot hugs the regression line.

Thus, we might estimate that ρ would equal −.45 if we measured the agility and age of all men. But, on the other hand, there is always the potential problem of sampling error. Maybe these variables are not really related in nature, but sampling error is misleading us to believe that they are. This leads to our statistical hypotheses. As usual, we can perform either a one- or two-tailed test.

REMEMBER

Never accept that a sample correlation coefficient reflects a real relationship in nature unless it is significant.

The two-tailed test is used when we do not predict the direction of the relationship, predicting that the correlation will be either positive or negative. First, the alternative hypothesis always says the predicted relationship exists. Thus, if the correlation in the population is either positive or negative, then ρ does not equal zero. Thus,

$$H_a: \rho \neq 0$$

As usual, the null hypothesis is that the predicted relationship does not exist, so the correlation in the population should be zero. Thus,

$$H_0: \rho = 0$$

These are the two-tailed hypotheses whenever you test that the sample either does or does not represent a relationship. This is the most common approach and the one we'll use. (You can also test the H_0 that your sample represents a nonzero ρ. Consult an advanced statistics book for the details.)

The Sampling Distribution of r Our H_0 implies that if r does not equal zero, it's because of sampling error. You can understand this by looking at the population scatter-plot in Figure 10.8. There is no relationship here, and so ρ is 0. However, the null hypothesis implies that, by chance, we obtained an elliptical *sample*

Figure 10.8

Scatterplot of a Population for which $\rho = 0$, as Described by H_0
Our r results from sampling error when selecting a sample from this scatterplot.

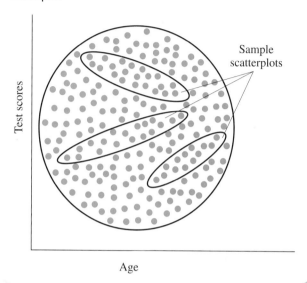

Sample scatterplots

Test scores

Age

scatterplot from this population that produced our r. Thus, our H_0 says that age and agility are not really related, but the scores in our sample happen to pair up so that it looks as though they're related. (On the other hand, H_a essentially says that the population's scatterplot would be similar to the sample's scatterplot.)

As usual, we test H_0, so here we determine the likelihood of obtaining our sample r from the population where ρ is zero. To do so, we envision the sampling distribution. To create this, it is as if we infinitely sample the population in Figure 10.8, each time computing r. The **sampling distribution of r** shows all possible values of the r that occur when samples are drawn from a population in which ρ is zero. Such a distribution is shown in Figure 10.9 on the next page. The only novelty here is that along the X axis are now different values of r. When $\rho = 0$, the most frequent sample r is also 0, so the mean of the sampling distribution—the average r— is 0. Because of sampling error, however, sometimes we might obtain a positive r and sometimes a negative r. Most often the r will be relatively small (close to 0), but less frequently we may obtain a larger r that falls more into the tails of the distribution. Thus, the larger

sampling distribution of r
A frequency distribution showing all possible values of r that occur when samples are drawn from a population in which ρ is zero

the *r* (whether positive or negative), the less likely it is to occur when the sample represents a population where $\rho = 0$.

To test H_0, we determine where our *r* is located in the sampling distribution. The size of our *r* directly communicates its location. For example, the mean of the sampling distribution is always 0, so our *r* of $-.45$ is a distance of .45 below the mean. Therefore, we test H_0 simply by examining our obtained *r*, which is r_{obt}. To determine whether r_{obt} is in the region of rejection, we compare it to r_{crit}.

Drawing Conclusions about r As with the *t*-distribution, the shape of the *r*-distribution is slightly different for each *df*, so there is a different value of r_{crit} for each *df*. Table 3 in Appendix B gives the critical values of the Pearson correlation coefficient. Use this "*r*-table" in the same way you've used the *t*-table: Find r_{crit} for either a one- or a two-tailed test at the appropriate α and *df*.

Figure 10.9

Sampling Distribution of *r* When $\rho = 0$

It is an approximately normal distribution, with values of *r* plotted along the *X* axis.

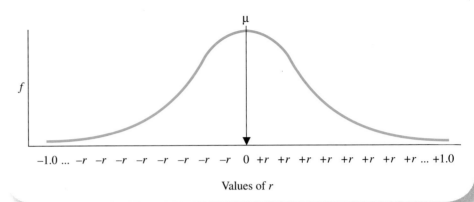

Figure 10.10

H_0 Sampling Distribution of *r* When H_0: $\rho = 0$

For the two-tailed test, there is a region of rejection for positive values of r_{obt} and for negative values of r_{obt}.

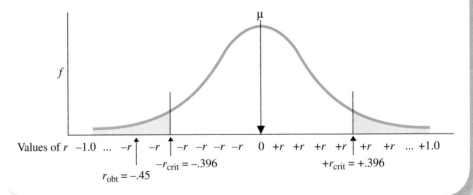

THE DEGREES OF FREEDOM IN THE PEARSON CORRELATION COEFFICIENT

$df = N - 2$, where *N* is the number of *X*-*Y* pairs in the data

For our example, *N* was 25, so $df = 23$. For $\alpha = .05$, the two-tailed r_{crit} is $\pm.396$. We set up the sampling distribution as in Figure 10.10. An r_{obt} of $-.45$ is beyond the r_{crit} of $\pm.396$, so it is in the region of rejection. Thus, our *r* is so unlikely to occur if we had been representing the population where ρ is 0, that we reject the H_0 that we were representing this population. We conclude that the r_{obt} is "significantly different from zero."

The rules for interpreting a significant result here are the same as with previous statistics. In particular, α is again the theoretical probability of a Type I error.

Here, a Type I error is rejecting the H_0 that the population correlation is zero, when in fact the correlation is zero. Also, as usual, rejecting H_0 does not *prove* anything. In particular, we have not proven that changes in age *cause* test scores to change. In fact, we have not even proven that there is a relationship in the population. We are simply more confident that the r_{obt} does not merely reflect some quirk of sampling error, and that instead, it represents a "real" relationship in nature.

Because the sample r_{obt} is $-.45$, our best estimate is that in the population, ρ equals $-.45$. However, recognizing that the sample may contain sampling error, we say that ρ is *around* $-.45$. (We could more precisely describe this ρ by computing a confidence interval for ρ. This confidence interval is computed using a different procedure from that discussed previously.)

If r_{obt} did not lie beyond r_{crit}, then we would retain H_0 and conclude that the sample *may* represent a population where $\rho = 0$. As usual, we have not proven there is no relationship in the population, we have simply failed to convincingly demonstrate that there *is* one. Therefore, we make no claims about the relationship that may or may not exist.

One-Tailed Tests of *r* If we had predicted only a positive correlation or only a negative correlation, then we would have performed a one-tailed test.

THE ONE-TAILED HYPOTHESES FOR
TESTING A CORRELATION COEFFICIENT

Predicting a positive correlation	Predicting a negative correlation
H_0: $\rho \leq 0$	H_0: $\rho \geq 0$
H_a: $\rho > 0$	H_a: $\rho < 0$

When predicting a positive relationship, we are saying that ρ will be greater than zero (in H_a) but H_0 says we are wrong. When predicting a negative relationship, we are saying that ρ will be less than zero (in H_a) but H_0 says we are wrong.

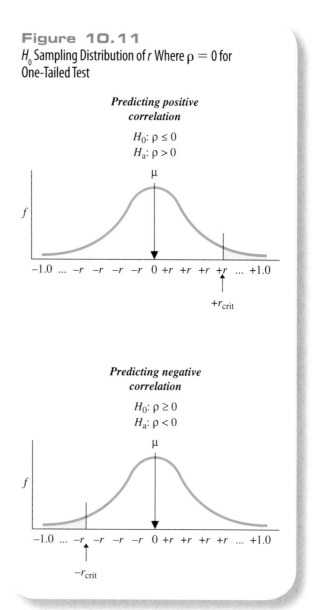

Figure 10.11
H_0 Sampling Distribution of *r* Where $\rho = 0$ for One-Tailed Test

We test each H_0 by again examining the sampling distribution for $\rho = 0$. From the *r*-table in Appendix B, find the one-tailed critical value for *df* and α, and set up one of the sampling distributions shown in Figure 10.11. When predicting a positive correlation, use the upper distribution: r_{obt} is significant if it falls beyond the positive r_{crit}. When predicting a negative correlation, use the lower distribution: r_{obt} is significant if it falls beyond the negative r_{crit}.

Summary of the Pearson Correlation Coefficient

The Pearson correlation coefficient describes the strength and direction of a linear relationship between normally distributed interval/ratio variables.

linear regression
The procedure used to predict Y scores based on correlated X scores

predictor variable
The X variable that has the known scores from which unknown Y scores are predicted when using the linear regression equation

criterion variable
The Y variable that has the unknown scores that are predicted based on a correlated X score when using the linear regression equation

1. Compute r_{obt}.

2. Create either the two-tailed or one-tailed H_0 and H_a.

3. Set up the sampling distribution and, using $df = N - 2$ (where N is the number of *pairs*), find r_{crit} in the r-table.

4. If r_{obt} is beyond r_{crit}, the results are significant, so describe the relationship and population correlation coefficient (ρ). If r_{obt} is not beyond r_{crit}, the results are not significant and we make no conclusion about the relationship.

10.3 STATISTICS IN THE RESEARCH LITERATURE: REPORTING r

Report the Pearson correlation coefficient using the same format as with previous statistics. Thus, in our agility study, the r_{obt} of $-.45$ was significant with 23 *df*. We report this as $r(23) = -.45, p < .05$. As usual, *df* is in parentheses and because $\alpha = .05$, the probability is less than .05 that we've made a Type I error.

Understand that, although theoretically a correlation coefficient may be as large as ± 1, in real research such values do not occur. Our scores reflect the behaviors of living organisms who always show variability. Therefore, adjust your expectations about real correlation coefficients: Typically, researchers obtain coefficients in the neighborhood of $\pm.30$ to $\pm.50$, so below $\pm.30$ is considered rather weak and above $\pm.50$ is considered extremely strong.

Finally, published research often involves a rather large N, producing a complex scatterplot that is diffi-cult to read. Therefore, instead, a graph showing only the regression line may be included, especially when the report focuses on predicting Y scores.

10.4 UNDER-STANDING LINEAR REGRESSION

Recall that we use a relationship and an individual's X score to predict his or her Y score. The regression line is the basis for making such predictions. **Linear regression** is the procedure for predicting unknown Y scores based on correlated X scores. To use regression, we first establish the relationship by computing r for our sample and determining that it *is* significant. Then, using the regression line, we identify the Y score that everyone in our sample scored *around* when they scored at a particular X. We predict that anyone else at that X would also score at around that Y. Therefore, we can measure the X scores of individuals who were not in our sample and we have a good idea of what their corresponding Y score would be. For example, the reason that students take the Scholastic Aptitude Test (SAT) when applying to some colleges is because researchers have previously established that SAT scores are positively correlated with college grades: We know the typical college grade average (Y) that is paired with a particular SAT score (X). Therefore, through regression techniques, the SAT scores of applying students are used to predict their future college performance. If the predicted grades are too low, the student is not admitted to the college.

Because we base our predictions on someone's X score, in statistical lingo the X variable is often referred to as the **predictor variable**. The Y variable is called the **criterion variable**. Thus, above, SAT score is the predictor variable and college grade average is the crite-rion variable.

You can understand the logic of how we use the regression line to predict scores by looking at Fig-ure 10.12. On the left is the scatterplot showing the relationship between scores on Test A and scores on Test B. The regression line passes through the center of the scatterplot so that the individuals at a particular X score have Y scores that are literally *around* (above and below) the line. For example, people scoring an X of 3 have Ys of 3, 4, and 5, so the regression line goes through the middle of these Y scores at 4. Thus, think of the regression line in the graph on the right as the summary of this relationship, composed of a series of data points itself, with each Y on the line located in the middle of the Y scores paired with that X. Thus, following the dotted lines, it shows that people at an X of 3 scored a Y around 4. Therefore, for anyone else scoring an X of 3, we'd predict a Y of 4 for them also. Notice that the symbol for this **predicted Y score** is Y' (pronounced Y *prime*).

Real data won't form such a perfectly balanced scatterplot as shown here, but each Y' will always be the central, summary Y score at an X based on the linear relationship in the data. Thus, even if no one at the X of 3 had actually scored a Y of 4, the line shows that participants scored *around* 4, so 4 is still our best prediction. Likewise, in the left-hand graph, people at X = 1 scored Ys around 2, so in the right-hand graph, for anyone scoring this X, we'd predict for them a Y' = 2. Thus, the Y' at any X is the value of Y falling *on* the regression line. The regression line, therefore, consists of the data points formed by pairing every possible value of X with its Y'.

Essentially, predicting Y' is the equivalent of traveling vertically from a particular X to the regression line and then traveling horizontally to determine the predicted Y' for that X. We actually accomplish this,

however, by using the linear regression equation. The **linear regression equation** is the equation that produces the value of Y' at each X and defines the straight line that summarizes a relationship. Thus, the equation allows us to do two things: First, we use the equation to produce the value of Y' at several Xs. When we plot the data points for these X-Y' pairs and connect them with a line, we have plotted the regression line. Second, because the equation allows us to determine the Y' at any X, we can use the equation to directly predict anyone's Y score.

The general form of the linear regression equation is $Y' = bX + a$. The b stands for the *slope* of the line, a number indicating the degree and direction the line slants. The X stands for the score on the X variable. The a stands for the Y *intercept,* the value of Y when the regression line intercepts, or crosses, the Y axis. Together, the slope and intercept describe how, starting at a particular value, the Y scores tend to change with each change in X. Thus, using our data we compute a and b (see Appendix A.3 for their formulas). Then we find the value of Y' for a particular X by multiplying b times that X and adding a.

The accuracy of our predictions depends on the strength of the relationship. (For example, SAT scores are not a very good predictor of college grades because they are

predicted Y score (Y') In linear regression, the best prediction of the Y scores at a particular X, based on the linear relationship summarized by the regression line

linear regression equation The equation that produces the value of Y' at each X and defines the straight line that summarizes a relationship

Figure 10.12

Graphs Showing the Actual Y Scores and the Predicted Y Scores from the Regression Line

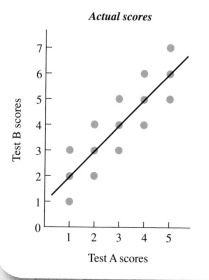

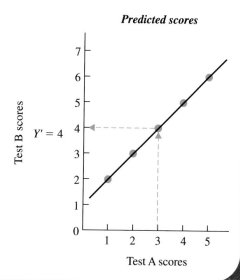

proportion
of variance
accounted for
The proportion of the
differences in *Y* scores
that is associated
with changes in the
X variable

only modestly correlated.) Recall that we said the larger the correlation coefficient, the more accurate will be our predictions. This is because in a stronger relationship, the data points at each *X* are *closer* to the regression line. Then the *Y* scores that participants actually obtain are closer to the *Y'* scores that we predict for them, so we have greater accuracy. Note, however, that although *r* indirectly communicates how accurate our predictions will be, an advanced statistic called the *standard error of the estimate* is used to directly measure our accuracy. (See Appendix A.3.) The standard error of the estimate is somewhat like the "average" difference between the actual *Y* scores that participants would obtain and the *Y'* scores we predict for them. Therefore, the larger it is, the greater our "average" error when using regression procedures to predict *Y* scores.

10.5 THE PROPORTION OF VARIANCE ACCOUNTED FOR: r^2

From the correlation coefficient we can compute one more piece of information about a relationship, called the *proportion of variance accounted for*. In the previous chapter we saw that, with experiments, this measured the "effect size." With correlational studies, we don't call it effect size, because we cannot confidently conclude that changing our *X* variable caused *Y* to change, so we can't say that *X* had an "effect." The logic, however, is the same. In any relationship, the **proportion of variance accounted for** describes the proportion of all differences in *Y* scores that are associated with changing the *X* variable. For example, consider the scores in this simple relationship:

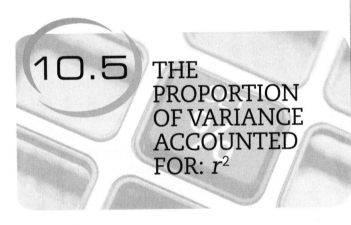

X	Y
1	1
1	2
1	3
2	5
2	6
2	7

When we change from an *X* of 1 to an *X* of 2, the *Y*s change from scores around 2 to scores around 6. These are differences in *Y* associated with changing *X*. However, we also see differences in *Y* when *X* does not change: At an *X* of 1, for example, one participant had a 1 while someone else had a 2. These are differences not associated with changing *X*. Thus, out of all the differences among these six *Y* scores, some differences are associated with changing *X* and some differences are not. The proportion of variance accounted for is the proportion of all differences in *Y* scores that are associated with changes in *X* scores.

With the Pearson correlation coefficient, the proportion of variance accounted for is easy to compute.

THE PROPORTION OF VARIANCE ACCOUNTED FOR

Proportion of variance accounted for $= r^2$

Not too tough! Compute *r* and then square it. For example, previously our age and agility scores produced $r = -.45$, so $r^2 = (-.45)^2 = .20$. Thus, out of all of the differences in the agility scores, .20 or 20% of them are associated with differences in our men's ages. (80% of the differences in agility scores do *not* occur along with changes in age.)

While *r* describes the consistency of the pairing of a particular *X* with a particular *Y* in a relationship, r^2 is slightly different: It indicates the extent to which the differences in *Y* occur along with changes in *X*. The r^2 can be as low as 0 (when $r = 0$), indicating that no differences in *Y* scores are associated with *X*, to as high as 1 (when $r = \pm 1$), indicating that 100% of the changes in *Y* occur when *X* changes, with no differences in *Y* occuring at the same *X*. However, we previously noted that in real research correlation coefficients tend to be between .30 and .50. Therefore, squaring these values indicates that the proportion of variance accounted for is usually between .09 and .25.

REMEMBER

r^2 *indicates the* proportion of variance accounted for *by the relationship. This is the proportion of all differences in* Y *that occur with changes in* X.

Notice that the proportion of variance accounted for also conveys how accurately the relationship allows us to predict Y scores. If 20% of the differences in agility scores are related to a man's age, then by using the resulting regression equation, we can predict these 20% of the differences. In other words, we are, on average, 20% closer to knowing a man's specific agility score when we use this relationship to predict scores, compared to if we did not use—or know about—the relationship. Therefore, this relationship improves our ability to predict agility scores by 20%.

The proportion of variance accounted for is used to judge the usefulness and scientific importance of a relationship. Although a relationship must be significant in order to be potentially important (and for us to even compute r^2), it can be significant and still be unimportant. The r^2 indicates the importance of a relationship because the larger it is, the closer the relationship gets us to our goal of understanding behavior. That is, by understanding the differences in Y scores that are associated with changes in X, we are actually describing the differences in *behaviors* that are associated with changes in X.

Further, we compare different relationships by comparing each r^2. Say that in addition to correlating age and agility, we also correlated a man's weight with his agility, finding a significant $r_{obt} = -.60$, so $r^2 = .36$. Thus, while a man's age accounts for only 20% of the differences in agility scores, his weight accounts for 36% of these differences. Therefore, the relationship involving a man's weight is the more useful and important relationship, because with it we are better able to predict and understand differences in physical agility.

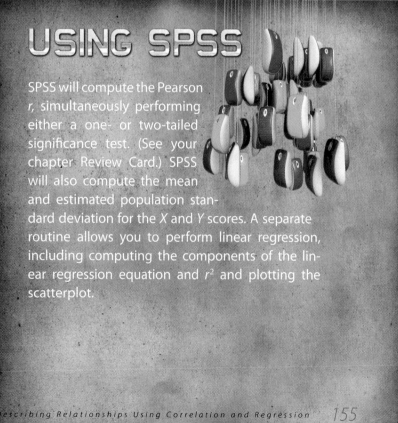

USING SPSS

SPSS will compute the Pearson r, simultaneously performing either a one- or two-tailed significance test. (See your chapter Review Card.) SPSS will also compute the mean and estimated population standard deviation for the X and Y scores. A separate routine allows you to perform linear regression, including computing the components of the linear regression equation and r^2 and plotting the scatterplot.

1. What distinguishes correlational research from experimental research?

2. When is it appropriate to calculate the Pearson *r*?

3. When is a scatterplot used and why is it used?

4. What is a regression line?

5. What two facts about a linear relationship are conveyed by a correlation coefficient?

6. What is meant by the strength of a linear relationship?

7. a. How does the strength of a relationship between two variables affect one's ability to make predictions about *Y* when *X* is known?

   b. Explain your answer.

8. If you have computed the correlation coefficient between two variables to be +1.7, should you think you have observed a very strong relationship? Explain your answer.

9. You correctly compute a correlation coefficient of +1.0 between the variables *X* and *Y*.

   a. What does this imply in terms of how the *X* and *Y* scores change, the scatterplot, and the degree of linear relationship?

   b. If, instead, you find the correlation coefficient to be 0, what would this mean?

   c. If, instead, you compute the coefficient as −.45, what would this mean?

10. What does ρ stand for, and how do we determine its value?

11. What does *Y'* represent?

12. Explain what each of the following correlation coefficients indicates about the direction in which *Y* scores change as *X* scores increase, the shape of the scatterplot, the variability of the *Y* scores at each *X*, and how closely the *Y* scores hug the regression line.

   a. −1.0   c. −.10

   b. +.32   d. −.71

13. A study of fitness habits produced the following interval/ratio scores.

Participants	Hours of Exercise (X)	Reported Life Satisfaction (Y)
1	2	6
2	0	2
3	5	13
4	6	15
5	1	3
6	2	6
7	4	10
8	4	12
9	3	8
10	4	10

   a. Calculate the correlation coefficient.

   b. How would you describe this relationship?

14. a. Generate a scatterplot for the data in problem 13.

   b. Does the scatterplot support your answer for part b in problem 13? Explain your answer.

15. If you have calculated the correlation coefficient for variables *X* and *Y* as −.63, what is the proportion of variance accounted for?

16. In a follow-up study to the one reported in problem 13, the following interval/ratio scores were recorded.

Participants	Hours of Television Watched (X)	Reported Life Satisfaction (Y)
1	10	4
2	8	2
3	3	16
4	6	11
5	2	20
6	2	16
7	12	3
8	8	10
9	6	14
10	4	13
11	7	8
12	12	2
13	5	14
14	7	6
15	9	8
16	3	18
17	10	3
18	6	10
19	8	15
20	4	18

   a. Generate a scatterplot for the data.

   b. Use the scatterplot to predict the direction and value of the correlation coefficient.

   c. Calculate the correlation coefficient.

   d. How would you describe this relationship?

   e. How close was your prediction to the actual value?

   f. What is the proportion of variance accounted for?

   g. Conduct a two-tailed inference test on *r* using α = .05.

Hypothesis Testing
Using the One-Way Analysis of Variance

In this chapter we return to analyzing experiments. We have only one more common type of inferential procedure to learn, and it is called the *analysis of variance*. It is used in experiments involving more than two conditions. This chapter will show you (1) the general logic behind the analysis of variance, (2) how to perform this procedure, and (3) an additional part to this analysis called *post hoc tests*.

11.1 AN OVERVIEW OF THE ANALYSIS OF VARIANCE

It is important to know about analysis of variance because it is *the* most common inferential statistical procedure used to analyze experiments. Why? Because there are actually many versions of this procedure, so it can be used with many different designs: It can be applied to independent or related samples, to an independent variable involving any number of conditions, and to a study involving any number of independent variables. Such complex designs are common, so you'll often encounter this statistic when conducting your own research or when reading that of others.

The analysis of variance has its own language that is also commonly used in research:

1. Analysis of variance is abbreviated as *ANOVA*.

Looking Back

- From Chapter 1, know what an independent variable, a dependent variable, and a condition are.
- From Chapter 4, understand that variance measures the differences between scores.
- From Chapter 7, know why we limit the probability of a Type I error to .05.
- From Chapter 9, know what independent samples and related samples are.

© Influx Productions/Photodisc/Jupiterimages

2. An independent variable is also called a **factor.**

3. Each condition of the independent variable is also called a **level** or a **treatment,** and differences produced by the independent variable are a **treatment effect.**

4. The symbol for the number of levels in a factor is k.

5. A **one-way ANOVA** is performed when one independent variable is tested. (A "two-way" ANOVA is used with two independent variables, and so on.)

6. When an independent variable is studied using independent samples, it is called a **between-subjects factor** and involves using the formulas from a **between-subjects ANOVA.**

7. When a factor is studied using related samples, it is called a **within-subjects factor** and requires the formulas for a **within-subjects ANOVA.**

In this chapter we discuss the one-way, between-subjects ANOVA. (The formulas for a one-way within-subjects ANOVA are presented in Appendix A.5.) As an example, let's examine how

ANOVA KEY TERMS

FACTOR In ANOVA, an independent variable

LEVELS (TREATMENTS) The conditions of the independent variable

TREATMENT EFFECT The result of changing the conditions of an independent variable so that different populations of scores having different μs are produced

ONE-WAY ANOVA The analysis of variance performed when an experiment has only one independent variable

BETWEEN-SUBJECTS FACTOR An independent variable that is studied using independent samples in all conditions

BETWEEN-SUBJECTS ANOVA The type of ANOVA that is performed when a study involves between-subjects factors

WITHIN-SUBJECTS FACTOR The type of factor created when an independent variable is studied using related samples in all conditions because participants are either matched or repeatedly measured

WITHIN-SUBJECTS ANOVA The type of ANOVA performed when a study involves within-subjects factors

---

people perform a task, depending on how difficult they believe the task will be (the "perceived difficulty" of the task). We'll create three conditions containing the unpowerful n of 5 participants each and provide them with the same easy 10 math problems. However, we will tell participants in Level 1 that the problems are easy, in Level 2 that the problems are of medium difficulty, and in Level 3 that the problems are difficult. Our dependent variable is the number of problems that participants solve within an allotted time.

The way to diagram a one-way ANOVA is shown in Table 11.1. Each column is a level of the factor, con-taining the scores of participants tested under that con-dition (here, symbolized by X). The mean of each level is the mean of the scores from that column. Because we have three levels, $k = 3$. (Notice that the general format is to label the factor as factor A, with levels A_1, A_2, A_3, and so on.)

As usual, the purpose here is to demonstrate a relationship between the independent variable and the dependent variable. Ideally, we'll find a different mean for each condition, suggesting that if we tested the entire population under each level of difficulty, we would find three different popu-lations of scores located at three different μs. But there's the usual problem: Perceived difficulty may not actu-ally produce differences in performance. Instead, the differences between our means may be due to sampling error, so that actu-ally we would find one population of scores, having the same μ, for all levels of difficulty. There-fore, we must test

Table 11.1

Diagram of a Study Having Three Levels of One Factor
Each column represents a condition of the independent variable.

Factor A: Independent Variable of Perceived Difficulty			
Level A_1: Easy	Level A_2: Medium	Level A_3: Difficult	Conditions ← $k = 3$
X	X	X	
X	X	X	
X	X	X	
X	X	X	
X	X	X	
$\overline{X}_1$	$\overline{X}_2$	$\overline{X}_3$	

LEVEL 3

whether the differences between our sample means reflect sampling error. The **analysis of variance (ANOVA)** is the parametric procedure for determining whether significant differences occur in an experiment containing two or more sample means. Don't be confused by this: When you have only two conditions, you can use either a two-sample *t*-test or the ANOVA. However, you *must* use ANOVA when you have more than two conditions.

The one-way between-subjects ANOVA requires that

1. all conditions contain independent samples

2. the dependent scores are normally distributed, interval or ratio scores

3. the variances of the populations are homogeneous.

Note: The *ns* in all conditions need not be equal, although they should not be massively different. However, these procedures are *much* easier to perform with equal *ns*.

Controlling the Experiment-Wise Error Rate

You might be wondering why we don't use the independent-samples *t*-test to find significant differences among the three means above. We might test whether $\overline{X}_1$ differs from $\overline{X}_2$, whether $\overline{X}_2$ differs from $\overline{X}_3$, and whether $\overline{X}_1$ differs from $\overline{X}_3$. The reason we cannot use this approach is because of the resulting probability of making a Type I error (rejecting a true H_0).

To understand this, we must distinguish between the probability of making a Type I error *when comparing a pair of means,* and the probability of making a Type I error *somewhere* in the experiment. With $\alpha = .05$, the probability of a Type I error when comparing a pair of means in a *single t*-test was .05. But now we can make a Type I error somewhere in the experiment when comparing $\overline{X}_1$ to $\overline{X}_2$, $\overline{X}_2$ to $\overline{X}_3$, or $\overline{X}_1$ to $\overline{X}_3$. The probability of making a Type I error somewhere in the experiment is called the **experiment-wise error rate.**

We can use the *t*-test when comparing only two means in an experiment because with only one comparison, the experiment-wise error rate equals α. But

REMEMBER

The reason for performing ANOVA is that it keeps the experiment-wise error rate *equal to* α.

analysis of variance (ANOVA) The parametric procedure for determining whether significant differences exist in an experiment containing two or more sample means

experiment-wise error rate The probability of making a Type I error when comparing all means in an experiment

with more than two means in the experiment, performing multiple *t*-tests would result in an experiment-wise error rate that is much larger than our α. Because of the importance of avoiding Type I errors, we do not want the error rate to be larger than we think it is, and it should never be larger than .05. Instead, we perform ANOVA: With it the experiment-wise error rate will equal the alpha we've chosen.

Statistical Hypotheses in ANOVA

ANOVA tests only two-tailed hypotheses. The null hypothesis is that there are no differences between the populations represented by the conditions. Thus, for our perceived difficulty study with the three levels of easy, medium, and difficult, we have

$$H_0: \mu_1 = \mu_2 = \mu_3$$

In general, when we perform ANOVA on a factor with k levels, the null hypothesis is $H_0: \mu_1 = \mu_2 = \ldots = \mu_k$. The " $\ldots = \mu_k$" indicates that there are as many μs as there are levels.

However, the alternative hypothesis is not $H_a: \mu_1 \neq \mu_2 \neq \mu_3$. A study may demonstrate differences between *some* but not *all* conditions. (Perhaps our data represent a difference between μ_1 and μ_2, but not between μ_1 and μ_3, or perhaps only μ_2 and μ_3 differ.) To communicate this idea, the alternative hypothesis is

$$H_a: \text{not all } \mu\text{s are equal}$$

H_a implies that a relationship is present because the population mean represented by one of the level means is different from the population mean represented by at least one other mean.

As usual, we test H_0, so ANOVA tests whether all sample means represent the same population mean.

Chapter 11: Hypothesis Testing Using the One-Way Analysis of Variance 161

The *F* Statistic and Post Hoc Comparisons

Completing an ANOVA requires two major steps. First, we compute the statistic called *F* to determine whether any of the means represent different μs. We calculate F_{obt} which we compare to F_{crit}.

When F_{obt} is not significant, it indicates that there are no significant differences between any means. Then, the experiment has failed to demonstrate a relationship and it's back to the drawing board.

When F_{obt} is significant, it indicates only that *somewhere* among the means *at least two* of them differ significantly. But, F_{obt} does not indicate *which* specific means differ significantly. If F_{obt} for the perceived difficulty study is significant, then we have one or more significant differences somewhere among the means of the easy, medium, and difficult levels, but we won't know where they are.

Therefore, when F_{obt} is significant we perform a second statistical procedure, called *post hoc comparisons*. **Post hoc comparisons** are like *t*-tests, in which we compare all possible *pairs* of means from a factor, one pair at a time, to determine which means differ significantly from each other. Thus, for the difficulty study we'll compare the means from easy and medium, from easy and difficult, and from medium and difficult. However, we perform post hoc comparisons *only* when F_{obt} is significant. A significant F_{obt} followed by post hoc comparisons ensures that the experiment-wise probability of a Type I error will equal our alpha.

The one exception to this rule is when you have only two levels in the factor. Then the significant difference indicated by F_{obt} must be between the only two means in the study, so it is unnecessary to perform post hoc comparisons.

---

post hoc comparisons In ANOVA, statistical procedures used to compare all possible pairs of sample means in a significant effect to determine which means differ significantly from each other

mean square (MS) In ANOVA, an estimated population variance

mean square within groups (MS_{wn}) In ANOVA, the variability in scores that occurs in the conditions

mean square between groups (MS_{bn}) In ANOVA, the variability that occurs between the levels in a factor

---

11.2 COMPONENTS OF THE ANOVA

The analysis of variance does just that—it analyzes variance. It involves computing variance from two perspectives in the sample data, so that we can estimate two components in the population. But, we do not *call* them an estimated variance. Instead, each is called a **mean square**. (This is shortened from "mean of the squared deviations," which is what variance actually is.) The symbol for a mean square is **MS**. The two mean squares that we compute are the *mean square within groups* and the *mean square between groups*.

The Mean Square within Groups

The **mean square within groups** describes the variability in scores *within* the conditions of an experiment. It is symbolized by MS_{wn}. Recall that variance is a way to measure the differences among the scores. Here, we find the differences among the scores within each condition and "pool" them (like we did in the independent-samples *t*-test). Thus, the MS_{wn} is the "average" variability of the scores within each condition. Because we look at scores within one condition at a time, MS_{wn} stays the same regardless of whether H_0 is true or false. Either way, the MS_{wn} is an estimate of the variability of *individual* scores in any of the populations being represented.

The Mean Square between Groups

The other variance we compute is the mean square between groups. The **mean square between groups** describes the differences *between* the levels in a factor. It is symbolized by MS_{bn}. Essentially, the computations of MS_{bn} measure how much the means of the conditions differ from each other. This MS_{bn} is used to estimate the differences between the sample means that would be found if we repeatedly sampled *one* population. That is, we are testing the H_0 that our data all come from the same single population. If so, sample means from that population will not

© iStockphoto.com/Andy Porter

necessarily equal μ or each other every time because of sampling error. Therefore, when H_0 is true, the differences between our means as measured by MS_{bn} will not be zero. Instead, MS_{bn} is an estimate of the "average" amount that sample means from that one population differ from each other due to chance, sampling error.

As we'll see, performing the ANOVA involves first using our data to compute the MS_{wn} and MS_{bn}. The final step is to then compare them by computing F_{obt}.

Comparing the Mean Squares: The Logic of the *F*-Ratio

The test of H_0 is based on the fact that statisticians have shown that when samples of scores are selected from *one* population, the size of the differences among the sample means will *equal* the size of the differences among individual scores. This makes sense because how much the sample means differ depends on how much the individual scores differ. Say that the variability in the population is small so that all scores are very close to each other. When we select samples we will have little variety in scores to choose from, so each sample will contain close to the same scores as the next and their means also will be close to each other. However, if the variability is very large, we have many different scores available. When we select samples of these scores, we will often encounter a very different batch each time, so the means also will be very different each time.

Here now is the key: the MS_{bn} estimates the variability of sample means in the population and the MS_{wn}

REMEMBER

In one population, the variability of sample means will equal the variability of individual scores.

estimates the variability of individual scores in the population. We've just seen that when we are dealing with only one population, sample means and individual scores will differ to the same degree. Therefore, when we are dealing with only one population, the MS_{bn} should *equal* MS_{wn}: the answer we compute for MS_{bn} should be the same answer as for MS_{wn}. *Our H_0 always says that we are dealing with only one population, so if H_0 is true for our study, then our MS_{bn} should equal our MS_{wn}.*

An easy way to determine if two numbers are equal is to make a fraction out of them, which is what we do when computing F_{obt}.

THE *F*-RATIO

$$F_{obt} = \frac{MS_{bn}}{MS_{wn}}$$

This fraction is referred to as the F-*ratio*. The **F-ratio** equals the MS_{bn} divided by the MS_{wn}. (The MS_{bn} is always on top!)

If we place the same number in the numerator as in the denominator, the ratio and our F_{obt} will equal 1. Thus, when H_0 is true and we are representing one population, the MS_{bn} should equal the MS_{wn}, so F_{obt} should equal 1.

Of course F_{obt} may not equal 1.0 *exactly* when H_0 is true, because we may have sampling error in either MS_{bn} or MS_{wn}. That is, either the differences among our individual scores and/or the differences among our means may inaccurately represent the corresponding differences in the population. Therefore, realistically, we expect that, if H_0 is true for our study, F_{obt} will equal 1 or will be close to 1. Also, if F_{obt} is less than 1, mathematically it can only be that H_0 is true and that we have sampling error in representing this. (Each MS is a variance, in which we square differences, so F_{obt} cannot be negative.)

It gets interesting, however, as F_{obt} becomes *larger* than 1. The H_0 implies that F_{obt} is "trying" to equal 1, and if it doesn't, it is because of sampling error. Let's think about that. If $F_{obt} = 2$, it is twice what H_0 "says" it should be, although according to H_0 we should

F-ratio In ANOVA, the ratio of the mean square between groups to the mean square within groups

sum of squares (SS) The sum of the squared deviations of a set of scores around the mean of those scores

conclude "no big deal—a little sampling error." Or, say that $F_{obt} = 4$, so that the MS_{bn} is *four times* the size of MS_{wn} (and F_{obt} is four times what H_0 says it should be.) Yet, H_0 claims that MS_{bn} would have equaled MS_{wn} except that, by chance, we happened to get some unrepresentative scores. If F_{obt} is, say, 10, then it and the MS_{bn} are *ten times* what H_0 says they should be! Still, H_0 would suggest that this is merely because we had sampling error in representing the population.

As this illustrates, the larger the F_{obt}, the more difficult it is to believe that *by luck* our data are poorly representing the situation where H_0 is true. Of course, if sampling error won't explain so large an F_{obt}, then we need something else that will. The answer is our independent variable. When H_a is true so that changing our conditions would involve more than one population of scores, MS_{bn} will be *larger* than MS_{wn}, and F_{obt} will be larger than 1. Further, the more that changing the levels of our factor changes scores, the larger will be the differences between our level means and so the larger will be MS_{bn}. (However, the MS_{wn} stays the same regardless of whether H_0 is true or not.) Thus, greater differences produced by our factor will produce a larger MS_{bn} which produces a larger F_{obt}. Turning this around, the larger the F_{obt}, the more it appears that H_a is true. Putting this all together:

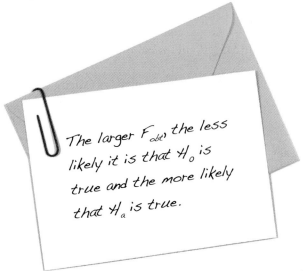

The larger F_{obt}, the less likely it is that H_0 is true and the more likely that H_a is true.

If our F_{obt} is large enough to be beyond F_{crit}, we will conclude that our F_{obt} is unlikely to occur if H_0 were true, so we will reject H_0 and accept H_a. (This is the logic of all ANOVAs, whether for a between- or within-subjects design.)

11.3 PERFORMING THE ANOVA

When we computed the estimated variance in Chapter 4, the quantity $\Sigma(X - \overline{X})^2$ was called the "sum of the squared deviations." In calculating the mean squares, we perform a similar operation, so this name is shortened to the **sum of squares.** The symbol for the sum of squares is **SS**. Then the general formula for a mean square is the fraction formed by the sum of squares divided by the degrees of freedom, or

$$MS = \frac{SS}{df}$$

Adding subscripts, we compute the mean square between groups (MS_{bn}) by computing the sum of squares between groups (SS_{bn}) and dividing by the degrees of freedom between groups (df_{bn}). Likewise, we compute the mean square within groups (MS_{wn}) by computing the sum of squares within groups (SS_{wn}) and dividing by the degrees of freedom within groups (df_{wn}). With MS_{bn} and MS_{wn}, we compute F_{obt}.

If all this strikes you as the most confusing thing ever devised, you'll find an ANOVA summary table very helpful. Below is the general format:

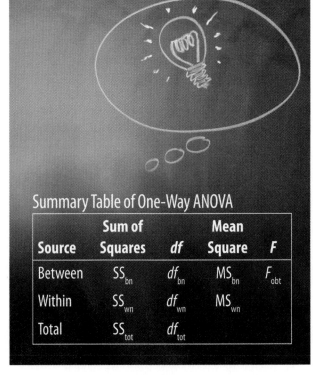

Summary Table of One-Way ANOVA

Source	Sum of Squares	df	Mean Square	F
Between	SS_{bn}	df_{bn}	MS_{bn}	F_{obt}
Within	SS_{wn}	df_{wn}	MS_{wn}	
Total	SS_{tot}	df_{tot}		

The "source" column identifies each source of variation, either *between* or *within*, and we also consider the *total*. Using the following formulas, we'll compute the components for the other columns.

Computing F_{obt}

Say that we performed the perceived difficulty study discussed earlier, telling participants that some math problems were easy, of medium difficulty, or difficult, and measuring the number of problems they solved. The data are presented in Table 11.2. As shown in the following sections, there are four parts to computing the F_{obt}: finding (1) the sum of squares, (2) the degrees of freedom, (3) the mean squares, and (4) F_{obt}.

Computing the Sums of Squares The computations here require 4 steps.

STEP 1: *Compute the sums and means.* As in Table 11.2, compute ΣX, ΣX^2, and $\overline{X}$ for each level (each column). Each n is the number of scores in the level. Then add together the ΣX from all levels to get the total, which is "ΣX_{tot}." Also, add together the ΣX^2 from all levels to get the total, which is "ΣX_{tot}^2." And, add the ns together to obtain N.

Table 11.2
Data from Perceived Difficulty Experiment

Factor A: Perceived Difficulty			
Level A_1: Easy	Level A_2: Medium	Level A_3: Difficult	
9	4	1	
12	6	3	
4	8	4	
8	2	5	
7	10	2	
			Totals
$\Sigma X = 40$	$\Sigma X = 30$	$\Sigma X = 15$	$\Sigma X_{tot} = 85$
$\Sigma X^2 = 354$	$\Sigma X^2 = 220$	$\Sigma X^2 = 55$	$\Sigma X_{tot}^2 = 629$
$n_1 = 5$	$n_2 = 5$	$n_3 = 5$	$N = 15$
$\overline{X}_1 = 8$	$\overline{X}_2 = 6$	$\overline{X}_3 = 3$	$k = 3$

STEP 2: *Compute the total sum of squares (SS_{tot}).*

THE TOTAL SUM OF SQUARES

$$SS_{tot} = \Sigma X_{tot}^2 - \frac{(\Sigma X_{tot})^2}{N}$$

Using the data from Table 11.2, $\Sigma X_{tot}^2 = 629$, $\Sigma X_{tot} = 85$, and $N = 15$, so

$$SS_{tot} = 629 - \frac{(85)^2}{15}$$

$$SS_{tot} = 629 - \frac{7225}{15}$$

$$SS_{tot} = 629 - 481.67$$

Thus, $SS_{tot} = 147.33$.

STEP 3: *Compute the sum of square between groups (SS_{bn}).*

SUM OF SQUARES BETWEEN GROUPS

$$SS_{bn} = \Sigma\left(\frac{(\Sigma X \text{ in column})^2}{n \text{ in column}}\right) - \frac{(\Sigma X_{tot})^2}{N}$$

In Table 11.2, each column represents a level of the factor. Thus, find the ΣX for a column, square that ΣX, then divide by the n in that level. After doing this for all levels, add the results together and subtract the quantity $(\Sigma X_{tot})^2/N$:

$$SS_{bn} = \left(\frac{(40)^2}{5} + \frac{(30)^2}{5} + \frac{(15)^2}{5}\right) - \frac{(85)^2}{5}$$

so

$$SS_{bn} = (320 + 180 + 45) - 481.67$$

and

$$SS_{bn} = 545 - 481.67$$

So, $SS_{bn} = 63.33$.

STEP 4: *Compute the sum of squares within groups (SS_{wn}).* Mathematically, SS_{tot} equals SS_{bn} plus SS_{wn}. Therefore, the total minus the between leaves the within.

SUM OF SQUARES WITHIN GROUPS

$$SS_{wn} = SS_{tot} - SS_{bn}$$

In the example, SS_{tot} is 147.33 and SS_{bn} is 63.33 so

$$SS_{wn} = 147.33 - 63.33$$

$$SS_{wn} = 84.00$$

Computing the Degrees of Freedom
Compute the df_{bn}, df_{wn}, and df_{tot}.

STEP 1: *The degrees of freedom between groups equals* k − 1, *where* k *is the number of levels in the factor. In the example there are three levels of perceived difficulty, so* $k = 3$. *Thus,* $df_{bn} = 2$.

STEP 2: *The degrees of freedom within groups equals* N − k, *where* N *is the total N in the experiment and* k *is the number of levels in the factor. In the example* N *is 15 and* k *is 3, so* $df_{wn} = 15 − 3 = 12$.

STEP 3: *The degrees of freedom total equals* N − 1, *where* N *is the total N in the experiment. In the example* N *is 15, so* $df_{tot} = 15 − 1 = 14$.

The df_{tot} must equal the df_{bn} plus the df_{wn}. At this point the ANOVA summary table looks like this:

Source	Sum of Squares	df	Mean Square	F
Between	63.33	2	MS_{bn}	F_{obt}
Within	84.00	12	MS_{wn}	
Total	147.33	14		

Computing the Mean Squares
You can work directly from the summary table to compute the mean squares.

STEP 1: *Compute the mean square between groups.*

THE MEAN SQUARE BETWEEN GROUPS

$$MS_{bn} = \frac{SS_{bn}}{df_{bn}}$$

From the summary table

$$MS_{bn} = \frac{63.33}{2} = 31.67$$

so MS_{bn} is 31.67.

STEP 2: *Compute the mean square within groups.*

THE MEAN SQUARE WITHIN GROUPS

$$MS_{wn} = \frac{SS_{wn}}{df_{wn}}$$

For the example

$$MS_{wn} = \frac{84}{12} = 7.00$$

so MS_{wn} is 7.00.

Do *not* compute the mean square for SS_{tot} because it has no use.

Computing the F
Finally, compute F_{obt}.

$$F_{obt} = \frac{MS_{bn}}{MS_{wn}}$$

In the example MS_{bn} is 31.67 and MS_{wn} is 7.00, so

$$F_{obt} = \frac{MS_{bn}}{MS_{wn}} = \frac{31.67}{7.00} = 4.52$$

Thus, F_{obt} is 4.52.

Now the completed ANOVA summary table is

Source	Sum of Squares	df	Mean Square	F
Between (difficulty)	63.33	2	31.67	4.52
Within	84.00	12	7.00	
Total	147.33	14		

The F_{obt} is always placed in the row labeled "Between." (Because this row reflects differences due to our treatment, we may also include the name of the independent variable here.)

Interpreting F_{obt}
The final step is to compare F_{obt} to F_{crit}, and for that we examine the F-distribution. The **F-distribution** is the sampling distribution showing the various values of F that occur when H_0 is true and all conditions represent one population. To create it, it is as if, using our ns and k, we select the scores for all of our conditions from one raw score population (like H_0 says we did in our experiment), and compute MS_{wn}, MS_{bn}, and then F_{obt}. We do this an infinite number of times, and plotting the Fs produces the sampling distribution, as shown in Figure 11.1.

© iStockphoto.com/malerapaso

The F-distribution is skewed because there is no limit to how large F_{obt} can be, but it cannot be less than zero. The mean of the distribution is 1 because, most often when H_0 is true, MS_{bn} will equal MS_{wn}, and so F will equal 1. The right-hand tail, however, shows that sometimes, by chance, F is greater than 1. Because our F_{obt} can reflect a relationship in the population only when it is greater than 1, the entire region of rejection is in this upper tail of the F-distribution. (That's right, ANOVA involves two-tailed hypotheses, but they are tested using only the upper tail of the sampling distribution.) If F_{obt} is larger than F_{crit}, then F_{obt}—and the differences between our level means—is so unlikely when H_0 is true that we reject H_0.

The F-distribution is actually a family of curves, each having a slightly different shape, depending on our degrees of freedom. However, two values of df determine the shape of an F-distribution: the df used when computing the mean square between groups (df_{bn}) and the df used when computing the mean square within groups (df_{wn}). Therefore, to obtain F_{crit}, turn to Table 4 in Appendix B, entitled "Critical Values of F." A portion of this "F-table" is presented in Table 11.3 below. Across the top of the table, the columns are labeled "df between groups." On the left-hand side, the rows are labeled "df within groups." Locate the appropriate column and row using the dfs from your study. The critical values in dark type are for $\alpha = .05$, and those in light type are for $\alpha = .01$. For our example, $df_{bn} = 2$ and $df_{wn} = 12$. For $\alpha = .05$, the F_{crit} is 3.88.

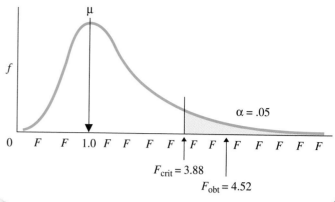

Figure 11.1
Sampling Distribution of F When H_0 Is True for $df_{bn} = 2$ and $df_{wn} = 12$

$\alpha = .05$

$F_{crit} = 3.88$

$F_{obt} = 4.52$

If your df_{bn} are not in the table, for df_{bn} between 30 and 50, your F_{crit} is the average of the two values shown for the df_{bn} that bracket your df_{bn}. For df_{bn} greater than 50, use the same approach as for t_{crit} in Chapter 8: Your results are significant if F_{obt} is larger than the larger bracketing F_{crit}. Your results are not significant if your F_{obt} is smaller than the smaller bracketing F_{crit}.

Thus, our F_{obt} is 4.52 and F_{crit} is 3.88, producing the complete sampling distribution in Figure 11.1 above. The H_0 says that our F_{obt} is greater than 1 because of sampling error: The differences between the means of our levels are due to sampling error, and all means poorly represent one population mean. However, our F_{obt} is in the region of rejection, telling us that such differences between $\overline{X}$s hardly ever occur when H_0 is true. Because F_{obt} is larger than F_{crit}, we reject H_0. Thus, we conclude that the F_{obt} is significant and that the factor of perceived difficulty produces a significant difference in performance. (Had F_{obt} been less than F_{crit}, then the corresponding differences between our means would *not* be unlikely to occur when H_0 is true, so we would not reject H_0.)

Because we rejected H_0 and accepted H_a, we return to the means from the levels:

Table 11.3
Portion of Table 4 in Appendix B, "Critical Values of F"

Degrees of Freedom Within Groups (degrees of freedom in denominator of F-ratio)	Degrees of Freedom Between Groups (degrees of freedom in numerator of F-ratio)					
	α	1	2	3	4	5
1	.05	**161**	**200**	**216**	**225**	**230**
	.01	4,052	4,999	5,403	5,625	5,764
—	—	—	—	—	—	—
—	—	—	—	—	—	—
11	.05	**4.84**	**3.98**	**3.59**	**3.36**	**3.20**
	.01	9.65	7.20	6.22	5.67	5.32
12	.05	**4.75**	**3.88**	**3.49**	**3.26**	**3.11**
	.01	9.33	6.93	5.95	5.41	5.06

Perceived Difficulty		
Easy	Medium	Difficult
$\overline{X}_1 = 8$	$\overline{X}_2 = 6$	$\overline{X}_3 = 3$

To see the *treatment effect*, look at the overall pattern: Because the

means change, the scores that produce them are changing, so a relationship is present; as perceived difficulty increases, performance scores decrease. However, we do not know if *every* increase in difficulty results in a *significant* decrease in scores. To determine that, we must perform the post hoc comparisons.

Tukey Post Hoc Comparisons

Remember that a significant F_{obt} indicates at least one significant difference somewhere among the level means. To determine which means differ, we perform post hoc comparisons. Statisticians have developed a variety of post hoc procedures that differ in how likely they are to produce Type I or Type II errors. One common procedure that has reasonably low error rates is the **Tukey HSD multiple comparisons test.** It is used only when the *n*s in all levels of the factor are equal. The *HSD* is a rearrangement of the *t*-test that computes the minimum difference between two means that is required for them to differ significantly. (*HSD* stands for the Honestly Significant Difference.) There are four steps to performing the *HSD* test.

STEP 1: *Find* q_k. Use Table 5 in Appendix B, entitled "Values of Studentized Range Statistic." Locate the column labeled with the *k* corresponding to the number of means in your factor. Find the row labeled with the df_{wn} used to compute your F_{obt}. (If your *df* is not in the table, use the *df* in the table that is closest to it.) Then find the q_k for the appropriate α. For our study above, $k = 3$, $df_{wn} = 12$, and $\alpha = .05$, so $q_k = 3.77$.

Tukey's HSD multiple comparisons test The post hoc procedure performed with ANOVA to compare means from a factor in which all levels have equal *n*s

STEP 2: *Compute the* HSD.

TUKEY'S *HSD*

$$HSD = (q_k)\left(\sqrt{\frac{MS_{wn}}{n}}\right)$$

MS_{wn} is the denominator from your significant *F*-ratio, and *n* is the number of scores in each level of the factor.

In the example MS_{wn} was 7.0, *n* was 5, and q_k is 3.77, so

$$HSD = (q_k)\left(\sqrt{\frac{MS_{wn}}{n}}\right) = (3.77)\left(\sqrt{\frac{7.0}{5}}\right)$$
$$= (3.77)(\sqrt{1.4}) = (3.77)(1.183) = 4.46$$

Thus, *HSD* is 4.46.

STEP 3: *Determine the differences between each pair of means.* Subtract each mean from every other mean. Ignore whether differences are positive or negative because this is a two-tailed test.

The differences for the perceived difficulty study can be diagramed as shown below.

Easy	Medium	Difficult
$\bar{X}_1 = 8$	$\bar{X}_2 = 6$	$\bar{X}_3 = 3$

2.0   3.0
5.0
HSD = 4.46

On the line connecting any two levels is the absolute difference between their means.

STEP 4: *Compare each difference to the* HSD. If the absolute difference between two means is *greater than* the HSD, then these means differ significantly. (It's as if you performed a *t*-test on them and t_{obt} was significant.) If the absolute difference between two means is less than or equal to the HSD, then it is *not* a significant difference (and would not produce a significant t_{obt}).

In the example, our *HSD* was 4.46. The means from the easy level (8) and the difficult level (3) differ by more than 4.46, so they differ significantly. The mean from the medium level (6), however, differs from the other means by less than 4.46, so it does not differ significantly from them.

Thus, our final conclusion is that we demonstrated a relationship between performance and perceived difficulty, but only when we changed from the easy to the difficult condition. If everyone in the population were tested under each of these conditions, we would expect to find two populations of scores, one for the easy condition at a μ around 8, and one for the difficult condition at a μ around 3. (In fact, recall from Chapter 8 that we can compute a *confidence interval* to describe the range of μs that is likely to be represented by the $\overline{X}$ of a condition.) However, we cannot say anything about the population produced by the medium condition, because it did not produce any significant differences.

3. Compare F_{obt} to F_{crit}. Find F_{crit} in the *F*-table using df_{bn} and df_{wn}. If F_{obt} is larger than F_{crit}, then F_{obt} is significant, indicating that the means in at least two conditions differ significantly.

4. With a significant F_{obt}, more than two levels, and equal *n*s, perform the Tukey *HSD* test.

a. Find q_k in Table 5, using *k* and df_{wn}.

b. Compute the *HSD*.

c. Find the difference between each pair of level means.

d. Any differences larger than the *HSD* are significant.

5. Draw conclusions about the influence of your independent variable by considering the significant means of your levels.

I can't say anything about the population!

11.4 SUMMARY OF THE ONE-WAY ANOVA

t's been a long haul, but after checking the assumptions, here is how to perform a one-way ANOVA:

1. The null hypothesis is H_0: $\mu_1 = \mu_2 = \ldots = \mu_k$, and the alternative hypothesis is H_a: not all μs are equal.

2. Compute F_{obt}.

a. Compute the sums of squares and the degrees of freedom.

b. Compute the mean squares.

c. Compute F_{obt}.

11.5 STATISTICS IN THE RESEARCH LITERATURE: REPORTING ANOVA

n research reports, an F_{obt} is reported using the same format as previous statistics, except that we include both the df_{bn} and the df_{wn}. In the perceived-difficulty study, the significant F_{obt} was 4.52, with $df_{bn} = 2$ and $df_{wn} = 12$. We report this as

$F(2, 12) = 4.52, p < .05$

Note: In the parentheses always report df_{bn} and then df_{wn}.

Usually the *HSD* value is not reported. Instead, indicate that the Tukey *HSD* was performed, the alpha level used, and identify which levels differ significantly. However, for completeness, the means and standard deviations from all levels are reported, even those that do not differ

USING SPSS

Your chapter Review Card describes how to use SPSS to perform the one-way *between-subjects* ANOVA. The program reports the significance level of F_{obt} and performs all steps of the *HSD* test. The program also computes the mean and unbiased standard deviation for each condition, determines the 95% confidence interval for each mean, and plots a line graph of the means. However, it does not compute η^2 (and the "partial eta squared" provided is not what we've discussed.)

Your chapter Review Card also describes how to perform the one-way *within-subjects* ANOVA, although we use a slightly convoluted route. The computations of this ANOVA are discussed in Appendix A.5. The program provides the same information as in the between-subjects ANOVA, except that it will not perform the *HSD* test.

significantly. Likewise, when graphing the results, the means from all levels are plotted.

11.6 EFFECT SIZE AND ETA²

Recall from Chapter 9 that in experiments we report the *effect size,* which indicates the amount of influence that changing the conditions of the independent variable had on dependent scores. The way to measure effect size in the ANOVA is to compute the "proportion of variance accounted for." Recall that this is the proportion of all differences in dependent scores that occur with—and presumably are produced by—changing the conditions of the independent variable.

In ANOVA, effect size is computed by squaring a new correlation coefficient symbolized by the Greek letter

eta squared
(η²) The proportion of variance in the dependent variable that is accounted for by changing the levels of a factor, and thus a measurement of effect size

"eta" (pronounced "ay-tah"). The symbol for eta squared is η^2. **Eta squared** indicates the proportion of variance in dependent scores that is accounted for by changing the levels of a factor. This is analogous to other measures we've discussed in previous chapters, except that η^2 can be used to describe any linear or nonlinear relationship containing two or more levels of a factor. In a particular experiment, η^2 will be a proportion between 0 and 1, indicating the extent to which dependent scores change as the independent variable changes.

ETA SQUARED

$$\eta^2 = \frac{SS_{bn}}{SS_{tot}}$$

The SS_{bn} reflects the differences that occur when we change the conditions. The SS_{tot} reflects the differences between all scores in the experiment. Thus, η^2 reflects the proportion of all differences in scores that are associated with changing the conditions.

For example, for the perceived-difficulty study, SS_{bn} was 63.33 and SS_{tot} was 147.33. So

$$\eta^2 = \frac{SS_{bn}}{SS_{tot}} = \frac{63.33}{147.33} = .43$$

The larger the η^2, the larger the role that factor plays in determining differences among participants' scores, so the more important the variable is for understanding differences in participants' corresponding behaviors. In the example above, 43% of all differences in scores were accounted for (produced) by changing the levels of perceived difficulty. Because 43% is a substantial amount, this factor is very important in determining participants' performance, so it is important for scientific study.

REMEMBER

η^2 (eta squared) *measures the* effect size *of a factor by indicating the proportion of all differences in dependent scores that is accounted for by changing the levels of a factor.*

using what you know

1. a. What does ANOVA stand for?
    b. What research design requires its use?
    c. What does this procedure enable us to decide?

2. a. What is a between-subjects factor?
    b. What is a within-subjects factor?
    c. What type of ANOVA is used in experiments with only one factor?

3. What are the assumptions of the one-way between-subjects ANOVA?

4. What do H_0 and H_a maintain in the ANOVA? (Provide a verbal description.)

5. What are the two types of mean squares and what does each describe?

6. a. Why is F_{obt} greater than 1 when H_0 is false?
    b. Why should F_{obt} equal 1 when H_0 is true?

7. Suppose we predict that the mean of one level will be smaller than the mean of another level. Why can't we perform a one-tailed test in ANOVA?

8. a. Why do we perform the ANOVA in experiments with more than two conditions of the independent variable rather than simply use multiple t-tests?
    b. What is meant by the experiment-wise error rate?

9. a. When is it necessary to perform post hoc comparisons and why?
    b. When is it unnecessary to perform post hoc comparisons and why?
    c. When is it appropriate to use the Tukey HSD test?

10. a. What does eta squared indicate?
    b. Give the symbol for this statistic.

11. Given the following information:

Source	Sum of Squares	df	Mean Square	F
Between groups	127.60	2		
Within groups	593.45	12		
Total	721.05	14		

    a. How many levels were involved?
    b. How many participants were involved?
    c. Complete the ANOVA summary table.
    d. Using $\alpha = .05$, what is F_{crit}?
    e. Is the F_{obt} significant?
    f. Based on your answer in part e, what else should you do?

12. Given the following information:

$\Sigma X_{tot} = 145$ $\Sigma X_1 = 32$

$\Sigma X_{tot}^2 = 881$ $\Sigma X_2 = 47$

$n_1 = n_2 = n_3 = 10$ $\Sigma X_3 = 66$

Source	Sum of Squares	df	Mean Square	F

    a. Complete the ANOVA summary table.
    b. Using $\alpha = .05$, what is F_{crit}?
    c. Is the F_{obt} significant?

13. Katie is studying aggression among adolescent girls. She believes there is a relationship between the level of interaction a girl has with her mother and the level of aggression. She has identified 5 girls who fall into each of 4 interaction levels and has measured their aggression scores. Her data are given below.

No Interaction	Low Interaction	Moderate Interaction	High Interaction
4	4	4	3
5	6	4	4
6	5	3	4
6	4	5	3
5	5	4	3

    a. What are the independent variable and the dependent variable?
    b. Should Katie conduct a between-subjects ANOVA or a within-subjects ANOVA? Explain your answer.
    c. How many factors are involved in Katie's study?
    d. How many levels are involved? Name the level(s).
    e. What are the H_0 and H_a?
    f. Complete the ANOVA summary table.
    g. If you use $\alpha = .05$, what is the appropriate F_{crit}?
    h. Is the F_{obt} significant? Why or why not?
    i. If the F_{obt} is significant, perform the Tukey's HSD.
    j. What should Katie conclude based on these findings?
    k. If appropriate, calculate the effect size.

Understanding the Two-Way Analysis of Variance

two-way ANOVA
The parametric inferential procedure performed when an experiment contains two independent variables

two-way, between-subjects ANOVA
The parametric inferential procedure performed when both factors are between-subjects factors

Researchers often conduct studies that simultaneously involve more than one independent variable. Therefore, this chapter briefly introduces the ANOVA used when experiments involve two factors. *The good news is that we will NOT focus on computations.* Nowadays, we usually analyze such experiments using a statistical computer program (although the formulas for the between-subjects version are presented in Appendix A.4). You will frequently encounter such designs in research publications, so you need to understand their basic logic and terminology. The following sections present (1) the general layout of a two-factor experiment, (2) the logic of the ANOVA, and (3) how to interpret such a study.

SECTIONS

12.1 UNDERSTANDING THE TWO-WAY ANOVA

Looking Back

- From Chapter 11, understand the terms *factor* and *level*, know what a significant *F* indicates, and know when to perform post hoc tests.

© Erik Dreyer/Stone+/Getty Images

The **two-way ANOVA** is the parametric inferential procedure that is applied to designs that involve two independent variables. When both factors involve independent samples, we perform the **two-way, between-subjects ANOVA.** When both factors involve related samples, we perform the **two-way, within-subjects ANOVA.** If one factor is tested using independent samples and the other factor involves related samples, we perform the **two-way, mixed-design ANOVA.** The logic of these ANOVAs is identical except for slight variations in the formulas.

A specific design is described using the number of levels in each factor. If, for example, factor A has two levels and factor B has two levels, we have a "two-by-two" ANOVA, which is written as 2×2. Or, with four levels of one factor and three levels of the other, we have a 4×3 ANOVA, and so on. Each factor can involve any number of levels.

Here's an example of a two-way, between-subjects design: Television commercials are often much louder than the programs themselves because increased volume may make the commercial more persuasive. To test this, we play a recording of an advertising message to participants at each of three volumes—soft, medium, or loud. We're also interested in

two-way, within-subjects ANOVA The parametric inferential procedure performed when both factors are within-subjects factors

two-way, mixed-design ANOVA The parametric inferential procedure performed when the design involves one within-subjects factor and one between-subjects factor

cell In a two-way ANOVA, the combination of one level of one factor with one level of the other factor

complete factorial design A two-way ANOVA design in which all levels of one factor are combined with all levels of the other factor

incomplete factorial design A two-way ANOVA design in which not all levels of the two factors are combined

main effect In a two-way ANOVA, the effect on the dependent scores of changing the levels of one factor; found by collapsing over the other factor

the differences between how males and females are persuaded, so our other factor is the gender of the listener. The dependent variable measures how persuasive each person believes the message to be, with higher scores indicating more persuasive.

The way to organize such a 3×2 design is shown in Table 12.1.

In the diagram:

1. Each *column* represents a level of the volume factor. (In general we'll call the column factor "factor A.") Thus, the scores in column A_1 are from participants tested under soft volume.

2. Each *row* represents a level of the gender factor. (In general we'll call the row factor "factor B.") Thus, scores in row B_1 are from male participants.

3. Each small square produced by combining a level of factor A with a level of factor B is called a **cell.** Here, we have six cells, each containing a sample of three participants who are one gender and given one volume. For example, the highlighted cell contains scores from 3 females presented with medium volume. (With 3 participants per cell, we have a total of 9 males and 9 females, so $N = 18$.)

4. Combining all levels of one factor with all levels of the other factor produces a **complete factorial design.** Here, all levels of gender are combined with all levels of our volume factor. On the other hand, in an **incomplete factorial design,** not all levels of the two factors are combined (e.g., if we had not tested females at the loud volume). Incomplete designs require procedures not discussed here.

We perform the two-way, between-subjects ANOVA if (1) each *cell* is an independent sample, and (2) we have normally distributed interval or ratio scores that have homogeneous variance. Then, as in the following sections, any two-way ANOVA involves examining three things: the two *main effects* and the *interaction effect*.

The Main Effect of Factor A

The **main effect** of a factor is the effect that changing the levels of that factor has on dependent scores while ignoring all other factors in the study. In the persuasiveness study, to find the main effect of factor A (volume), we will ignore the levels of factor B (gender). To do this we literally erase the horizontal line that separates the rows of males and females in Table 12.1.

Table 12.1

A 3 × 2 Design for the Factors of Volume and Gender

Each column represents a level of the volume factor; each row represents a level of the gender factor; each cell contains the scores of participants tested under a particular combination of a level of volume and gender.

		Factor A: Volume		
		Level A_1: Soft	Level A_2: Medium	Level A_3: Loud
Factor B: Gender	Level B_1: Male	9 4 11	8 12 13	18 17 15
	Level B_2: Female	2 6 4	9 10 17	6 8 4

One of the six cells ⬏ $N = 18$

Once we erase that horizontal line, we treat the experiment as if it were a one-way design:

Factor A: Volume		
Level A$_1$: Soft	Level A$_2$: Medium	Level A$_3$: Loud
9	8	18
4	12	17
11	13	15
2	9	6
6	10	8
4	17	4
$\overline{X}_{A_1} = 6$	$\overline{X}_{A_2} = 11.5$	$\overline{X}_{A_3} = 11.33$

We ignore the distinction between males and females, so we simply have six *people* in each column (so $n = 6$). Thus, we have one factor with three levels of volume. The number of levels in factor A is called k_A, so here $k_A = 3$. Averaging the scores in each column yields the **main effect means** for the column factor. Here, we have the "main effect means for volume."

In statistical terminology, we created the main effect means for volume by *collapsing* across the factor of gender. **Collapsing** across a factor refers to averaging together all scores from all levels of that factor. When we collapse across one factor, we have the main effect means for the remaining factor.

Once we have the main effect means, we can see the *main effect* of a factor by looking at the overall pattern in the means. For the main effect of volume, we see how persuasiveness scores change as volume increases: Scores go up from around 6 (at soft) to around 11.5 (at medium), but then scores drop slightly to around 11.3 (at high). To determine if these are significant differences—if there is a *significant main effect of the volume factor*—we essentially perform a one-way ANOVA that compares these main effect means. In general, H_0 says there is no difference between the levels of factor A in the population, so H_0: $\mu_{A_1} = \mu_{A_2} = \mu_{A_3}$. The H_a is

that at least two of the main effect means reflect different populations, so H_a: not all μ_A are equal.

We test this H_0 by computing an F_{obt}, which, in general, is F_A. **Approach this exactly as you did the one-way ANOVA in the previous chapter.** First, we compare F_A to F_{crit}, and if it is significant, it indicates that at least two means from factor A differ significantly. Then, if needed, we determine which specific conditions differ by performing *post hoc tests* (such as the Tukey *HSD*). We also compute the factor's *effect size* and graph the mean effect means. Then we describe and interpret the relationship (here describing how, by changing volume, we influence persuasiveness scores).

The Main Effect of Factor B

After analyzing the main effect of factor A, we examine the main effect of factor B. Here, we *collapse* across the levels of factor A (volume), so we erase the vertical lines separating the levels of volume back in Table 12.1. Then we have this:

Factor B: Gender	Level B$_1$: Male	9	8	18	$\overline{X}_{B_1} = 11.89$
		4	12	17	
		11	13	15	
	Level B$_2$: Female	2	9	6	$\overline{X}_{B_2} = 7.33$
		6	10	8	
		4	17	4	

We simply have the persuasiveness scores of 9 males and 9 females, ignoring the fact that some of each heard the message at different volumes. Thus, we have one factor with two levels, so $k_B = 2$. Notice with 9 scores in each row now, $n = 9$. Averaging the scores in each row yields the mean persuasiveness score for each gender, which are our *main effect means for factor B*.

To see the main effect of this factor, we again look at the pattern of the means: Apparently, changing from males to females leads to a drop in scores from around 11.89 to around 7.33. To determine if this is a significant difference—if there is a significant *main effect of the gender factor*—we perform essentially another one-way

main effect means The mean of the level of one factor after collapsing across the levels of the other factor

collapsing In a two-way ANOVA, averaging together all scores from all levels of one factor in order to calculate the main effect means for the other factor

REMEMBER

When we examine the main effect of factor A, *we look at the overall mean of each level of A, examining the column means.*

ANOVA that compares these main effect means. Our H_0 says there is no difference between the levels of factor B in the population, so $H_0: \mu_{B_1} = \mu_{B_2}$. Our H_a is that at least two of the main effect means reflect different populations, so H_a: not all μ_B are equal.

We test this H_0 by computing another F_{obt}, which, in general, is F_B. We compare this to F_{crit}, and if F_B is significant, it indicates that at least two main effect means from factor B differ significantly. Then, if needed, we perform *post hoc tests* (such as the Tukey *HSD*) to determine which means differ, we compute the effect size, and we describe and interpret this relationship (here describing how gender influences persuasiveness scores).

The Interaction Effect

After we have examined the main effects of factors A and B, we examine the effect of their interaction. The interaction of two factors is called a *two-way interaction*. It is the influence on scores created by combining each level of factor A with each level of factor B. In the example, it is combining each volume with each gender. An interaction is identified as A × B. Here, factor A has 3 levels and factor B has 2 levels, so it is a 3 × 2 (say "3 *by* 2") interaction. Because an interaction examines the influence of combining the levels of the factors, we do not collapse (ignore) either factor. Instead, we examine the cell means. A **cell mean** is the mean of the scores from one cell. The cell means for the interaction between volume and gender are shown in Table 12.2. Here we will ask whether the scores in male–soft are different from those in male–medium or from the scores in female–soft, and so on. (Notice that with six cells, $k_{A \times B} = 6$, and with 3 scores per cell, $n = 3$.)

cell mean The mean of the scores from one cell in a two-way design

However, examining an interaction is not as simple as saying that the cell means differ significantly. Interpreting an interaction is difficult because both independent variables are changing, as are the dependent scores.

To simplify the process, look at the influence of changing the levels of factor A under *one* level of factor B. Then see if this effect—this pattern—for factor A is *different* when you look at the other levels of factor B. For example, here is the first row of Table 12.2, showing the relationship between volume and scores for *males.* As volume increases, mean persuasiveness scores also increase in a positive, linear relationship.

	Factor A: Volume		
	Soft	Medium	Loud
B_1: Male	$\bar{X} = 8$	$\bar{X} = 11$	$\bar{X} = 16.67$

However, now look at the relationship between volume and scores for *females,* using the cell means from the bottom row of Table 12.2.

	Factor A: Volume		
	Soft	Medium	Loud
B_2: Female	$\bar{X} = 4$	$\bar{X} = 12$	$\bar{X} = 6$

Here, as volume increases, mean persuasiveness scores first increase but then decrease, producing a nonlinear relationship.

Thus, there is a different relationship between volume and persuasiveness scores for each gender level.

REMEMBER: For the interaction effect, you compare the cell means. For a main effect, you compare the level means.

Table 12.2
The Volume by Gender Interaction
Each mean is a cell mean.

		Factor A: Volume		
		Soft	Medium	Loud
Factor B: Gender	Male	$\bar{X} = 8$	$\bar{X} = 11$	$\bar{X} = 16.67$
	Female	$\bar{X} = 4$	$\bar{X} = 12$	$\bar{X} = 6$

A **two-way interaction effect** is present when the relationship between one factor and the dependent scores changes with, or depends on, the level of the other factor that is present. Thus, an interaction effect occurs when the influence of changing one factor is not the same for each level of the other factor. Here, for example, increasing volume does not have the same effect for males as it does for females. Or, in other words, an interaction is present when you must use the word "depends" to describe the influence of either factor. Here, whether increasing volume always increases scores *depends on* whether we're talking about males or females.

You can also see the interaction by looking at the difference between males and females at each level of volume. Sometimes the males score higher, and sometimes the females do; it *depends* on which level of volume we're talking about.

Conversely, an interaction effect would not be present if the cell means formed the *same* pattern for males and females. For example, say the cell means had been as follows:

		Factor A: Volume		
		Soft	Medium	Loud
Factor B: Gender	Male	$\bar{X} = 5$	$\bar{X} = 10$	$\bar{X} = 15$
	Female	$\bar{X} = 20$	$\bar{X} = 25$	$\bar{X} = 30$

Increasing the volume increases scores by about 5 points, *regardless* of whether it's for males or females. Or, females always score higher, regardless of volume. Therefore, an interaction effect is not present when the influence of changing the levels of one factor does not depend on which level of the other variable we are talking about. In other words, there's no interaction when we see the same relationship between the dependent scores and one factor for each level of the other factor.

Here is an example of an interaction effect from a different study. Say that for one factor we measure whether participants are in a sad or happy mood. Our second factor involves having participants learn a list of 15 happy words (e.g., *love, beauty,* etc.) or a list of 15 sad words (e.g., *death, pain,* etc.). Each group then recalls its list. Research suggests we would obtain mean recall scores forming a pattern like this:

		Words	
		Sad	Happy
Mood	Sad	$\bar{X} = 10$	$\bar{X} = 5$
	Happy	$\bar{X} = 5$	$\bar{X} = 10$

This is an interaction because when in a sad mood, people recall more sad words than happy words, but when in a happy mood, they recall more happy words than sad words. So, what is the influence on recall when we change the factor from sad to happy words? It *depends* on what mood participants are in. Or, in which mood will people recall words best? It depends on whether they are recalling happy or sad words.

To determine if we have a *significant interaction effect,* we perform essentially another one-way ANOVA that compares the cell means. To write the H_0 and H_a in symbols is complicated, but in words, H_0 is that the cell means do not represent an interaction effect in the population, and H_a is that at least some of the cell means do represent an interaction effect in the population.

To test H_0, we compute another F_{obt}, called $F_{A\times B}$. If $F_{A\times B}$ is significant, it indicates that at least two of the cell means differ significantly in a way that produces an interaction effect. Then we perform *post hoc tests* (such as the Tukey *HSD*) to determine which specific cell means differ, we compute the effect size, and we describe and interpret the relationship (in our commercial persuasiveness example, describing how the different combinations of volume and gender influence persuasiveness scores).

two-way interaction effect In a two-way ANOVA, the effect in which the relationship between one factor and the dependent scores depends on the level of the other factor that is present

12.2 INTERPRETING THE TWO-WAY EXPERIMENT

As you've seen, in the two-way ANOVA we compute three *F*s: one for the main effect of factor A, one for the main effect of factor B, and one for the interaction of A × B. The formulas for the two-way, between-subjects ANOVA applied to our persuasiveness data are presented in Appendix A.4. The completed ANOVA Summary Table is shown in Table 12.3 on the next page.

The row labeled *Factor A* reflects differences between groups due to the main effect

© Thinkstock/Jupiterimages

Table 12.3
Completed Summary Table of Two-Way ANOVA

Source	Sum of Squares	df	Mean Square	F
Between				
Factor A (volume)	117.45	2	58.73	7.14
Factor B (gender)	93.39	1	93.39	11.36
Interaction (vol × gen)	102.77	2	51.39	6.25
Within	98.67	12	8.22	
Total	412.28	17		

of changing volume. The row labeled *Factor B* reflects differences between groups due to the main effect of gender. The row labeled *Interaction* reflects the differences between the groups (cells) formed by combining volume and gender. The row labeled *Within* reflects differences among scores within each cell, which we then pool and use as our estimate of the variability of scores in the population.

The logic and calculations for our *F*s here are the same as in the one-way ANOVA. For each row we first compute the sum of squares and then divide by the appropriate *df* to obtain the mean square. As usual, an *F* equals the mean square between-groups divided by the mean square within groups. Thus, the F_{obt} for Factor A (volume) of 7.14 is produced by dividing 58.73 by 8.22. The F_{obt} for Factor B (gender) of 11.36 is produced by dividing 93.39 by 8.22. And the F_{obt} for the interaction (volume × gender) of 6.25 is produced by dividing 51.39 by 8.22.

Each F_{obt} is tested by comparing it to F_{crit} from the *F*-table in Appendix B. The *df*s for a particular F_{crit} are those used in computing that particular F_{obt}. Because factor A and B may have different *df*s, usually you will have a different F_{crit} for each main effect and for the interaction.

Note: Whether one F_{obt} is significant does not influence whether any other one is significant, so any combination of significant main effects and/or the interaction is possible.

We interpret the two-way ANOVA by examining the means from the significant main effects and/or interaction. As

shown in Appendix A.4, the persuasiveness data produced significant *F*s for A (volume), for B (gender), and for their interaction. Table 12.4 shows the means for the persuasiveness study.

Completing the Main Effects

Approach each main effect as the one-way ANOVA that we originally diagrammed. As usual, a significant F_{obt} merely indicates differences *somewhere* among the means. Therefore, if the *n*s are equal and we have more than two levels in the factor, we determine which specific means differ by performing Tukey's *HSD* test. Use the formula and procedure described in the previous chapter.

Note: As in our example, the *n* and *k* may be different for each factor. If so, this requires computing a separate *HSD* for each factor.

As shown in Appendix A.4, the *HSD* test for our main effect (column) means for the volume factor indicates that the soft condition (6) differs significantly from both the medium (11.5) and loud (11.33) levels. However, medium and loud do not differ significantly. For the gender factor, the *HSD* test is not necessary: It must be that the mean for males (11.89) differs significantly from the mean for females (7.33). If factor B had more than two levels, however, we would compute a new *HSD* and proceed as usual.

We report each F_{obt} in the usual format. Also, it is appropriate to compute the effect size (η^2) for each significant main effect. Further, researchers may produce a separate graph for each main effect. As you saw in Chapter 3, we show the relationship between the levels

© Samantha Everton/Photolibrary

Table 12.4
Main Effect and Interaction Means from the Persuasiveness Study

		Factor A: Volume			
		Level A$_1$: Soft	Level A$_2$: Medium	Level A$_3$: Loud	
Factor B: Gender	Level B$_1$: Male	8	11	16.67	$\overline{X}_{male} = 11.89$
	Level B$_2$: Female	4	12	6	$\overline{X}_{fem} = 7.33$
		$\overline{X}_{soft} = 6$	$\overline{X}_{med} = 11.5$	$\overline{X}_{loud} = 11.33$	

The way to read the graph is to look at one line at a time. For males (the dashed line), as volume increases, mean persuasiveness scores increase. However, for females (the solid line), as volume increases, persuasiveness scores first increase but then decrease.

Thus, we see one relationship for males and a different relationship for females. Therefore, the graph shows an interaction effect by showing that the effect that increasing volume has on persuasiveness scores depends on whether the participants are male or female.

of the factor (independent variable) on the *X* axis, and the main effect means (dependent variable) on the *Y* axis. Include all means, even those that do not differ significantly.

We examine interaction effects in the same ways, but they involve slightly different procedures.

Graphing the Interaction Effect

An interaction can be a beast to interpret, so always graph it! As usual, label the *Y* axis with the mean of the dependent variable. To produce the simplest graph, place on the *X* axis the factor with the most levels. You'll show the other factor by drawing a separate line for each of its levels.

Thus, for the persuasiveness study the *X* axis is labeled with the three volume levels. Then we plot the *cell* means. The resulting graph is shown in Figure 12.1. Remember that any graph shows the relationship between the *X* variable and the *Y* variable. Here, we show the relationship between the *X* factor and the dependent scores, but we do this separately for each level of the other factor. Thus, approach this in the same way that we examined the means for the interaction effect. There we first looked at the relationship between volume and persuasiveness scores for only *males*: From Table 12.4 their cell means are $\overline{X}_{soft} = 8$, $\overline{X}_{medium} = 11$, and $\overline{X}_{loud} = 16.67$. Plot these three means and connect the data points with straight lines. Then look at the relationship between volume and scores for *females*: Their cell means are $\overline{X}_{soft} = 4$, $\overline{X}_{medium} = 12$, and $\overline{X}_{loud} = 6$. Plot these means and connect their adjacent data points with straight lines. (*Notice:* Always provide a key to identify each line.)

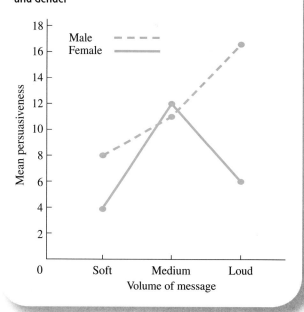

Figure 12.1
Graph of Cell Means, Showing the Interaction of Volume and Gender

REMEMBER: Graph the interaction by drawing a separate line that shows the relationship between the factor on the X axis and the dependent (Y) scores for each level of the other factor.

confounded comparison In a two-way ANOVA, a comparison of two cells that differ along more than one factor

As another example, let's say that a 2 × 2 experiment produces the cell means shown below. To produce the graph of the interaction, we plot data points at 2 and 6 (the cell means at A_1 and A_2 for B_1) and connect them with the solid line. We plot data points at 10 and 4 (the cell means at A_1 and A_2 for B_2) and connect them with the dashed line.

	A_1	A_2
B_1	$\bar{X} = 2$	$\bar{X} = 6$
B_2	$\bar{X} = 10$	$\bar{X} = 4$

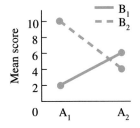

Take note that an interaction effect can produce an infinite variety of different graphs, but when it is present, it produces lines that are *not parallel*. Each line summarizes the relationship, and a line that is shaped or oriented differently from another line indicates a different relationship. Therefore, lines that are not parallel indicate that the relationship between *X* and *Y* changes depending on the level of the second factor, so that an interaction effect is present. Conversely, when an interaction effect is not present, the lines will be essentially parallel, with each line depicting essentially the same relationship.

REMEMBER

An interaction effect *is present when its graph produces lines that are* not parallel.

Performing Unconfounded Post Hoc Comparisons on the Interaction

The Tukey *HSD* is also used with a significant interaction effect so that we can determine which of the *cell* means differ significantly. However, we do not compare every cell mean to every other cell mean. Look at Table 12.5. We would not, for example, compare the mean for males at soft volume to the mean for females at medium volume. Because the two cells differ both in terms of gender and volume, we cannot determine which of these variables caused the difference. Therefore, we are confused or "confounded." A **confounded comparison** occurs when two cells differ along more than one factor. Other examples of confounded comparisons would involve comparing cells connected by the dashed lines in Table 12.5. Instead, we perform only

Table 12.5
Interaction Means for Persuasiveness Study
Any horizontal or vertical comparison is unconfounded; any diagonal comparison is confounded.

		Factor A: Volume		
		Soft	Medium	Loud
Factor B: Gender	B_1: Male	$\bar{X} = 8$	$\bar{X} = 11$	$\bar{X} = 16.67$
	B_2: Female	$\bar{X} = 4$	$\bar{X} = 12$	$\bar{X} = 6$

unconfounded comparisons, in which two cells differ along only one factor. The cells connected by solid lines in Table 12.5 are examples of unconfounded comparisons. Thus, we compare only cell means within the same column because these differences result from factor B. We also compare means within the same row because these differences result from factor A. We do not, however, make any diagonal comparisons, because these are confounded comparisons.

Thus, we find the difference between all pairs of cell means in the same column or in the same row to compare to the *HSD*. (*Note:* As shown in Appendix A.4, the *HSD* is computed slightly differently for an interaction than for a main effect.) For the persuasiveness study, only three significant differences occur between the cell means: (1) between females at soft volume and females at medium volume, (2) between males at soft volume and males at loud volume, and (3) between males at loud volume and females at loud volume.

Interpreting the Experiment

The way to interpret an experiment is to look at the significant differences between means from the post hoc comparisons for all significant main effects and interaction effects. All of the differences found in the persuasiveness study are summarized in Table 12.6. Each line connecting two means indicates that they differ significantly.

Usually the interpretation of a two-way study rests with the significant interaction, even when main effects are significant. This is because the conclusions about main effects are contradicted by the interaction. For example, our main effect means for gender suggest that males scored higher than females. However, the cell means of the interaction show that gender differences *depend* on volume: Only in the loud condition is there a significant difference between males and females. Therefore, we cannot make an overall, general conclusion about differences between males and females.

Likewise, the main effect means for volume showed that increasing volume from soft to medium and from soft to loud produced significant differences in scores. However, the interaction indicates that increasing volume from soft to medium actually produced a significant difference only for females; increasing the volume from soft to loud produced a significant difference only for males.

As above, usually we cannot draw any conclusions about significant main effects when the interaction is significant. After all, the interaction indicates that the influence of one factor *depends* on the levels of the other factor and vice versa, so we should not act like either factor has a consistent effect by itself. When the interaction is not significant, then we focus on any significant main effects. For completeness, however, always perform the entire ANOVA and report all results.

Thus, we conclude our persuasiveness study by saying that increasing the volume of a message beyond soft tends to increase persuasiveness scores in the population,

unconfounded comparison In a two-way ANOVA, a comparison of two cells that differ along only one factor

REMEMBER

The primary interpretation of a two-way ANOVA rests on the interaction when it is significant.

Table 12.6
Summary of Significant Differences in the Persuasiveness Study
Each line connects two means that differ significantly.

		Factor A: Volume			
		Level A$_1$: Soft	Level A$_2$: Medium	Level A$_3$: Loud	
Factor B: Gender	Level B$_1$: Male	8.0	11	16.67	$\bar{X}_{male} = 11.89$
	Level B$_2$: Female	4.0	12	6	$\bar{X}_{fem} = 7.33$
		$\bar{X}_{soft} = 6$	$\bar{X}_{med} = 11.5$	$\bar{X}_{loud} = 11.33$	

but this increase occurs for females with medium volume and for males with loud volume. Further, differences in persuasiveness scores occur between males and females in the population, but only if the volume of the message is loud. (And, after all of the above shenanigans, for all of these conclusions together, the probability of a Type I error in the study—the *experiment-wise error rate*—is still $p < .05$.)

© Comstock Images/Jupiterimages

using what you know

1. a. What are the characteristics of an experiment calling for a two-way, complete factorial, between-subjects ANOVA?

   b. What are the other assumptions of the two-way, between-subjects ANOVA?

2. What is a cell?

3. a. What is the main effect of a factor?

   b. How do we collapse across factor B to compute the main effect means of A?

   c. What does a significant main effect indicate about these means?

4. a. What information can be obtained from a two-way ANOVA that cannot be obtained from two one-way designs using the same variables?

   b. How do we obtain the means examined in an interaction effect?

   c. What does a significant interaction effect indicate about these means?

5. Describe the null and alternative hypotheses for the main effects and the interaction effect in a two-way ANOVA.

6. Identify the *F*-ratios computed in a two-way ANOVA.

7. a. What must be done for each significant effect in a two-way ANOVA before attempting to interpret the results of an experiment?

   b. Explain why this must be done.

8. Why is the interpretation of the results from a two-way ANOVA limited to the interaction when it is significant?

9. a. When is a two-way, within-subjects ANOVA performed?

   b. When is a two-way, mixed-design ANOVA performed?

10. a. What is an unconfounded comparison?

    b. What is a confounded comparison?

11. After performing a 3 × 3 ANOVA with equal *n*s, you find all *F*s are significant.

    a. What is your next step?

    b. What other procedure should you perform?

12. Complete the following ANOVA summary table.

Source	Sum of Squares	df	Mean Square	F
Between	150	5		
Factor A	80	2		
Factor B	30	1		
A × B Interaction	40	2		
Within	350	24	14.58	
Total	500	29		

13. a. Graph the following cell means for two experiments. Label the *X* axis with Factor A.

Study 1				Study 2		
	A_1	A_2	A_3		A_1	A_2	A_3
B_1	10	8	7	B_1	2	6	9
B_2	9	13	18	B_2	3	8	12

   b. Using the graphs you developed in part a, what would you conclude about a possible A × B interaction?

14. In a study involving the effect of amount of fat in the diet and amount of exercise on the mental acuity of middle-aged men, Dr. Parks is using three treatment levels for diet (<30% fat, 30%–60% fat, and >60% fat) and two levels for exercise (<60 minutes per week and 60 or more minutes per week). The data from her study follow:

		Factor A		
		<30% fat	30%–60% fat	>60% fat
Factor B	<60 minutes	4 4 2 4 3	3 1 2 2 3	2 2 2 2 1
	60 minutes or more	6 5 4 4 5	8 8 7 8 6	5 7 5 5 6

   a. What is (are) the independent variable(s)?
   b. What is (are) the dependent variable(s)?
   c. How many levels does factor A have?
   d. How many levels does factor B have?
   e. How many participants received each level of factor A?
   f. How many participants received each level of factor B?
   g. How many participants received each treatment?

15. Using the data given in problem 14:

Source	Sum of Squares	df	Mean Square	F
Between	114.800			
Factor A	6.200			
Factor B	90.133			
A × B Interaction	18.467			
Within	16.000	24		
Total	130.800	29		

a. Complete the summary table.
b. Determine which effects are significant.
c. Based on this information, what can you conclude?

16. A researcher examines the effects of attendance in class and completion of homework assignments on statistics test grades. Half his subjects complete 5 homework assignments, and the other half complete 15. He determines the number of absences for each participant and assigns each to either the low- or high-absenteeism condition. From the 10 people per cell, he obtains the following mean test scores and sums of squares:

		Factor A: Absenteeism		
		Low	High	
Factor B: Homework	5	8.7	6.4	7.6
	15	9.4	8.6	9.0
		9.1	7.5	

Summary Table of Two-Way ANOVA				
Source	Sum of Squares	df	Mean Square	F
Between				
Factor A	25.60			
Factor B	22.50			
A × B Interaction	5.63			
Within	35.25			
Total	88.98			

a. Complete the summary table.
b. Determine which effects are significant.
c. What conclusions can be drawn from this study?

Chi Square
and Nonparametric
Procedures

Previous chapters have discussed the category of inferential statistics called *parametric procedures*. Now we'll turn to the other category called *nonparametric statistics*. Nonparametric procedures are still for deciding whether the relationships in the sample accurately represent the relationship in the population. Therefore, H_0 and H_a, sampling distributions, Type I and Type II errors, alpha, critical values, and significance all apply. Although a number of different nonparametric procedures are available, we'll focus on the most common ones. This chapter presents (1) the one-way chi square, (2) the two-way chi square, and (3) a brief review of the procedures for ordinal scores.

13.1 PARAMETRIC VERSUS NONPARAMETRIC STATISTICS

Previous parametric procedures have required that dependent scores involve an interval or ratio scale, that the scores are normally distributed, and that the population variances are homogeneous. But sometimes researchers obtain data that do not fit these requirements. Some dependent variables are nominal variables (e.g., whether someone is male or female). Sometimes we can measure a dependent variable only by assigning ordinal scores (e.g., judging this participant as showing the most of an attribute, this one the second-most, and so on). And sometimes

Looking Back

- From Chapter 1, know the four types of measurement scales (nominal, ordinal, interval, and ratio).
- From Chapter 7, understand the concepts of Type I errors and power.

© 2010 Masterfile Corporation

a variable involves an interval or ratio scale, but the populations are severely skewed and/or do not have homogeneous variance (e.g., we saw that yearly income forms a positively skewed distribution).

It is better to design a study that allows us to use parametric procedures, because they are more *powerful* than nonparametric procedures: we are less likely to miss a relationship and make a Type II error. In fact, parametric procedures will tolerate some violation of their assumptions, so usually we can use them. But if the data severely violate the rules, then we increase the probability of making a Type I error (rejecting H_0 when it's true), so that the actual probability of a Type I error will be larger than the alpha level we've set. Therefore, when data clearly do not fit a parametric procedure, we turn to nonparametric procedures. **Nonparametric statistics** do not assume a normal distribution or homogeneous variance, and the data may be nominal or ordinal. Using nonparametric procedures with such data will keep the probability of a Type I error equal to our alpha.

nonparametric statistics Inferential procedures that do not require stringent assumptions about the raw score population represented by the sample data

REMEMBER

Use nonparametric statistics when dependent scores form very nonnormal distributions, when the population variances are not homogeneous, or when scores are measured using ordinal or nominal scales.

Nonparametric procedures fall into one of two groups: those that are used with nominal data and those that are used with ordinal data. With nominal data, we use the *chi square procedure*.

13.2 CHI SQUARE PROCEDURES

chi square procedure (χ^2) The nonparametric inferential procedure for testing whether the frequencies of category membership in the sample represent the predicted frequencies in the population

one-way chi square The chi square procedure for testing whether the sample frequencies of category membership on one variable represent the predicted distribution of frequencies in the population

Chi square procedures are used when participants are measured using a nominal variable. With nominal variables we do not measure an amount, but rather we indicate the *category* that participants fall into, and then count the number—the *frequency*—of individuals in each category. Thus, we have nominal variables when counting how many individuals answer yes or no to a question; how many claim to vote Republican, Democratic, or Socialist; how many say that they were or were not abused children; and so on.

The next step is to determine what the data represent. For example, we might find that out of 100 people, 40 say yes to a question and 60 say no. These numbers indicate how the *frequencies are distributed* across the categories of yes/no. As usual, we want to draw inferences about the population: Can we infer that if we asked the entire population this question, 40% of the population would say yes and 60% would say no? Or would the frequencies be distributed in a different manner? To make inferences about the frequencies in the population, we perform the chi square procedure (pronounced "kigh square"). The **chi square procedure** is the nonparametric inferential procedure for testing whether the frequencies in each category in sample data represent specified frequencies in the population. The symbol for the chi square statistic is χ^2.

Theoretically, there is no limit to the number of categories—levels—you may have in a variable and no limit to the number of variables you may have. Therefore, we describe a chi square design in the same way we described ANOVA: When a study has only one variable, we use the *one-way chi square*; when a study has two variables, we use the *two-way chi square*.

REMEMBER

Use the chi square procedure (χ^2) when you measure the number of participants falling into different categories.

13.3 ONE-WAY CHI SQUARE: THE GOODNESS OF FIT TEST

The **one-way chi square** is used when data consist of the frequencies with which participants belong to the different categories of *one* variable. Here we examine the

© iStockphoto.com/ Hamza Türkkol

No ☐   ☐ Yes

relationship between the different categories and the frequency with which participants fall into each. We ask "As the categories change, do the frequencies in the categories also change?"

Here is an example. Being right-handed or left-handed is apparently related to brain organization and many of history's great geniuses were left-handed. Therefore, using an IQ test, we select a sample of 50 geniuses. Then we ask them whether they are left- or right-handed (ambidextrous is not an option). The total numbers of left- and right-handers are the frequencies in the two categories. The results are shown here:

Handedness	
Left-Handers	Right-Handers
$f_o = 10$	$f_o = 40$
$k = 2$	
$N = $ total $f_o = 50$	

Each column contains the frequency with which participants are in that category. We call this the **observed frequency,** symbolized by f_o. The sum of the f_os from all categories equals N, the total number of participants in the study. Notice that k stands for the number of categories, or levels, and here $k = 2$.

Above, 10 of the 50 geniuses (20%) are left-handers, and 40 of them (80%) are right-handers. Therefore, we might argue that the same distribution of 20% left-handers and 80% right-handers would occur in the population of all geniuses. But there is the usual problem: sampling error. Maybe our sample is unrepresentative, so that in the population of all geniuses we would not find this distribution of right- and left-handers. Maybe our results poorly represent some *other* distribution. As usual, this is the null hypothesis, implying that we are being misled by sampling error.

Usually, researchers test the H_0 that there is no difference among the frequencies in the categories in the population, meaning that there is no relationship in the population. In the handedness study, for the moment we'll ignore that there are generally more right-handers than left-handers in the world. Therefore, if there is no relationship in the population, then our H_0 is that the frequencies of left- and right-handed

geniuses in the population are equal. There is no conventional way to write this in symbols, so simply write H_0: *all frequencies in the population are equal.* This implies that if the observed frequencies (f_o) in the sample are not equal, it is because of sampling error.

The alternative hypothesis always implies that the study did demonstrate the predicted relationship, so H_a: *not all frequencies in the population are equal.* For our handedness study, H_a implies that our observed frequencies represent the frequencies of left- and right-handers found in the population of geniuses. We can test only whether the sample frequencies are or are not equal, so the one-way χ^2 tests *only* two-tailed hypotheses.

The one-way χ^2 has five assumptions: (1) Participants are categorized along one variable having two or more categories, and we count the frequency in each category. (2) Each participant can be in only one category (i.e., you cannot have repeated measures). (3) Category membership is independent: The fact that an individual is in one category does not influence the probability that another participant will be in any category. (4) We include the responses of all participants in the study (i.e., you would not count only the number of right-handers, or in a different study, you would count both those who agree and disagree with a statement). (5) The theoretical basis requires that our "expected frequencies" be at least 5 per category.

Computing the One-Way χ^2

The first step in computing χ^2 is to translate H_0 into the expected frequency for each category. The **expected frequency** is the frequency we expect in a category if the sample data perfectly represent the distribution of frequencies in the population described by H_0. The symbol for an expected frequency is f_e. Our H_0 is that

observed frequency (f_o) The frequency with which participants fall into a category of a variable

expected frequency (f_e) The frequency expected in a category if the sample data perfectly represent the distribution of frequencies in the population described by the null hypothesis

χ^2-distribution

The sampling distribution of all possible values of χ^2 that occur when the samples represent the distribution of frequencies described by the null hypothesis

the frequencies of left- and right-handedness are equal. We translate this into the expected frequency in each group based on our N. If our samples perfectly represented equal frequencies, then out of our 50 participants, 25 should be right-handed and 25 should be left-handed. Thus, the expected frequency in each category is $f_e = 25$.

With an H_0 that the frequencies in the categories are equal, the f_e will be the same in all categories, and there's a shortcut for computing it:

THE FORMULA FOR EACH EXPECTED FREQUENCY WHEN TESTING AN H_0 OF NO DIFFERENCE

$$f_e \text{ in each category} = \frac{N}{k}$$

Thus, in our study, with $N = 50$ and $k = 2$, the f_e in each category $= 50/2 = 25$. (Sometimes f_e may contain a decimal. For example, if we included a third category, ambidextrous, then $k = 3$, and each f_e would be 16.67.)

For the handedness study we have these frequencies:

Handedness	
Left-Handers	Right-Handers
$f_o = 10$	$f_o = 40$
$f_e = 25$	$f_e = 25$

The χ^2 compares the difference between our observed frequencies and the expected frequencies. We compute an obtained χ^2, which we call χ^2_{obt}.

CHI SQUARE

$$\chi^2_{obt} = \Sigma\left(\frac{(f_o - f_e)^2}{f_e}\right)$$

This says to find the difference between f_o and f_e in each category, square that difference, and then divide it by the f_e for that category. After doing this for all categories, sum the quantities, and the answer is χ^2_{obt}. Thus:

STEP 1: *Compute the f_e for each category.* We computed our f_e to be 25 per category.

STEP 2: *Create the fraction $\frac{(f_o - f_e)^2}{f_e}$ for each category.* Thus, the formula becomes

$$\chi^2_{obt} = \Sigma\left(\frac{(f_o - f_e)^2}{f_e}\right) = \left(\frac{(10 - 25)^2}{25}\right) + \left(\frac{(40 - 25)^2}{25}\right)$$

STEP 3: *Perform the subtraction in the numerator of each fraction.* After subtracting,

$$\chi^2_{obt} = \left(\frac{(-15)^2}{25}\right) + \left(\frac{(15)^2}{25}\right)$$

STEP 4: *Square the numerator in each fraction.* This gives

$$\chi^2_{obt} = \left(\frac{225}{25}\right) + \left(\frac{225}{25}\right)$$

STEP 5: *Perform the division in each fraction and then sum the results.*

$$\chi^2_{obt} = 9 + 9 = 18$$

so

$$\chi^2_{obt} = 18$$

STEP 6: *Compare χ^2_{obt} to χ^2_{crit}. This is discussed below.*

Interpreting the One-Way χ^2

We interpret χ^2_{obt} by determining its location on the χ^2 sampling distribution. The χ^2-**distribution** contains all possible values of χ^2 that occur when H_0 is true. Thus, for the handedness study, the χ^2-distribution is the distribution of all possible values of χ^2 when the frequencies in the two categories in the population are equal. You can envision the χ^2-distribution as shown in Figure 13.1.

Even though the χ^2-distribution is not at all normal, it is used in the same way as previous sampling distributions. When the data perfectly represent the H_0 situation, so that each f_o equals its f_e, then χ^2 is zero. However, sometimes by chance the observed frequencies differ from the expected frequencies, producing a χ^2 greater than zero. The larger the χ^2, the larger the differences, and then the less likely they are to occur when H_0 is true.

Figure 13.1
Sampling Distribution of χ^2 When H_0 is True

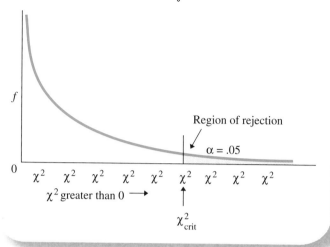

f

Region of rejection

$\alpha = .05$

0   χ^2 χ^2 χ^2 χ^2 χ^2 χ^2 χ^2 χ^2 χ^2

χ^2 greater than 0 $\longrightarrow$

χ^2_{crit}

With chi square we again have two-tailed hypotheses but one region of rejection. If χ^2_{obt} is larger than the critical value, then it is in the region of rejection and so χ^2_{obt} is significant: The observed frequencies are unlikely to represent the distribution of frequencies in the population described by H_0.

Therefore, to determine if χ^2_{obt} is significant, we compare it to the critical value, symbolized by χ^2_{crit}. As with previous statistics, the χ^2-distribution changes shape as the degrees of freedom change, so to find the appropriate value of χ^2_{crit}, you must first determine the degrees of freedom.

DEGREES OF FREEDOM IN A ONE-WAY CHI SQUARE

$df = k - 1$

Remember that k is the number of categories.

Find the critical value of χ^2 in Table 6 in Appendix B, entitled "Critical Values of Chi Square." For the handedness study, $k = 2$ so $df = 1$, and with $\alpha = .05$, the $\chi^2_{crit} = 3.84$. Our χ^2_{obt} of 18 is larger than this χ^2_{crit}, so the results are significant: We reject the H_0 that our categories are poorly representing a distribution of equal frequencies in the population (rejecting that we are poorly representing that geniuses are equally left- or right-handed).

When our χ^2_{obt} is significant, we accept the H_a that the sample represents frequencies in the population that are not equal. In fact, as in our samples, we would expect to find about 20% left-handers and 80% right-handers in the population of geniuses. We conclude that we have evidence of a relationship between the categories of handedness and the frequency with which geniuses fall into each. Then, as usual, we interpret the relationship, here attempting to explain what aspects of being left-handed and being a genius are related.

Unlike ANOVA, a significant one-way chi square that involves more than two conditions is *not* followed by post hoc comparisons: A significant χ^2_{obt} indicates that all categories are significantly different. Then we use the observed frequency in each category to estimate the frequencies that would be found in the population. Also, there is no measure of effect size here.

If χ^2_{obt} had not been significant, we would not reject H_0 and would have no evidence—one way or the other—regarding how handedness is distributed among geniuses.

Notice that the one-way chi square procedure is also called the **goodness of fit test**: Essentially, it tests how "good" the "fit" is between our data and the frequencies we expect if H_0 is true. This is simply another way of asking whether sample data are likely to represent the distribution of frequencies in the population described by H_0. Notice too that our H_0 may be that the frequencies are distributed in some way other than equally across the categories. For example, only about 10% of

goodness of fit test A name for the one-way chi square, because it tests how "good" the "fit" is between the data and H_0

the general population is left-handed, so we should have tested whether our geniuses fit this model. Now H_0 is that our geniuses are like the general population, being 10% left-handed and 90% right-handed. Our H_a is that the data represent a population of geniuses that does not have this distribution. Each f_e is again based on H_0, so left-handed geniuses should occur 10% of the time: 10% of our 50 geniuses is 5, so $f_e = 5$. Right-handed geniuses should occur 90% of the time: 90% of 50 is 45, so $f_e = 45$. We then compute χ^2_{obt}, find χ^2_{crit}, and interpret the results as we did above.

13.4 THE TWO-WAY CHI SQUARE: THE TEST OF INDEPENDENCE

The **two-way chi square** is used when we count the frequency of category membership along *two* variables. (This is similar to the factorial ANOVA discussed in the previous chapter.) The procedure for computing χ^2 is the same regardless of the number of categories in each variable. The assumptions of the two-way chi square are the same as for the one-way chi square.

Logic of the Two-Way Chi Square

Here is a study that calls for a two-way chi square. At one time psychologists claimed that someone with

a "Type A" personality tends to be very pressured and never seems to have enough time. The "Type B" personality, however, tends not to be so time pressured, and is more relaxed and mellow. A controversy developed over whether people with Type A personalities are less healthy, especially when it comes to having heart attacks. Therefore, say that we select a sample of 80 people and determine how many are Type A and how many are Type B. We then count the frequency of heart attacks in each type. We must also count how many in each type have *not* had heart attacks. Therefore, we have two categorical variables: personality type (A or B) and health (heart attack or no heart attack). Table 13.1 shows the layout of this study. Notice, with two rows and two columns, this is a 2×2 ("2 *by* 2") matrix, so we have a 2×2 design. With different variables, the design might be a 2×3, a 3×4, etc.

Table 13.1

A Two-Way Chi Square Design Comparing Participants' Personality Type and Health

		Personality Type	
		Type A	Type B
Health	Heart Attack	f_o	f_o
	No Heart Attack	f_o	f_o

TYPE B

TYPE A

two-way chi square The chi square procedure for testing whether, in the population, frequency of category membership on one variable is independent of frequency of category membership on the other variable

Although this looks like a two-way ANOVA, it is not analyzed like one. The two-way χ^2 is also called the **test of independence**: It tests whether the frequency of participants falling into the categories of one variable is independent of, or unrelated to, the frequency of their falling into the categories of the other variable. Thus, our example study will test whether the frequencies of having or not having a heart attack are independent of the frequencies of being Type A or Type B. Essentially, the two-way χ^2 tests the *interaction* which, as in the two-way ANOVA, tests whether the influence of one factor *depends on* the level of the other factor that is present. Thus, we'll ask, "Does the frequency of people having heart attacks depend on their frequency of being Type A or B?"

To understand "independence," Table 13.2 shows an example where category membership is perfectly independent. Here, the frequency of having or not having a heart attack does not depend on the frequency of being Type A or Type B. Another way to view the two-way χ^2 is as a test of whether a *correlation* exists between the two variables. When variables are independent, there is no correlation, and using the categories from one variable is no help in predicting the frequencies for the other variable. Here, knowing if people are Type A or Type B does not help to predict if they have heart attacks (and heart attacks do not help in predicting personality type).

However, Table 13.3 shows a pattern we might see when the variables are totally dependent. Here, the frequency of a heart attack or no heart attack *depends* on personality type. Likewise, a perfect correlation exists because whether people are Type A or Type B is a perfect predictor of whether or not they have had a heart attack (and vice versa).

Say that our actual data are shown in Table 13.4. There is a *degree* of dependence here because a heart attack tends to be more frequent for Type A, while no heart attack is more frequent for Type B. Therefore, there is some degree of correlation between the variables.

On the one hand, we'd like to conclude that this relationship occurs in the population. On the other hand, perhaps there really is no correlation in the population, but by chance we obtained frequencies that poorly represent this. The above translate into our null and alternative hypotheses. In the two-way chi square:

1. H_0 is that category membership on one variable is independent of (not correlated with) category membership on the other variable. If the sample data look correlated, this is due to sampling error.

2. H_a is that category membership on the two variables in the population is dependent (correlated).

> **test of independence** A name for the two-way chi square, because it tests whether the frequency of category membership on one variable is independent of frequency of category membership on the other variable

Table 13.2
Example of Independence
Personality type and heart attacks are perfectly independent.

Health		Personality Type	
		Type A	Type B
	Heart Attack	$f_o = 20$	$f_o = 20$
	No Heart Attack	$f_o = 20$	$f_o = 20$

Table 13.3
Example of Dependence
Personality type and heart attacks are perfectly dependent.

Health		Personality Type	
		Type A	Type B
	Heart Attack	$f_o = 40$	$f_o = 0$
	No Heart Attack	$f_o = 0$	$f_o = 40$

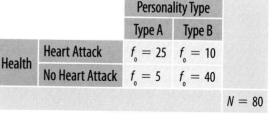

Table 13.4
Observed Frequencies as a Function of Personality Type and Health

Health		Personality Type	
		Type A	Type B
	Heart Attack	$f_o = 25$	$f_o = 10$
	No Heart Attack	$f_o = 5$	$f_o = 40$
			$N = 80$

Computing the Two-Way Chi Square

Again the first step is to compute the expected frequencies. To do so, first compute the total of the observed frequencies in each column and the total of the observed frequencies in each row. This is shown in Table 13.5. Also, note N, the total of all observed frequencies. Now we will compute the expected frequency in each *cell*. (A cell is a square in the diagram in Table 13.5 formed by a combination of personality type and heart attack/no heart attack.)

Each f_e is based on the probability of a participant falling into the cell if the two variables are independent. For example, for the cell of Type A and heart attack, we'd determine the probability of someone in our study being Type A and the probability of someone in our study reporting a heart attack, when these variables are independent. The expected frequency in this cell then equals this probability multiplied by N. Luckily, the steps involved in this can be combined to produce this formula:

COMPUTING THE EXPECTED FREQUENCY IN EACH CELL OF A TWO-WAY CHI SQUARE

$$f_e = \frac{(\text{Cell's row total } f_o)(\text{Cell's column total } f_o)}{N}$$

This says that, for each cell, multiply the total observed frequency for the row containing the cell times the total observed frequency for the column containing the cell. Then divide by the N of the study.

Table 13.5

Diagram Containing f_o and f_e for Each Cell

Each f_e equals the row total times the column total, divided by N.

		Personality Type		
		Type A	Type B	
Health	Heart Attack	$f_o = 25$ $f_e = 13.125$ (35)(30)/80	$f_o = 10$ $f_e = 21.875$ (35)(50)/80	Row Total = 35
	No Heart Attack	$f_o = 5$ $f_e = 16.875$ (45)(30)/80	$f_o = 40$ $f_e = 28.125$ (45)(50)/80	Row Total = 45
		Column total = 30	Column total = 50	$N = 80$

Thus, to compute the two-way χ^2:

STEP 1: *Compute the f_e for each cell.* Table 13.5 shows the completed f_e for the example.

STEP 2: *Compute χ^2_{obt}.* Use the same formula used in the one-way design, which is

$$\chi^2_{obt} = \Sigma \left(\frac{(f_o - f_e)^2}{f_e} \right)$$

Here, the formula says to first form a fraction for each *cell*: In the numerator of each, square the difference between the f_e and f_o for the cell. In the denominator is the f_e for the cell. Thus, with the data in Table 13.5 we have

$$\chi^2_{obt} = \left(\frac{(25 - 13.125)^2}{13.125} \right) + \left(\frac{(10 - 21.875)^2}{21.875} \right) + \left(\frac{(5 - 16.875)^2}{16.875} \right) + \left(\frac{(40 - 28.125)^2}{28.125} \right)$$

STEP 3: *Perform the subtraction in the numerator of each fraction.* After subtracting, we have

$$\chi^2_{obt} = \frac{(11.875)^2}{13.125} + \frac{(-11.875)^2}{21.875} + \frac{(-11.875)^2}{16.875} + \frac{(11.875)^2}{28.125}$$

STEP 4: *Square the numerator in each fraction.* This gives

$$\chi^2_{obt} = \frac{141.016}{13.125} + \frac{141.016}{21.875} + \frac{141.016}{16.875} + \frac{141.016}{28.125}$$

STEP 5: *Perform the division in each fraction and then sum the results.*

$$\chi^2_{obt} = 10.74 + 6.45 + 8.36 + 5.01$$

so

$$\chi^2_{obt} = 30.56$$

STEP 6:

Compare χ^2_{obt} to χ^2_{crit}. First, determine the degrees of freedom. Look at your diagram of your study and count the number of rows and columns in your matrix. Then:

DEGREES OF FREEDOM IN A TWO-WAY CHI SQUARE

$$df = (\text{Number of rows} - 1)(\text{Number of columns} - 1)$$

For our study, df is $(2 - 1)(2 - 1)$, which is 1. Find the critical value of χ^2 in Table 6 in Appendix B. At $\alpha = .05$ and $df = 1$, the χ^2_{crit} is 3.84.

Our χ^2_{obt} of 30.56 is larger than χ^2_{crit}, so it is significant. Therefore, we reject H_0 that the variables are independent and accept the alternative hypothesis: We are confident that the sample represents frequencies from two variables that are dependent in the population. In other words, the correlation is significant such that the frequency of having or not having a heart attack depends on the frequency of being Type A or Type B (and vice versa).

If χ^2_{obt} is not larger than the critical value, do not reject H_0. Then we cannot say whether these variables are independent or not.

Describing the Relationship in a Two-Way Chi Square

A significant two-way chi square indicates a significant correlation between the two variables. To determine the size of this correlation, we compute one of two new correlation coefficients, either the *phi coefficient* or the *contingency coefficient*.

REMEMBER

A significant two-way χ^2 indicates that the sample data are likely to represent two variables that are dependent (correlated) in the population.

If you have performed a 2×2 chi square and it is significant, compute the **phi coefficient**. The symbol for the phi coefficient is ϕ, and its value can be between 0 and 1. Think of phi as comparing your data to the ideal situations that were illustrated back in Tables 13.2 and 13.3 when the variables are or are not dependent. A value of 0 indicates that the data are perfectly independent. The larger the value of phi, however, the closer the data come to being perfectly dependent. (Real research tends to find values in the range of .20 to .50.)

PHI COEFFICIENT

$$\phi = \sqrt{\frac{\chi^2_{obt}}{N}}$$

The formula says to divide the χ^2_{obt} by N (the total number of participants) and then find the square root.

For the heart attack study, χ^2_{obt} was 30.56 and N was 80, so

$$\phi = \sqrt{\frac{\chi^2_{obt}}{N}} = \sqrt{\frac{30.56}{80}} = \sqrt{.382} = .62$$

Thus, on a scale of 0 to 1, where 1 indicates perfect dependence, the correlation is .62 between the frequency of heart attacks and the frequency of personality types.

Further, recall that by squaring a correlation coefficient we obtain the *proportion of variance accounted for*, the proportion of differences in one variable that is associated with the other variable. By not computing the square root in the formula above, we have ϕ^2. For our study $\phi^2 = .38$. This is analogous to r^2, indicating that about 38% of the differences in whether people have heart attacks are associated with differences in their personality type—and vice versa.

> **phi coefficient (ϕ)**
> The statistic that describes the strength of the relationship in a two-way chi square when there are only two categories for each variable

contingency coefficient (C) The statistic that describes the strength of the relationship in a two-way chi square when there are more than two categories for either variable

The other correlation coefficient is the **contingency coefficient,** symbolized by *C*. This is used to describe a significant two-way chi square that is *not* a 2 × 2 design (it's a 2 × 3, a 3 × 3, etc.).

CONTINGENCY COEFFICIENT

$$C = \sqrt{\frac{\chi^2_{\text{obt}}}{N + \chi^2_{\text{obt}}}}$$

This says to first add N to the χ^2_{obt} in the denominator. Then divide that quantity into χ^2_{obt}, and then find the square root. Interpret C in the same way as ϕ. (Likewise, C^2 is analogous to ϕ^2.)

13.5 STATISTICS IN THE RESEARCH LITERATURE: REPORTING χ^2

A chi square is reported like previous results, except that in addition to *df*, we also include the *N*. For example, in our one-way design involving geniuses and handedness, we tested an *N* of 50, *df* was 1, and the significant χ^2_{obt} was 18. We report these results as $\chi^2(1, N = 50) = 18.00, p < .05$. We report a two-way χ^2 using the same format.

As usual, a graph is useful for summarizing the data. For a one-way design, label the *Y* axis with frequency and the *X* axis with the categories, and then plot the f_o in each category. Because the *X* variable is nominal, we create a *bar graph*. The bar graph on the left in Figure 13.2 shows the results of our handedness study.

The right-hand graph in Figure 13.2 shows the bar graph for our heart attack study. To graph the data from a two-way design, place frequency on the *Y* axis and one of the nominal variables on the *X* axis. The levels of the other variable are indicated in the body of the graph. (This is similar to the way a two-way interaction

was plotted in the previous chapter, except that here we create bar graphs.)

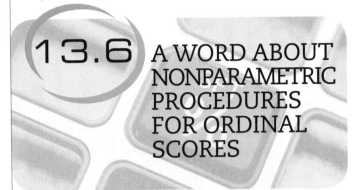

13.6 A WORD ABOUT NONPARAMETRIC PROCEDURES FOR ORDINAL SCORES

We also have nonparametric procedures that are used with ordinal (rank-ordered) scores. We obtain ranked scores in a study for one of two reasons. First, sometimes we directly measure participants using ranked scores (directly assigning participants a score of 1st, 2nd, etc.). Second, sometimes we initially measure interval or ratio scores, but they violate the assumptions of parametric procedures by not being normally distributed or not having homogeneous variance. Then we transform these scores to ranks (the highest raw score is ranked 1, the next highest score is ranked 2, and so on). Regardless of how we obtained the ranked scores, we then compute the same nonparametric inferential statistics.

Although the computations for nonparametric procedures are different from those of parametric procedures, their logic and rules are the same. A relationship occurs here when the ordinal scores consistently change. For example, we might see the scores change from predominantly low ranks around 1st and 2nd in one condition to predominantly higher ranks in another condition, say around 6th or 7th. The null hypothesis says that our data show this because of sampling error, and we are poorly representing that no relationship occurs in the population. The alternative hypothesis is that the pattern in our data represents a similar relationship that would be found in the population. As usual, we test H_0 by computing an obtained statistic that describes our data. By comparing it to a critical value, we determine whether the sample relationship is significant. If it is, we conclude that the predicted relationship exists in the population (in nature). If the data are not significant, we retain H_0.

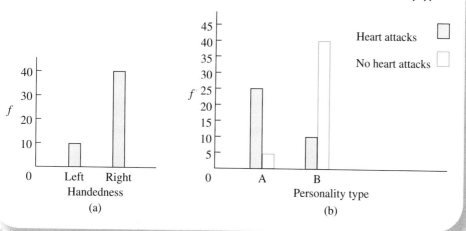

REMEMBER

Perform nonparametric procedures for ranked data *when the dependent variable is measured in, or transformed to, ordinal scores.*

In the literature you will encounter a number of nonparametric procedures. The computations for each are found in more advanced textbooks (or you can use SPSS). We won't dwell on their computations because you are now experienced enough to compute and understand them if you encounter them. However, you should know when we use the most common procedures.

Common Nonparametric Procedures for Ranked Scores

1. The **Spearman rank-order correlation coefficient** is analogous to the Pearson correlation coefficient for ranked data. Its symbol is r_s. It produces a number between ± 1 that describes the strength and type of linear relationship that is present when data consist of pairs of X-Y scores that are both ordinal scores. If significant, r_s estimates the corresponding population coefficient.

2. The **Mann–Whitney U test** is analogous to the independent-samples t-test. It is performed when a study contains two independent samples of ordinal scores.

3. The **Wilcoxon T test** is analogous to the related-samples t-test. It is performed when a study has two related samples of ordinal scores. Recall that related samples occur either through *matching*

Spearman rank-order correlation coefficient (r_s) The correlation coefficient that describes the linear relationship between pairs of ranked scores

Mann–Whitney U test The nonparametric version of the independent-samples t-test for ranked scores when there are two independent samples of ranked scores

Wilcoxon T test The nonparametric version of the related-samples t-test for ranked scores

Kruskal–Wallis H test The nonparametric version of the one-way, between-subjects ANOVA for ranked scores

Friedman χ^2 test The nonparametric version of the one-way, within-subjects ANOVA for ranked scores

pairs of participants or through *repeated measures* of the same participants.

4. The **Kruskal–Wallis H test** is analogous to a one-way, between-subjects ANOVA. It is performed when a study has one factor with at least three conditions, and each involves independent samples of ordinal scores.

5. The **Friedman χ^2 test** is analogous to a one-way, within-subjects ANOVA. It is performed when a study has one factor with at least three levels, and each involves related samples of ordinal scores.

USING SPSS

Your chapter Review Cards detail how to use SPSS to perform the one-way or the two-way chi square procedure. The program counts the f_o in each group, computes χ^2, and reports its significance level. In the two-way design, you can also compute ϕ or C, and plot a bar graph of the data.

The chapter Review Card for this chapter also describes how to use SPSS to perform the five common nonparametric procedures for ranked scores listed here.

using what you know

1. a. What do we determine when we perform nonparametric statistics?
    b. When are they used?

2. a. When is the one-way χ^2 test used?
    b. When is the two-way χ^2 test used?

3. a. What are the symbol for and meaning of observed frequency?
    b. What are the symbol for and meaning of expected frequency?

4. a. What does a significant one-way χ^2 indicate?
    b. What does a significant two-way χ^2 indicate?

5. a. What is the phi coefficient, and when is it used?
    b. What is the contingency coefficient, and when is it used?

6. Which nonparametric test for ranks corresponds to each of the following parametric tests?
    a. The related samples *t*-test
    b. Pearson correlation coefficient
    c. The *t*-test for independent samples
    d. The one-way, between-subjects ANOVA
    e. The one-way, within-subjects ANOVA

7. Assume each of the following studies cannot be analyzed using a parametric procedure. For each

of the following studies, indicate the appropriate nonparametric procedure:

a. One group of mothers has played classical music to their babies every day for 6 months, while another group of mothers has not. The babies are tested to determine if there is a significant difference in the motor coordination skills of these two groups.

b. College freshmen with high test anxiety are assigned to three groups. One group receives free individualized counseling, one group participates in a class to teach them to reduce their anxiety, and the third group receives no treatment. They are tested to determine if there is a significant difference in the test anxiety levels of these three groups.

c. Elementary school children with attention deficit disorder are tested for their ability to read a 5-page age-appropriate document. Each child is then assigned to a reading tutor. After 3 weeks of working with the tutor, each child's reading ability is tested again.

d. A survey of local residents has been conducted to determine the level of support for a new shopping plaza. Those surveyed were allowed to respond in one of five categories—strongly support, support, no opinion, do not support, strongly do not support. The number of people in each category was recorded.

8. Which parametric tests correspond to the following nonparametric tests?

    a. The Mann-Whitney U test
    b. The Wilcoxon T test
    c. The Kruskal-Wallis H test
    d. The Friedman χ^2 test

9. A newspaper article claims there are many more newborn females than males, although you had assumed there was a fifty-fifty chance of being born male or female. Through the local hospital, you determine 628 baby boys and 718 baby girls were born over the past two months.

    a. What are H_0 and H_a?
    b. Compute the appropriate statistic.
    c. Determine whether the results are significant at $\alpha = .05$.
    d. Report your results in the correct format and interpret them.

10. A researcher collected the following handedness data from 60 students. Determine whether gender and handedness are unrelated variables.

	Left-handed	Ambidextrous	Right-handed
Men	29	3	8
Women	1	9	10

    a. What are the null and alternative hypotheses?
    b. Compute the appropriate statistic.
    c. Determine whether the results are significant at $\alpha = .05$.
    d. Describe the strength of this relationship. How useful a relationship is it?
    e. Report and interpret the results.

11. In a recent report, Charlene's company claimed they hire as many female as male engineers. Using the company's hiring information for the past year, Charlene (who is the Human Resources Manager) decides to test this claim.

	Male	Female
Number of Engineers Hired	23	17

    a. What test should Charlene use?
    b. What are her H_0 and H_a?
    c. Conduct the appropriate test. Use $\alpha = .05$
    d. What should Charlene conclude about her company's claim?

12. Michael wants to know if support for a local referendum is related to political party affiliation. He conducts a survey of 200 local residents and asks them their political party affiliation and whether they intend to vote for or against the referendum. The number of individuals in each cell is reported in the following table.

	For	Against
Republican	16	54
Democrat	72	28
Other	27	3

    a. What test should Michael use?
    b. What are H_0 and H_a?
    c. Conduct the appropriate test. Use $\alpha = .05$.
    d. What should Michael conclude?

13. An experiment was conducted to determine whether it is easier to recall 3-letter or 5-letter nonsense syllables. Each participant learned both a 3-letter and a 5-letter nonsense-syllable list, after which the number of errors in recall was determined. Suppose, however, you discover the scores cannot be assumed to have homogeneity of variance.

    a. Under these circumstances, which statistical test would be appropriate for analyzing these data?
    b. If the data were homogeneous, what would be the more powerful test?

Math Review
and Additional Formulas

A.1 REVIEW OF BASIC MATH

T he following is a review of the math used in performing statistical procedures. There are accepted systems for identifying mathematical operations, for rounding answers, for computing a proportion and a percent, and for creating graphs.

Identifying Mathematical Operations

Here are the mathematical operations you'll use in statistics, and they are simple ones. Addition is indicated by the plus sign, so for example, $4 + 2$ is 6 (I said this was simple!). Subtraction is indicated by the minus sign. We read from left to right, so $X - Y$ is read as "X minus Y." This order is important because $10 - 4$, for example, is $+6$, but $4 - 10$ is -6. With subtraction, pay attention to what is subtracted from what and whether the answer is positive or negative. Adding two negative numbers together gives a larger negative number, so $-4 + -3 = -7$. Adding a negative number to a positive number is the same as subtracting by the negative's amount, so $5 + -2 = 5 - 2 = 3$. When subtracting a negative number, a double negative produces a positive. Thus, in $4 - -3$, the minus -3 becomes $+3$, so we have $4 + 3 = 7$.

We indicate division by forming a fraction, such as X/Y. The number above the dividing line is called the numerator, and the number below the line is called the denominator. *Always express fractions as decimals,* dividing the denominator *into* the numerator. (After all, 1/2 equals .5, not 2!)

Multiplication is indicated in one of two ways. We may place two components next to each other: XY means "X times Y." Or we may indicate multiplication using parentheses: $4(2)$ and $(4)(2)$ both mean "4 times 2."

The symbol X^2 means square the score, so if X is 4, X^2 is 16. Conversely, $\sqrt{X}$ means "find the square root of X," so $\sqrt{4}$ is 2. (The symbol $\sqrt{}$ also means "use your calculator.")

Determining the Order of Mathematical Operations

Statistical formulas often call for a series of mathematical steps. Sometimes the steps are set apart by parentheses. Parentheses mean "the quantity," so always find the quantity inside the parentheses first and then perform the operations outside of the parentheses on that quantity. For example, $(2)(4 + 3)$ indicates to multiply 2 times "the quantity 4 plus 3." So first add, which gives $(2)(7)$, and then multiply to get 14.

A square root sign also operates on "the quantity," so always compute the quantity inside the square root sign first. Thus, $\sqrt{2 + 7}$ means find the square root of the quantity $2 + 7$; so $\sqrt{2 + 7}$ becomes $\sqrt{9}$, which is 3.

Most formulas are giant fractions. Pay attention to how far the dividing line is drawn because the length of a dividing line determines the quantity that is in the numerator and the denominator. For example, you might see a formula that looks like this:

$$X = \frac{\frac{6}{3} + 14}{\sqrt{64}}$$

The longest dividing line means you should divide the square root of 64 into the quantity in the numerator, so first work in the numerator. Before you can add 6/3 to 14 you must reduce 6/3 by dividing $6/3 = 2$. Then you have

$$X = \frac{2 + 14}{\sqrt{64}}$$

Now adding $2 + 14$ gives 16, so

$$X = \frac{16}{\sqrt{64}}$$

Before we can divide we must find the square root of 64, which is 8, so we have

$$X = \frac{16}{8} = 2$$

When working with complex formulas, perform one step at a time and then *rewrite* the formula. Trying to do several steps in your head is a good way to make mistakes.

If you become confused in reading a formula, remember that there is a rule for the order of mathematical operations. Often this is summarized with PEMDAS or you may recall the phrase "Please Excuse My Dear Aunt Sally." Either way the letters indicate that, unless otherwise indicated, first compute inside any **P**arentheses, then compute **E**xponents (squaring and square roots), then **M**ultiply or **D**ivide, and finally, **A**dd or **S**ubtract. Thus, for $(2)(4) + 5$, first multiply 2 times 4 and then add 5. For $2^2 + 3^2$, first square each number, resulting in $4 + 9$, which is then 13. Finally, an important distinction is whether a squared sign is inside or outside of parentheses. Thus, in $(2^2 + 3^2)$ we square first, giving $(4 + 9)$ so the answer is 13. But! For $(2 + 3)^2$ we add first, so $2 + 3 = 5$ and then squaring gives $(5)^2$, so the answer is 25!

Working with Formulas

We use a formula to find an answer, and we have symbols that stand for that answer. For example, in the formula $B = AX + K$, the B stands for the answer we will obtain. The symbol for the unknown answer is always isolated on one side of the equal sign, but we will know the numbers to substitute for the symbols on the other side of the equal sign. For example, to find B, say that $A = 4$, $X = 11$, and $K = 3$. In working any formula, the first step is to copy the formula and then rewrite it, replacing the symbols with their known values. Thus, start with

$$B = AX + K$$

Filling in the numbers gives

$$B = 4(11) + 3$$

Rewrite the formula after performing each mathematical operation. Above, multiplication takes precedence over addition, so multiply and then rewrite the formula as

$$B = 44 + 3$$

After adding,

$$B = 47$$

Do not skip rewriting the formula after each step!

Rounding Numbers

Close counts in statistics, so you must carry out calculations to the appropriate number of decimal places. Usually, you must "round off" your answer. The rule is this: *Always carry out calculations so that your final answer after rounding has two more decimal places than the original scores.* Usually, we have whole-number scores (e.g., 2 and 11) so the final answer contains two decimal places. But say the original scores contain one decimal place (e.g., 1.4 and 12.3). Here the final answer should contain *three* decimal places.

So when beginning a problem, first decide the number of decimals to be in your final answer. However, if there are intermediate computing steps, do not round off to this number of decimals at each step. This will produce substantial error in your final answer. Instead, carry out each intermediate step to more decimals than you'll ultimately need before rounding. Then the error introduced will be smaller. If the final answer is to contain two decimal places, round off your intermediate answers to at least three decimal places. Then after you've completed all calculations, round off the final answer to two decimal places.

To round off a calculation use the following rules:

If the number in the next decimal place is 5 or greater, round up. For example, to round to two decimal places, 2.366 is rounded to 2.370, which becomes 2.37.

If the number in the next decimal place is less than 5, round down: 3.524 is rounded to 3.520, which becomes 3.52.

We add zeroes to the right of the decimal point to indicate the level of precision we are using. For example, rounding 4.966 to two decimal places produces 5, but to show we used the precision of two decimal places, we report it as 5.00.

Computing Proportions and Percents

Sometimes we will transform an individual's original score into a proportion. A *proportion* is a decimal number between 0 and 1 that indicates a fraction of the total. To transform a number to a proportion, divide the number by the total. If 4 out of 10 people pass an exam, then the proportion of people passing the exam is 4/10, which equals .4. Or, if you score 6 correct on a test out of a possible 12, the proportion you have correct is 6/12, which is .5.

We can also work in the opposite direction from a known proportion to find the number out of the total it represents. Here, multiply the proportion times the total. Thus, to find how many questions out of 12 you must answer correctly to get .5 correct, multiply .5 times 12, and voilà, the answer is 6.

We can also transform a proportion into a percent. A *percent* (or percentage) is a proportion multiplied by 100. Above, your proportion correct was .5, so you had (.5)(100) or 50% correct. Altogether, to transform the original test score of 6 out of 12 to a percent, first divide the score by the total to find the proportion and then multiply by 100. Thus (6/12)(100) equals 50%.

To transform a percent back into a proportion, divide the percent by 100 (above, 50/100 equals .5). Altogether, to find the test score that corresponds to a certain percent, transform the percent to a proportion and then multiply the proportion times the total number possible. Thus, to find the score that corresponds to 50% of 12, transform 50% to the proportion, which is .5 and then multiply .5 times 12. So, 50% of 12 is equal to (50/100)(12), which is 6.

Recognize that a percent is a whole unit: Think of 50% as 50 of those things called percents. On the other hand, a decimal in a percent is a proportion of *one* percent. Thus, .2% is .2 or two-tenths, of one percent, which is .002 of the total.

Creating Graphs

Recall that the horizontal line across the bottom of a graph is the X axis, and the vertical line at the left-hand side is the Y axis. (Draw the Y axis so that it is about 60 to 75% of the length of the X axis.) Where the two axes intersect is always labeled as a score of zero on X and a score of zero on Y. On the X axis, scores become larger positive scores as you move to the *right*. On the Y axis, scores become larger positive scores as you move *upward*.

Say that we measured the height and weight of several people. We decide to place weight on the Y axis and height on the X axis. (How to decide this is discussed later.) We plot the scores as shown in Figure A.1. Notice that because the lowest height score is 63, the lowest label on the X axis is also 63. The symbol // in the axis indicates that we cut out the part between 0 and 63. We do this with either axis when there is a large gap between 0 and the lowest score we are plotting.

In the body of the graph, we plot the scores from the table on the left. Jane is 63 inches tall and weighs 130 pounds, so we place a dot above the height of 63 and opposite the weight of 130. And so on. As mentioned in Chapter 2, each dot on a graph is called a data point. Notice that you read the graph by using the scores on one axis and the data points. For example, to find the weight of the person who has a height of 67, travel vertically from 67 to the data point and then horizontally back to the Y axis: 165 is the corresponding weight.

Always label the X and Y axes to indicate what the scores measure (not just X and Y), and always give your graph a title indicating what it describes.

Practice these concepts with the following practice problems.

Figure A.1
Plot of Height and Weight Scores

Person	Height	Weight
Jane	63	130
Bob	64	140
Mary	65	155
Tony	66	160
Sue	67	165
Mike	68	170

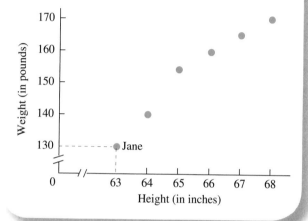

For Practice

1. Round off the following numbers to two decimal places:
   (a) 13.7462 (b) 10.043 (c) 10.047 (d) .079 (e) 1.004

2. The intermediate answers in a formula based on whole-number scores are $X = 4.3467892$ and $Y = 3.3333$. What values of X and Y do we use when performing the next step in the calculations?

3. For $Q = (X + Y)(X^2 + Y^2)$ find Q when $X = 8$ and $Y = -2$.

4. Below find D when $X = 14$ and $Y = 3$.

$$D = \left(\frac{X - Y}{Y}\right)\left(\sqrt{X}\right)$$

5. Using the formula in problem 4, find D when $X = 9$ and $Y = -4$.

6. (a) What proportion is 5 out of 15? (b) What proportion of 50 is 10? (c) One in a thousand equals what proportion?

7. Transform each answer in problem 6 to a percent.

8. Of the 40 students in a gym class, 35% played volleyball and 27.5% ran track. (a) What proportion of the class played volleyball? (b) How many students played volleyball? (c) How many ran track?

9. You can earn a total of 135 points in your statistics course. To pass you need 60% of these points. (a) How many points must you earn to pass the course? (b) You actually earned a total of 115 points. What percent of the total did you earn?

10. Create a graph showing the data points for the following scores:

X Score Student's Age	Y Score Student's Test Score
20	10
25	30
35	20
45	60
25	55
40	70
45	3

Answers

1. (a) 13.75 (b) 10.04 (c) 10.05 (d) .08 (e) 1.00
2. Carry at least three places, so $X = 4.347$ and $Y = 3.333$.
3. $Q = (8 + -2)(64 + 4) = (6)(68) = 408$
4. $D = (3.667)(3.742) = 13.72$
5. $D = \left(\frac{13}{-4}\right)(3) = (-3.250)(3) = -9.75$
6. (a) $5/15 = .33$ (b) $10/50 = .20$ (c) $1/1000 = .001$
7. (a) 33% (b) 20% (c) .1%
8. (a) 35%/100 = .35 (b) (.35)(40) = 14
   (c) (27.5%/100)(40) = 11
9. (a) 60% of 135 is (60%/100)(135) = 81
   (b) (115/135)(100) = 85%
10.

Plot of Students' Age and Test Scores

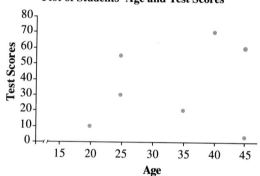

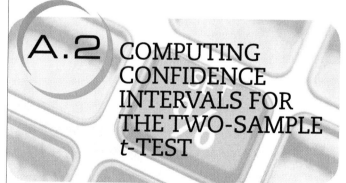

A.2 COMPUTING CONFIDENCE INTERVALS FOR THE TWO-SAMPLE t-TEST

Two versions of a confidence interval can be used to describe the results from the two-sample *t*-test described in Chapter 9. For the independent-samples *t*-test we compute the *confidence interval for the difference between two μs*; for the related-samples *t*-test we compute the *confidence interval for* μ_D.

Confidence Interval for the Difference between Two μs

The **confidence interval for the difference between two μs** describes a range of *differences* between two μs, any one of which is likely to be represented by the *difference* between our two sample means. This procedure is appropriate when the sample means are from *independent* samples. For example, in Chapter 9 we discussed the experiment that compared recall scores under the conditions of Hypnosis and No-hypnosis. We found a difference of +3 between the sample means ($\overline{X}_1 - \overline{X}_2$). If we could examine the corresponding μ_1 and μ_2, we'd expect that their difference ($\mu_1 - \mu_2$) would be *around* +3. We say "around" because we may have sampling error, so the actual difference between μ_1 and μ_2 might be 2 or 4. The confidence interval contains the highest and lowest values around +3 that the difference between our sample means is likely to represent.

THE CONFIDENCE INTERVAL FOR THE DIFFERENCE BETWEEN TWO μs

$$(s_{\overline{X}_1 - \overline{X}_2})(-t_{crit}) + (\overline{X}_1 - \overline{X}_2) \leq \mu_1 - \mu_2 \leq$$
$$(s_{\overline{X}_1 - \overline{X}_2})(+t_{crit}) + (\overline{X}_1 - \overline{X}_2)$$

Here, $\mu_1 - \mu_2$ stands for the unknown difference we are estimating. The t_{crit} is the *two-tailed* value found for the appropriate α at $df = (n_1 - 1) + (n_2 - 1)$. The values of $s_{\overline{X}_1 - \overline{X}_2}$ and ($\overline{X}_1 - \overline{X}_2$) are computed in the independent-samples *t*-test.

REMEMBER

The confidence interval for the difference between two μs describes the difference between the population means represented by the difference between our sample means in the independent-samples t-test.

In the hypnosis study, the two-tailed t_{crit} for $df = 30$ and α = .05 is ±2.042, $s_{\overline{X}_1 - \overline{X}_2}$ is 1.023, and $\overline{X}_1 - \overline{X}_2$ is +3. Filling in the formula gives

$$(1.023)(-2.042) + (+3) \leq \mu_1 - \mu_2 \leq$$
$$(1.023)(+2.042) + (+3)$$

Multiplying 1.023 times ±2.042 gives

$$-2.089 + (+3) \leq \mu_1 - \mu_2 \leq +2.089 + (+3)$$

So finally,

$$.911 \leq \mu_1 - \mu_2 \leq 5.089$$

Because α = .05, this is the 95% confidence interval: We are 95% confident that the interval between .911 and 5.089 contains the difference we'd find between the μs for no-hypnosis and hypnosis. In essence, if someone asked how big is the average difference between when the population recalls under hypnosis and when it recalls under no-hypnosis, we'd be 95% confident the difference is between .91 and 5.09.

For Practice

1. In problem 13 on your Chapter 9 Review Card, $\overline{X}_1 = 43$, $\overline{X}_2 = 39$, and $s_{\overline{X}_1 - \overline{X}_2} = 1.78$. At α = .05 and $df = 28$, $t_{crit} = ±2.048$. What is the confidence interval for the difference between the μs?

Answer

1. $(1.78)(-2.048) + 4 \leq \mu_1 - \mu_2 \leq (1.78)(+2.048) + 4 =$
    $.35 \leq \mu_1 - \mu_2 \leq 7.65$

Computing the Confidence Interval for μ_D

The other confidence interval is used with the related-samples *t*-test to describe the μ of the population of difference scores (μ_D) that is represented by our sample of difference scores ($\overline{D}$). The **confidence interval for μ_D** describes a range of values of μ_D, one of which our sample mean is likely to represent. The interval contains the highest and lowest values of μ_D that are not significantly different from $\overline{D}$.

THE CONFIDENCE INTERVAL FOR μ_D

$$(s_{\overline{D}})(-t_{crit}) + \overline{D} \leq \mu_D \leq (s_{\overline{D}})(+t_{crit}) + \overline{D}$$

The t_{crit} is the *two-tailed* value for $df = N - 1$, where N is the number of difference scores, $s_{\overline{D}}$ is the standard error of the mean difference computed in the *t*-test, and $\overline{D}$ is the mean of the difference scores.

For example, in Chapter 9 we compared the fear scores of participants who had or had not received our phobia therapy. We found the mean difference score in the sample was $\overline{D} = +3.6$, $s_{\overline{D}} = 1.25$, and with $\alpha = .05$ and $df = 4$, t_{crit} is ± 2.776. Filling in the formula gives

$$(1.25)(-2.776) + 3.6 \leq \mu_D \leq (1.25)(+2.776) + 3.6$$

which becomes

$$(-3.47) + 3.6 \leq \mu_D \leq (+3.47) + 3.6$$

and so

$$.13 \leq \mu_D \leq 7.07$$

Thus, we are 95% confident that our sample mean of differences represents a population μ_D within this interval. In other words, if we performed this study on the entire population, we would expect the average difference in before- and after-therapy scores to be between .13 and 7.07.

For Practice

1. In problem 17 on your Chapter 9 Review Card, $\overline{D} = +2.63$ and $s_{\overline{D}} = .75$. At $\alpha = .05$ and $df = 7$, $t_{crit} = \pm 2.365$. What is the confidence interval for μ_D?

2. In problem 19 on your Chapter 9 Review Card, $\overline{D} = 1.2$ and $s_{\overline{D}} = .359$. At $\alpha = .05$ and $df = 9$, $t_{crit} = \pm 2.262$. What is the confidence interval for the difference between the μs?

Answers

1. $(.75)(-2.365) + 2.63 \leq \mu_D \leq (.75)(+2.365) + 2.63 = .86 \leq \mu_D \leq 4.40$

2. $(.359)(-2.262) + 1.2 \leq \mu_D \leq (.359)(+2.262) + 1.2 = .39 \leq \mu_D \leq 2.01$

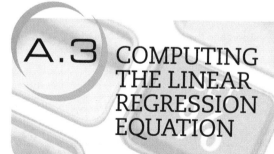

A.3 COMPUTING THE LINEAR REGRESSION EQUATION

As discussed in Chapter 10, computing the linear regression equation involves computing two components: the *slope* and the *Y intercept*.

First we compute the *slope* of the regression line, which is symbolized by b. This is a number that mathematically conveys the degree and slant of the regression line. A negative number indicates a negative linear relationship; a positive number indicates a positive relationship. A slope of 0 indicates no relationship.

THE SLOPE OF THE LINEAR REGRESSION LINE

$$b = \frac{N(\Sigma XY) - (\Sigma X)(\Sigma Y)}{N(\Sigma X^2) - (\Sigma X)^2}$$

N is the number of pairs of scores in the sample, and X and Y are the scores in our data. Notice that the numerator of the formula here is the same as the numerator of the formula for r, and the denominator of the formula here is the left-hand quantity in the denominator of the formula for r. [An alternative formula is $b = (r)(S_Y/S_X)$.]

For example, in Chapter 10 we examined the relationship between daily juice consumption and yearly doctor visits and found $r = -.95$. In the data (in Table 10.1) we found $\Sigma X = 17$, $(\Sigma X)^2 = 289$, $\Sigma X^2 = 45$, $\Sigma Y = 47$, $\Sigma XY = 52$, and $N = 10$. Filling in the above formula for b gives:

$$b = \frac{10(52) - (17)(47)}{10(45) - 289} = \frac{520 - 799}{450 - 289}$$

$$= \frac{-279}{161} = -1.733$$

Thus, in this negative relationship we have a negative slope of $b = -1.733$.

Next we compute the Y intercept, symbolized by a. This is the value of Y when the regression line crosses the Y axis. (Notice that a negative number is also possible here if the regression line crosses the Y axis at a point below the X axis.)

THE Y INTERCEPT OF THE LINEAR REGRESSION LINE

$$a = \overline{Y} - (b)(\overline{X})$$

Here we first multiply the mean of the X scores times the slope of the regression line. Then subtract that quantity from the mean of the Y scores. For our example, $\overline{X} = 1.70$, $\overline{Y} = 4.70$, and $b = -1.733$, so

$$a = 4.70 - (-1.733)(1.70) = 4.70 - (-2.946)$$

Subtracting a negative number is the same as adding its positive value so,

$$a = 4.70 + 2.946 = 7.646$$

Thus, when we plot the regression line, it will cross the Y axis at the Y of 7.646.

Applying the Linear Regression Equation

We apply the slope and the Y intercept in the linear regression equation.

LINEAR REGRESSION EQUATION

$$Y' = bX + a$$

This says to obtain the Y' for a particular X, multiply the X by b and then add a.

For our example data, substituting our values of b and a we have

$$Y' = (-1.733)X + 7.646$$

This is the equation for the regression line that summarizes our juice–doctor visits data.

To plot the regression line: We need at least two data points to plot a line. Therefore, choose a low value of X, insert it into the completed regression equation, and calculate the value of Y'. Choose a higher value of X and calculate Y' for that X. (Do not select values of X that are above or below those found in your original data.) Plot your values of X-Y' and connect the data points with a straight line.

To predict a Y score: To predict an individual's Y score, enter his/her X score into the completed regression equation and compute the corresponding Y'. This is the Y score we predict for anyone who scores at that X.

The Standard Error of the Estimate

We compute the standard error of the estimate to determine the amount of error we expect to have when we use a relationship to predict Y scores. Its symbol is $S_{Y'}$.

THE STANDARD ERROR OF THE ESTIMATE

$$S_{Y'} = (S_Y)\left(\sqrt{1 - r^2}\right)$$

This says to find the square root of the quantity $1 - r^2$ and then multiply it times the standard deviation of all Y scores (S_Y). In our juice–doctor visits data in Table 10.1, the $\Sigma Y = 47$, $\Sigma Y^2 = 275$, and $N = 10$. Thus, first we compute S_Y.

$$S_Y = \sqrt{\frac{\Sigma Y^2 - \frac{(\Sigma Y)^2}{N}}{N}} = \sqrt{\frac{275 - \frac{(47)^2}{10}}{10}}$$

$$= \sqrt{5.41} = 2.326$$

Our r in these data was $-.95$, so the standard error of the estimate is

$$S_{Y'} = (S_Y)\left(\sqrt{1 - r^2}\right) = (2.326)\left(\sqrt{1 - (-.95)^2}\right)$$

$$= (2.326)\left(\sqrt{1 - .9025}\right) = (2.326)(.312) = .73$$

Thus, in this relationship, $S_{Y'} = .73$. It indicates that we expect the "average error" in our predictions to be .73 when we use this regression equation to predict Y scores. For example, if someone's Y' score is 2, we expect to be "off" by about $\pm.73$, so we expect his or her actual Y score will be between 1.27 and 2.73.

For Practice

1. Compute the regression equation for each set of scores.

a.

X	Y
1	3
1	2
2	4
2	5
3	5
3	6

b.

X	Y
1	5
1	3
2	4
2	3
3	2
4	1

2. What will the standard error of the estimate indicate for each set of scores?

Answers

1. a. $b = \dfrac{6(56) - (12)(25)}{6(28) - (12)^2} = 1.5$;
   $a = 4.167 - (+1.5)(2) = 1.17$;

   $Y' = +1.5X + 1.17$.

   b. $b = \dfrac{6(32) - (13)(18)}{6(35) - (13)^2} = -1.024$;
   $a = 3 - (-1.024)(2.167) = +5.219$;

   $Y' = -1.042X + 5.219$

2. It indicates the "average error" we expect between the actual Y scores and the predicted Y' scores.

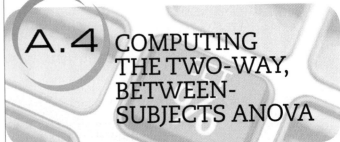

A.4 COMPUTING THE TWO-WAY, BETWEEN-SUBJECTS ANOVA

As discussed in Chapter 12, the following presents the formulas for computing the two-way, between subjects ANOVA, the Tukey *HSD* test for main effects and interactions, and η^2.

Table A.1
Summary of Data for 3 × 2 ANOVA

		Factor A: Volume			
		A_1: Soft	A_2: Medium	A_3: Loud	
Factor B: Gender	B_1: Male	4 9 11 $\bar{X} = 8$ $\Sigma X = 24$ $\Sigma X^2 = 218$ $n = 3$	8 12 13 $\bar{X} = 11$ $\Sigma X = 33$ $\Sigma X^2 = 377$ $n = 3$	18 17 15 $\bar{X} = 16.67$ $\Sigma X = 50$ $\Sigma X^2 = 838$ $n = 3$	$\bar{X}_{male} = 11.89$ $\Sigma X = 107$ $n = 9$
	B_2: Female	2 6 4 $\bar{X} = 4$ $\Sigma X = 12$ $\Sigma X^2 = 56$ $n = 3$	9 10 17 $\bar{X} = 12$ $\Sigma X = 36$ $\Sigma X^2 = 470$ $n = 3$	6 8 4 $\bar{X} = 6$ $\Sigma X = 18$ $\Sigma X^2 = 116$ $n = 3$	$\bar{X}_{fem} = 7.33$ $\Sigma X = 66$ $n = 9$
		$\bar{X}_{soft} = 6$ $\Sigma X = 36$ $n = 6$	$\bar{X}_{med} = 11.5$ $\Sigma X = 69$ $n = 6$	$\bar{X}_{loud} = 11.33$ $\Sigma X = 68$ $n = 6$	$\Sigma X_{tot} = 173$ $\Sigma X^2_{tot} = 2075$ $N = 18$

Computing the ANOVA

Chapter 12 discusses a 3 × 2 design for the factors of volume of a message and participants' gender, and the dependent variable of persuasiveness. Organize the data as shown in Table A.1. The ANOVA involves five parts: computing (1) the sums and means, (2) the sums of squares, (3) the degrees of freedom, (4) the mean squares, and (5) the Fs.

Computing the Sums and Means

STEP 1: *Compute ΣX and ΣX^2 in each cell.* Note the n of the cell. For example, in the male–soft cell, $\Sigma X = 4 + 9 + 11 = 24$; $\Sigma X^2 = 4^2 + 9^2 + 11^2 = 218$; $n = 3$. Also, compute the mean in each cell (for the male–soft cell, $\bar{X} = 8$). These are the interaction means.

STEP 2: *Compute ΣX vertically in each column of the study's diagram.* Add the ΣXs from the cells in a column (e.g., for soft, $\Sigma X = 24 + 12$). Note the n in each column (here, $n = 6$) and compute the mean for each column (e.g., $\bar{X}_{soft} = 6$). These are the main effect means for factor A. (*Note:* If the ns are equal, averaging the cell means in a column also gives the overall mean of the column.)

STEP 3: *Compute ΣX horizontally in each row of the diagram.* Add the ΣXs from the cells in a row (for males, $\Sigma X = 24 + 33 + 50 = 107$). Note the n in each row (here, $n = 9$). Compute the mean for each row (e.g., $\bar{X}_{male} = 11.89$). These are the main effect means for factor B. (If the ns are equal, averaging the cell means in a row also gives the overall mean of the row.)

STEP 4: *Compute ΣX_{tot}. Add the ΣX from the levels (columns) of factor A, so $\Sigma X_{tot} = 36 + 69 + 68 = 173$. (Or, add the ΣX from the levels of factor B.)*

STEP 5: *Compute ΣX^2_{tot}. Add the ΣX^2 from all cells, so $\Sigma X^2_{tot} = 218 + 377 + 838 + 56 + 470 + 116 = 2075$. Note N (18).*

Computing the Sums of Squares

STEP 1: *Compute the total sum of squares.*

THE TOTAL SUM OF SQUARES

$$SS_{tot} = \Sigma X^2_{tot} - \left[\frac{(\Sigma X_{tot})^2}{N} \right]$$

This says to divide $(\Sigma X_{tot})^2$ by N and then subtract the answer from ΣX^2_{tot}.

From Table A.1, $\Sigma X_{tot} = 173$, $\Sigma X^2_{tot} = 2075$, and $N = 18$. Filling in the formula gives

$$SS_{tot} = 2075 - \left[\frac{(173)^2}{18} \right] = 2075 - 1662.72$$

$$= 412.28$$

Note: $(\Sigma X_{tot})^2/N$ above is also used later and is called the *correction* (here, the correction equals 1662.72).

STEP 2: *Compute the sum of squares for factor A. Always have factor A form your columns.*

THE SUM OF SQUARES BETWEEN GROUPS FOR COLUMN FACTOR A

$$SS_A = \Sigma \left[\frac{(\Sigma X \text{ in the column})^2}{n \text{ of scores in the column}} \right] - \left[\frac{(\Sigma X_{tot})^2}{N} \right]$$

This says to square the ΣX in each column of the study's diagram, divide by the n in the column, add the answers together, and subtract the correction.

From Table A.1 the column sums are 36, 69, and 68, and n was 6, so

$$SS_A = \left(\frac{36^2}{6} + \frac{69^2}{6} + \frac{68^2}{6} \right) - \left(\frac{173^2}{18} \right)$$

$$= (216 + 793.5 + 770.67) - 1662.72$$

$$SS_A = 1780.17 - 1662.72 = 117.45$$

STEP 3: *Compute the sum of squares between groups for factor B. Factor B should form the rows.*

THE SUM OF SQUARES BETWEEN GROUPS FOR ROW FACTOR B

$$SS_B = \Sigma \left[\frac{(\Sigma X \text{ in the row})^2}{n \text{ of scores in the row}} \right] - \left[\frac{(\Sigma X_{tot})^2}{N} \right]$$

This says to square the ΣX for each row of the diagram and divide by the n in the level. Then add the answers and subtract the correction.

In Table A.1, the row sums are 107 and 66, and n was 9, so

$$SS_B = \left(\frac{107^2}{9} + \frac{66^2}{9} \right) - 1662.72$$

$$= 1756.11 - 1662.72 = 93.39$$

STEP 4: *Compute the sum of squares between groups for the interaction. First, compute the overall sum of squares between groups, SS_{bn}.*

THE OVERALL SUM OF SQUARES BETWEEN GROUPS

$$SS_{bn} = \Sigma \left[\frac{(\Sigma X \text{ in the cell})^2}{n \text{ of scores in the cell}} \right] - \left[\frac{(\Sigma X_{tot})^2}{N} \right]$$

Find $(\Sigma X)^2$ for each cell and divide by the n of the cell. Then add the answers together and subtract the correction.

From Table A.1

$$SS_{bn} = \left(\frac{24^2}{3} + \frac{33^2}{3} + \frac{50^2}{3} + \frac{12^2}{3} + \right.$$

$$\left. \frac{36^2}{3} + \frac{18^2}{3} \right) - 1662.72$$

$$SS_{bn} = 1976.33 - 1662.72 = 313.61$$

To find $SS_{A \times B}$, subtract the sum of squares for both main effects (in Steps 2 and 3) from the overall SS_{bn}. Thus,

THE SUM OF SQUARES BETWEEN GROUPS FOR THE INTERACTION

$$SS_{A \times B} = SS_{bn} - SS_A - SS_B$$

In our example $SS_{bn} = 313.61$, $SS_A = 117.45$, and $SS_B = 93.39$, so

$$SS_{A \times B} = 313.61 - 117.45 - 93.39 = 102.77$$

STEP 5: *Compute the sum of squares within groups.* Subtract the overall SS_{bn} in Step 4 from the SS_{tot} in Step 1 to obtain the SS_{wn}.

THE SUM OF SQUARES WITHIN GROUPS

$$SS_{wn} = SS_{tot} - SS_{bn}$$

Above, $SS_{tot} = 412.28$ and $SS_{bn} = 313.61$, so

$$SS_{wn} = 412.28 - 313.61 = 98.67$$

Computing the Degrees of Freedom

STEP 1: *The degrees of freedom between groups for factor A is* $k_A - 1$, where k_A is the number of levels in factor A. (In our example, k_A is the three levels of volume, so $df_A = 2$.)

STEP 2: *The degrees of freedom between groups for factor B is* $k_B - 1$, where k_B is the number of levels in factor B. (In our example k_B is the two levels of gender, so $df_B = 1$.)

STEP 3: *The degrees of freedom between groups for the interaction is the* df *for factor A multiplied by the* df *for factor B.* (In our example $df_A = 2$ and $df_B = 1$, so $df_{A \times B} = 2$.)

STEP 4: *The degrees of freedom within groups equals* $N - k_{A \times B}$, where N is the total N of the study and $k_{A \times B}$ is the number of cells in the study. (In our example N is 18 and we have six cells, so $df_{wn} = 18 - 6 = 12$.)

STEP 5: *The degrees of freedom total equals* $N - 1$. Use this to check your previous calculations, because the sum of the above dfs should equal df_{tot}. (In our example $df_{tot} = 17$.)

Place each SS and df in the ANOVA summary table as shown in Table A.2. Perform the remainder of the computations using this table.

Computing the Mean Squares

STEP 1: *Compute the mean square between groups for factor A.*

THE MEAN SQUARE BETWEEN GROUPS FOR FACTOR A

$$MS_A = \frac{SS_A}{df_A}$$

From Table A.2,

$$MS_A = \frac{117.45}{2} = 58.73$$

STEP 2: *Compute the mean square between groups for factor B.*

THE MEAN SQUARE BETWEEN GROUPS FOR FACTOR B

$$MS_B = \frac{SS_B}{df_B}$$

In our example

$$MS_B = \frac{93.39}{1} = 93.39$$

STEP 3: *Compute the mean square between groups for the interaction.*

THE MEAN SQUARE BETWEEN GROUPS FOR THE INTERACTION

$$MS_{A \times B} = \frac{SS_{A \times B}}{df_{A \times B}}$$

Thus, we have

$$MS_{A \times B} = \frac{102.77}{2} = 51.39$$

Table A.2

Summary Table of Two-Way ANOVA with df and Sums of Squares

Source	Sum of Squares	df	Mean Square	F
Between				
Factor A (volume)	117.45	2	MS_A	F_A
Factor B (gender)	93.39	1	MS_B	F_B
Interaction (vol × gen)	102.77	2	$MS_{A \times B}$	$F_{A \times B}$
Within	98.67	12	MS_{wn}	
Total	412.28	17		

STEP 4: *Compute the mean square within groups.*

THE MEAN SQUARE WITHIN GROUPS

$$MS_{wn} = \frac{SS_{wn}}{df_{wn}}$$

Thus, we have

$$MS_{wn} = \frac{98.67}{12} = 8.22$$

Computing *F*

STEP 1: *Compute the F_{obt} for factor A.*

THE MAIN EFFECT OF FACTOR A

$$F_A = \frac{MS_A}{MS_{wn}}$$

In our example we have

$$F_A = \frac{58.73}{8.22} = 7.14$$

STEP 2: *Compute the F_{obt} for factor B.*

THE MAIN EFFECT OF FACTOR B

$$F_B = \frac{MS_B}{MS_{wn}}$$

Thus,

$$F_B = \frac{93.39}{8.22} = 11.36$$

STEP 3: *Compute the F_{obt} for the interaction.*

THE INTERACTION EFFECT

$$F_{A \times B} = \frac{MS_{A \times B}}{MS_{wn}}$$

Thus, we have

$$F_{A \times B} = \frac{51.39}{8.22} = 6.25$$

And now the finished summary table is in Table A.3.

Interpreting Each *F* Determine whether each F_{obt} is significant by comparing it to the appropriate F_{crit}. To find each F_{crit} in the *F*-table (Table 4 in Appendix B), use the df_{bn} and the df_{wn} used in computing the corresponding F_{obt}.

Table A.3
Completed Summary Table of Two-Way ANOVA

Source	Sum of Squares	df	Mean Square	F
Between				
Factor A (volume)	117.45	2	58.73	7.14
Factor B (gender)	93.39	1	93.39	11.36
Interaction (vol × gen)	102.77	2	51.39	6.25
Within	98.67	12	8.22	
Total	412.28	17		

1. To find F_{crit} for testing F_A, use df_A as the *df* between groups and df_{wn}. In our example $df_A = 2$ and $df_{wn} = 12$. So for $\alpha = .05$, the F_{crit} is 3.88.

2. To find F_{crit} for testing F_B, use df_B as the *df* between groups and df_{wn}. In our example $df_B = 1$ and $df_{wn} = 12$. So at $\alpha = .05$, the F_{crit} is 4.75.

3. To find F_{crit} for the interaction, use $df_{A \times B}$ as the *df* between groups and df_{wn}. In our example $df_{A \times B} = 2$ and $df_{wn} = 12$. Thus, at $\alpha = .05$, the F_{crit} is 3.88.

Interpret each F_{obt} as you have previously: If an F_{obt} is larger than its F_{crit}, the corresponding main effect or interaction effect is significant. For the example all three effects are significant: F_A (7.14) is larger than its F_{crit} (3.88), F_B (11.36) is larger than its F_{crit} (4.75), and $F_{A \times B}$ (6.25) is larger than its F_{crit} (3.88).

Performing the Tukey *HSD* Test

Perform post hoc comparisons on any significant F_{obt}. If the *n*s in all levels are equal, perform Tukey's *HSD* procedure. However, the procedure is computed differently for an interaction than for a main effect.

Performing Tukey's *HSD* Test on Main Effects Perform the *HSD* on each main effect, using the procedure described in Chapter 11 for a one-way design. Recall that the formula for the *HSD* is

$$HSD = (q_k)\left(\sqrt{\frac{MS_{wn}}{n}}\right)$$

The MS_{wn} is from the two-way ANOVA, and q_k is found in Table 5 of Appendix B for df_{wn} and *k* (where *k* is the number of levels in the factor). The *n* in the formula is the number of scores in a level. Be careful here: For each factor there may be a different value of n and k. In the example, six scores went into each mean for a level of volume (each column), but nine scores went into each mean for a level of gender (each row). The n is the n in each group that you are presently comparing! Also, because q_k depends on k, when factors have a different k, they have different values of q_k.

After computing the HSD for a factor, find the difference between each pair of its main effect means. Any difference that is larger than the HSD is significant. In the example, for volume, $n = 6$, $MS_{wn} = 8.22$, and with $\alpha = .05$, $k = 3$, and $df_{wn} = 12$, the q_k is 3.77. Thus, the HSD is 4.41. The main effect mean for soft (6) differs from the means for medium (11.5) and loud (11.33) by more than 4.41, so these are significant differences. The means for medium and loud, however, differ by less than 4.41, so they do not differ significantly. No HSD is needed for the gender factor.

Performing Tukey's *HSD* Test on Interaction Effects

The post hoc comparisons for a significant interaction involve the cell means. However, as discussed in Chapter 12, we perform only *unconfounded* comparisons, in which two cells differ along only one factor. Therefore, we find the differences only between the cell means within the same column or within the same row. Then we compare the differences to the HSD. However, when computing the HSD for an interaction, we find q_k using a slightly different procedure.

Previously, we found q_k in Table 5 using k, the number of means being compared. For an interaction first determine the *adjusted k*. This value "adjusts" for

Table A.4

Values of Adjusted *k*

Design of Study	Number of Cell Means in Study	Adjusted Value of *k*
2 × 2	4	3
2 × 3	6	5
2 × 4	8	6
3 × 3	9	7
3 × 4	12	8
4 × 4	16	10
4 × 5	20	12

Table A.5

Table of the Interaction Cells Showing the Difference between Unconfounded Means

the actual number of unconfounded comparisons you will make. Obtain the *adjusted k* from Table A.4 (or at the beginning of Table 5 of Appendix B). In the left-hand column locate the design of your study. Do not be concerned about the order of the numbers. We called our persuasiveness study a 3 × 2 design, so look at the row labeled "2 × 3." Reading across that row, confirm that the middle column contains the number of cell means in the interaction (we have 6). In the right-hand column is the *adjusted k* (for our study it is 5).

The *adjusted k* is the value of k to use to obtain q_k from Table 5. Thus, for the persuasiveness study with $\alpha = .05$, $df_{wn} = 12$, and in the column labeled $k = 5$, the q_k is 4.51. Now compute the HSD using the same formula used previously. In each cell are 3 scores, so

$$HSD = (q_k)\left(\sqrt{\frac{MS_{wn}}{n}}\right) = (4.51)\left(\sqrt{\frac{8.22}{3}}\right) = 7.47$$

The HSD for the interaction is 7.47.

The differences between our cell means are shown in Table A.5. On the line connecting any two cells is the absolute difference between their means. Any difference between two means that is larger than the HSD is a significant difference. Only three differences are significant: (1) between the mean for females at the soft volume and the mean for females at the medium volume, (2) between the mean for males at the soft volume and the mean for males at the loud volume, and (3) between the mean for males at the loud volume and the mean for females at the loud volume.

Computing η^2

In the two-way ANOVA we again compute *eta squared* (η^2) to describe effect size—the proportion of variance in dependent scores that is accounted for by a relationship. Compute a separate η^2 for each significant main and interaction effect. Recall that the formula was

$$\eta^2 = \frac{SS_{bn}}{SS_{tot}}$$

Here, we divide the SS_{tot} into the sum of squares between groups for each significant effect, either SS_A, SS_B, or $SS_{A \times B}$. For example, for our factor A (volume), SS_A was 117.45 and SS_{tot} was 412.28. Therefore, the η^2 is .28. Thus, the main effect of changing the volume of a message accounts for 28% of our differences in persuasiveness scores. For the gender factor, SS_B is 93.39, so η^2 is .23: The conditions of male or female account for an additional 23% of the variance in scores. Finally, for the interaction, $SS_{A \times B}$ is 102.77, so $\eta^2 = .25$: The particular combination of gender and volume we created in our cells accounts for an additional 25% of the differences in persuasiveness scores.

For Practice

1. A study compared the performance of males and females tested by either a male or a female experimenter. Here are the data:

		Factor A: Participants	
		Level A₁: Males	Level A₂: Females
Factor B: Experimenter	Level B₁: Male	6	8
		11	14
		9	17
		10	16
		9	19
	Level B₂: Female	8	4
		10	6
		9	5
		7	5
		10	7

a. Using $\alpha = .05$, perform an ANOVA and complete the summary table.

b. Compute the main effect means and interaction means.

c. Perform the appropriate post hoc comparisons.

d. What do you conclude about the relationships that this study demonstrates?

e. Compute the effect size where appropriate.

Answers

1. a.

Source	Sum of Squares	df	Mean Square	F
Between groups				
Factor A	7.20	1	7.20	1.19
Factor B	115.20	1	115.20	19.04
Interaction	105.80	1	105.80	17.49
Within groups	96.80	16	6.05	
Total	325.00			

For each factor, $df = 1$ and 16, so $F_{crit} = 4.49$: Factor B and the interaction are significant, $p < .05$.

b. For factor A, $\overline{X}_1 = 8.9$, $\overline{X}_2 = 10.1$; for factor B, $\overline{X}_1 = 11.9$, $\overline{X}_2 = 7.1$; for the interaction, $\overline{X}_{A_1B_2} = 9.0$, $\overline{X}_{A_1B_2} = 8.8$, $\overline{X}_{A_2B_1} = 14.8$, $\overline{X}_{A_2B_1} = 5.4$.

c. Because factor A is not significant and factor B contains only two levels, such tests are unnecessary for them. For A × B, *adjusted k* = 3, so $q_k = 3.65$, $HSD = (3.65)(\sqrt{6.05/5}) = 4.02$; the only significant differences are between males and females tested by a male, and between females tested by a male and females tested by a female.

d. Conclude that a relationship exists between gender and test scores when testing is done by a male, and that male versus female experimenters produce a relationship when testing females, $p < .05$.

e. For B, $\eta^2 = \frac{115.2}{325} = .35$; for A × B, $\eta^2 = \frac{105.8}{325} = .33$.

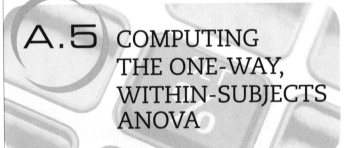

A.5 COMPUTING THE ONE-WAY, WITHIN-SUBJECTS ANOVA

This section contains formulas for the one-way, within-subjects ANOVA discussed in Chapter 11. (However, it involves the concept of an interaction described in Chapter 12.) This ANOVA is used when either the same participants are measured repeatedly or different participants are matched under all levels of one factor. (Statistical terminology still uses the old fashioned term *subjects* instead of the more modern *participants*.) The other assumptions are (1) the dependent variable is a normally distributed ratio or interval variable and (2) the population variances are homogeneous.

Logic of the One-Way, Within-Subjects ANOVA

As an example, say we're interested in whether one's form of dress influences how comfortable people feel in a social setting. On three consecutive days, we ask participants to "greet" people participating in a different experiment. On day one, participants dress casually, on another day, they dress semiformally, on another day, they dress formally. Each day participants complete a questionnaire measuring the dependent variable of their comfort level. Our data are shown in Table A.6.

As usual, we test whether the means from the levels represent different μs. Therefore, H_0: $\mu_1 = \mu_2 = \mu_3$, and H_a: Not all μs are equal.

Notice that this one-way ANOVA can be viewed as a two-way ANOVA: Factor A (the columns) is one factor, and the different participants or subjects (the rows) are a second factor, here with five levels. The interaction is between subjects and type of dress.

In Chapters 11 and 12, we compute F by dividing by the mean square within groups (MS_{wn}). This was an estimate of the variability in the population. We computed MS_{wn} using the differences between the scores in each *cell* and the mean of the cell. However, in Table A.4, each cell contains only one score. Therefore, the mean of each cell *is* the score in the cell, and the differences within a cell are always zero. So, we cannot compute MS_{wn} in the usual way.

Instead, the mean square for the interaction between factor A and subjects (abbreviated $MS_{A \times subs}$) reflects the variability of scores. It is because of the variability among people that the effect of type of dress will change as we change the "levels" of which participant we test. Therefore, $MS_{A \times subs}$ is our estimate of the variance in the scores, and it is used as the denominator of the F-ratio.

Computing the One-Way, Within-Subjects ANOVA

STEP 1: *Compute the ΣX, the $\overline{X}$, and the ΣX^2 for each level of factor A (each coumn).* Then compute ΣX_{tot} and ΣX_{tot}^2. Also, compute ΣX_{sub}, which is the ΣX for each participant's scores (each row).

Then follow these steps.

STEP 2: *Compute the total sum of squares.*

THE TOTAL SUMS OF SQUARES

$$SS_{tot} = \Sigma X_{tot}^2 - \left(\frac{(\Sigma X_{tot})^2}{N} \right)$$

From the example, we have

$$SS_{tot} = 629 - \left(\frac{(85)^2}{15} \right)$$

$$SS_{tot} = 629 - 481.67$$

$$= 147.33$$

Note that the quantity $(\Sigma X_{tot})^2/N$ is the *correction* in the following computations. (Here, the correction is 481.67.)

Table A.6
One-Way Repeated Measures Study of the Factor of Type of Dress

		Factor A: Type of Dress			
		Level A_1: Casual	Level A_2: Semiformal	Level A_3: Formal	
Subjects	1	4	9	1	$\Sigma X_{sub_1} = 14$
	2	6	12	3	$\Sigma X_{sub_2} = 21$
	3	8	4	4	$\Sigma X_{sub_3} = 16$
	4	2	8	5	$\Sigma X_{sub_4} = 15$
	5	10	7	2	$\Sigma X_{sub_5} = 19$
					Total:
		$\Sigma X = 30$	$\Sigma X = 40$	$\Sigma X = 15$	$\Sigma X_{tot} = 30 + 40 + 15 = 85$
		$\Sigma X^2 = 220$	$\Sigma X^2 = 354$	$\Sigma X^2 = 55$	$\Sigma X_{tot}^2 = 220 + 354 + 55 = 629$
		$n_1 = 5$	$n_2 = 5$	$n_3 = 5$	$N = 15$
		$\overline{X}_1 = 6$	$\overline{X}_2 = 8$	$\overline{X}_3 = 3$	$k = 3$

3: *Compute the sum of squares for the column factor, factor A.*

THE SUM OF SQUARES BETWEEN GROUPS FOR FACTOR A

$$SS_A = \Sigma\left[\frac{(\text{Sum of scores in the column})^2}{n \text{ of scores in the column}}\right] - \left[\frac{(\Sigma X_{tot})^2}{N}\right]$$

Find ΣX in each level (column) of factor A, square the sum and divide by the n of the level. After doing this for all levels, add the results together and subtract the correction.

In the example

$$SS_A = \left(\frac{30^2}{5} + \frac{40^2}{5} + \frac{15^2}{5}\right) - 481.67$$

$$SS_A = 545 - 481.67 = 63.33$$

 4: *Find the sum of squares for the row factor of subjects.*

THE SUM OF SQUARES FOR SUBJECTS

$$SS_{subs} = \frac{(\Sigma X_{sub1})^2 + (\Sigma X_{sub2})^2 + \cdots + (\Sigma X_n)^2}{k_a} - \frac{(\Sigma X_{tot})^2}{N}$$

Square the sum for each subject (ΣX_{sub}). Then add the squared sums together. Next, divide by k_a, the number of levels of factor A. Finally, subtract the correction. In the example,

$$SS_{subs} = \frac{14^2 + 21^2 + 16^2 + 15^2 + 19^2}{3} - 481.67$$

$$SS_{subs} = 493 - 481.67 = 11.33$$

 5: *Find the sum of squares for the interaction by subtracting the sums of squares for the other factors from the total.*

THE INTERACTION OF FACTOR A BY SUBJECTS

$$SS_{A \times subs} = SS_{tot} - SS_A - SS_{subs}$$

In the example

$$SS_{A \times subs} = 147.33 - 63.33 - 11.33 = 72.67$$

STEP 6: *Determine the degrees of freedom.*

THE DEGREES OF FREEDOM BETWEEN GROUPS FOR FACTOR A

$$df_A = k_A - 1$$

k_A is the number of levels of factor A. In the example, $k_A = 3$, so df_A is 2.

THE DEGREES OF FREEDOM FOR THE INTERACTION

$$df_{A \times subs} = (k_A - 1)(k_{subs} - 1)$$

k_A is the number of levels of factor A, and k_{subs} is the number of participants. In the example with three levels of factor A and five subjects, $df_{A \times subs} = (2)(4) = 8$.

STEP 7: *Find the mean squares for factor A and the interaction.*

THE MEAN SQUARE FOR FACTOR A

$$MS_A = \frac{SS_A}{df_A}$$

In our example,

$$MS_A = \frac{SS_A}{df_A} = \frac{63.33}{2} = 31.67$$

THE MEAN SQUARE FOR THE INTERACTION BETWEEN FACTOR A AND SUBJECTS

$$MS_{A \times subs} = \frac{SS_{A \times subs}}{df_{A \times subs}}$$

In our example,

$$MS_{A \times sub} = \frac{SS_{A \times subs}}{df_{A \times subs}} = \frac{72.67}{8} = 9.08$$

STEP 8: *Find F_{obt}.*

THE WITHIN-SUBJECTS F-RATIO

$$F_{obt} = \frac{MS_A}{MS_{A \times subs}}$$

In the example,

$$F_{obt} = \frac{MS_A}{MS_{A \times subs}} = \frac{31.67}{9.08} = 3.49$$

The finished summary table is

Source	Sum of Squares	df	Mean Square	F
Subjects	11.33	4		
Factor A (dress)	63.33	2	31.67	3.49
Interaction (A × subjects)	72.67	8	9.08	
Total	147.33	14		

STEP 9: *Find the critical value of F in Table 5 of Appendix B. Use df_A as the degrees of freedom between groups and $df_{A \times subs}$ as the degrees of freedom within groups. In the example for $\alpha = .05$, $df_A = 2$, and $df_{A \times subs} = 8$, the F_{crit} is 4.46.*

Interpret the above F_{obt} the same way you did in Chapter 11. Our F_{obt} is *not* larger than F_{crit}, so it is not significant. Had F_{obt} been significant, then at least two of the level means differ significantly. Then, for post hoc comparisons, graphing, eta squared, and confidence intervals, follow the procedures discussed in Chapter 11. However, in any of those formulas, in place of the term MS_{wn} use $MS_{A \times subs}$.

For Practice

1. We study the influence of practice on eye–hand coordination. We test people with no practice, 1 hour, or 2 hours of practice.
   a. What are H_0 and H_a?
   b. Complete the ANOVA summary table.
   c. With $\alpha = .05$, what do you conclude about F_{obt}?
   d. Perform the post hoc comparisons.
   e. What is the effect size in this study?

Subjects	Amount of Practice		
	None	1 Hour	2 Hours
1	4	3	6
2	3	5	5
3	1	4	3
4	3	4	6
5	1	5	6
6	2	6	7
7	2	4	5
8	1	3	8

2. You measure 21 students' degree of positive attitude toward statistics at four equally spaced intervals during the semester. The mean score for each level is: time 1, 62.50; time 2, 64.68; time 3, 69.32; and time 4, 72.00. You obtain the following sums of squares:

Source	Sum of Squares	df	Mean Square	F
Subjects	402.79			
Factor A	189.30			
A × subjects	688.32			
Total	1280.41			

a. What are H_0 and H_a?
b. Complete the ANOVA summary table
c. With $\alpha = .05$, what do you conclude about F_{obt}?
d. Perform the appropriate post hoc comparisons.
e. What is the effect size in this study?
f. What should you conclude about this relationship?

Answers

1. a. $H_0: \mu_1 = \mu_2 = \mu_3$; H_a: Not all μs are equal.

   b. $SS_{tot} = 477 - 392.04$; $SS_A = 445.125 - 392.04$; and
   $$SS_{subs} = \left(\frac{1205}{3}\right) - 392.04$$

Source	Sum of Squares	df	Mean Square	F
Subjects	9.63	7		
Factor A	53.08	2	26.54	16.69
A × subjects	22.25	14	1.59	
Total	84.96	23		

   c. With $df_A = 2$ and $df_{A \times subs} = 14$, the F_{crit} is 3.74. The F_{obt} is significant.

   d. The $q_k = 3.70$ and $HSD = 1.65$. The means for 0, 1, and 2 hours are 2.13, 4.25, and 5.75, respectively. Significant differences occurred between 0 and 1 hour and between 0 and 2 hours, but not between 1 and 2 hours.

   e. Eta squared $(\eta^2) = \frac{53.08}{84.96} = .62$.

2. a. $H_0: \mu_1 = \mu_2 = \mu_3 = \mu_4$; H_a: not all μs are equal.
   b.

Source	Sum of Squares	df	Mean Square	F
Subjects	402.79	20		
Factor A	189.30	3	63.10	5.50
A × subjects	688.32	60	11.47	
Total	1280.41	83		

   c. With $df_A = 3$ and $df_{A \times subs} = 60$, the F_{crit} is 2.76. The F_{obt} is significant.

   d. The $q_k = 3.74$ and $HSD = 2.76$. The means at time 1 and time 2 differ from those times 3 and 4, but time 1 and time 2 do not differ significantly, and neither do times 3 and 4.

   e. Eta squared $(\eta^2) = 189.30/1280.41 = .15$.

Statistical Tables

Table 1

Proportions of Area under the Standard Normal Curve: The z-Table

Column (A) lists z-score values. Column (B) lists the proportion of the area between the mean and the z-score value. Column (C) lists the proportion of the area beyond the z-score in the tail of the distribution. (Note: Because the normal distribution is symmetrical, areas for negative z-scores are the same as those for positive z-scores.)

A z	B Area between Mean and z	C Area beyond z in Tail	A z	B Area between Mean and z	C Area beyond z in Tail	A z	B Area between Mean and z	C Area beyond z in Tail
0.00	.0000	.5000	0.30	.1179	.3821	0.60	.2257	.2743
0.01	.0040	.4960	0.31	.1217	.3783	0.61	.2291	.2709
0.02	.0080	.4920	0.32	.1255	.3745	0.62	.2324	.2676
0.03	.0120	.4880	0.33	.1293	.3707	0.63	.2357	.2643
0.04	.0160	.4840	0.34	.1331	.3669	0.64	.2389	.2611
0.05	.0199	.4801	0.35	.1368	.3632	0.65	.2422	.2578
0.06	.0239	.4761	0.36	.1406	.3594	0.66	.2454	.2546
0.07	.0279	.4721	0.37	.1443	.3557	0.67	.2486	.2514
0.08	.0319	.4681	0.38	.1480	.3520	0.68	.2517	.2483
0.09	.0359	.4641	0.39	.1517	.3483	0.69	.2549	.2451
0.10	.0398	.4602	0.40	.1554	.3446	0.70	.2580	.2420
0.11	.0438	.4562	0.41	.1591	.3409	0.71	.2611	.2389
0.12	.0478	.4522	0.42	.1628	.3372	0.72	.2642	.2358
0.13	.0517	.4483	0.43	.1664	.3336	0.73	.2673	.2327
0.14	.0557	.4443	0.44	.1700	.3300	0.74	.2704	.2296
0.15	.0596	.4404	0.45	.1736	.3264	0.75	.2734	.2266
0.16	.0636	.4364	0.46	.1772	.3228	0.76	.2764	.2236
0.17	.0675	.4325	0.47	.1808	.3192	0.77	.2794	.2206
0.18	.0714	.4286	0.48	.1844	.3156	0.78	.2823	.2177
0.19	.0753	.4247	0.49	.1879	.3121	0.79	.2852	.2148
0.20	.0793	.4207	0.50	.1915	.3085	0.80	.2881	.2119
0.21	.0832	.4168	0.51	.1950	.3050	0.81	.2910	.2090
0.22	.0871	.4129	0.52	.1985	.3015	0.82	.2939	.2061
0.23	.0910	.4090	0.53	.2019	.2981	0.83	.2967	.2033
0.24	.0948	.4052	0.54	.2054	.2946	0.84	.2995	.2005
0.25	.0987	.4013	0.55	.2088	.2912	0.85	.3023	.1977
0.26	.1026	.3974	0.56	.2123	.2877	0.86	.3051	.1949
0.27	.1064	.3936	0.57	.2157	.2843	0.87	.3078	.1922
0.28	.1103	.3897	0.58	.2190	.2810	0.88	.3106	.1894
0.29	.1141	.3859	0.59	.2224	.2776	0.89	.3133	.1867

(continued)

Table 1 (cont.)

Proportions of Area under the Standard Normal Curve: The z-Table

A	B	C	A	B	C	A	B	C
	Area	Area		Area	Area		Area	Area
	between	beyond z		between	beyond z		between	beyond z
z	Mean and z	in Tail	z	Mean and z	in Tail	z	Mean and z	in Tail
0.90	.3159	.1841	1.25	.3944	.1056	1.60	.4452	.0548
0.91	.3186	.1814	1.26	.3962	.1038	1.61	.4463	.0537
0.92	.3212	.1788	1.27	.3980	.1020	1.62	.4474	.0526
0.93	.3238	.1762	1.28	.3997	.1003	1.63	.4484	.0516
0.94	.3264	.1736	1.29	.4015	.0985	1.64	.4495	.0505
0.95	.3289	.1711	1.30	.4032	.0968	1.65	.4505	.0495
0.96	.3315	.1685	1.31	.4049	.0951	1.66	.4515	.0485
0.97	.3340	.1660	1.32	.4066	.0934	1.67	.4525	.0475
0.98	.3365	.1635	1.33	.4082	.0918	1.68	.4535	.0465
0.99	.3389	.1611	1.34	.4099	.0901	1.69	.4545	.0455
1.00	.3413	.1587	1.35	.4115	.0885	1.70	.4554	.0446
1.01	.3438	.1562	1.36	.4131	.0869	1.71	.4564	.0436
1.02	.3461	.1539	1.37	.4147	.0853	1.72	.4573	.0427
1.03	.3485	.1515	1.38	.4162	.0838	1.73	.4582	.0418
1.04	.3508	.1492	1.39	.4177	.0823	1.74	.4591	.0409
1.05	.3531	.1469	1.40	.4192	.0808	1.75	.4599	.0401
1.06	.3554	.1446	1.41	.4207	.0793	1.76	.4608	.0392
1.07	.3577	.1423	1.42	.4222	.0778	1.77	.4616	.0384
1.08	.3599	.1401	1.43	.4236	.0764	1.78	.4625	.0375
1.09	.3621	.1379	1.44	.4251	.0749	1.79	.4633	.0367
1.10	.3643	.1357	1.45	.4265	.0735	1.80	.4641	.0359
1.11	.3665	.1335	1.46	.4279	.0721	1.81	.4649	.0351
1.12	.3686	.1314	1.47	.4292	.0708	1.82	.4656	.0344
1.13	.3708	.1292	1.48	.4306	.0694	1.83	.4664	.0336
1.14	.3729	.1271	1.49	.4319	.0681	1.84	.4671	.0329
1.15	.3749	.1251	1.50	.4332	.0668	1.85	.4678	.0322
1.16	.3770	.1230	1.51	.4345	.0655	1.86	.4686	.0314
1.17	.3790	.1210	1.52	.4357	.0643	1.87	.4693	.0307
1.18	.3810	.1190	1.53	.4370	.0630	1.88	.4699	.0301
1.19	.3830	.1170	1.54	.4382	.0618	1.89	.4706	.0294
1.20	.3849	.1151	1.55	.4394	.0606	1.90	.4713	.0287
1.21	.3869	.1131	1.56	.4406	.0594	1.91	.4719	.0281
1.22	.3888	.1112	1.57	.4418	.0582	1.92	.4726	.0274
1.23	.3907	.1093	1.58	.4429	.0571	1.93	.4732	.0268
1.24	.3925	.1075	1.59	.4441	.0559	1.94	.4738	.0262

Table 1 (cont.)

Proportions of Area under the Standard Normal Curve: The z-Table

A	B	C	A	B	C	A	B	C
	Area	Area		Area	Area		Area	Area
	between	beyond z		between	beyond z		between	beyond z
z	Mean and z	in Tail	z	Mean and z	in Tail	z	Mean and z	in Tail
1.95	.4744	.0256	2.30	.4893	.0107	2.65	.4960	.0040
1.96	.4750	.0250	2.31	.4896	.0104	2.66	.4961	.0039
1.97	.4756	.0244	2.32	.4898	.0102	2.67	.4962	.0038
1.98	.4761	.0239	2.33	.4901	.0099	2.68	.4963	.0037
1.99	.4767	.0233	2.34	.4904	.0096	2.69	.4964	.0036
2.00	.4772	.0228	2.35	.4906	.0094	2.70	.4965	.0035
2.01	.4778	.0222	2.36	.4909	.0091	2.71	.4966	.0034
2.02	.4783	.0217	2.37	.4911	.0089	2.72	.4967	.0033
2.03	.4788	.0212	2.38	.4913	.0087	2.73	.4968	.0032
2.04	.4793	.0207	2.39	.4916	.0084	2.74	.4969	.0031
2.05	.4798	.0202	2.40	.4918	.0082	2.75	.4970	.0030
2.06	.4803	.0197	2.41	.4920	.0080	2.76	.4971	.0029
2.07	.4808	.0192	2.42	.4922	.0078	2.77	.4972	.0028
2.08	.4812	.0188	2.43	.4925	.0075	2.78	.4973	.0027
2.09	.4817	.0183	2.44	.4927	.0073	2.79	.4974	.0026
2.10	.4821	.0179	2.45	.4929	.0071	2.80	.4974	.0026
2.11	.4826	.0174	2.46	.4931	.0069	2.81	.4975	.0025
2.12	.4830	.0170	2.47	.4932	.0068	2.82	.4976	.0024
2.13	.4834	.0166	2.48	.4934	.0066	2.83	.4977	.0023
2.14	.4838	.0162	2.49	.4936	.0064	2.84	.4977	.0023
2.15	.4842	.0158	2.50	.4938	.0062	2.85	.4978	.0022
2.16	.4846	.0154	2.51	.4940	.0060	2.86	.4979	.0021
2.17	.4850	.0150	2.52	.4941	.0059	2.87	.4979	.0021
2.18	.4854	.0146	2.53	.4943	.0057	2.88	.4980	.0020
2.19	.4857	.0143	2.54	.4945	.0055	2.89	.4981	.0019
2.20	.4861	.0139	2.55	.4946	.0054	2.90	.4981	.0019
2.21	.4864	.0136	2.56	.4948	.0052	2.91	.4982	.0018
2.22	.4868	.0132	2.57	.4949	.0051	2.92	.4982	.0018
2.23	.4871	.0129	2.58	.4951	.0049	2.93	.4983	.0017
2.24	.4875	.0125	2.59	.4952	.0048	2.94	.4984	.0016
2.25	.4878	.0122	2.60	.4953	.0047	2.95	.4984	.0016
2.26	.4881	.0119	2.61	.4955	.0045	2.96	.4985	.0015
2.27	.4884	.0116	2.62	.4956	.0044	2.97	.4985	.0015
2.28	.4887	.0113	2.63	.4957	.0043	2.98	.4986	.0014
2.29	.4890	.0110	2.64	.4959	.0041	2.99	.4986	.0014

(continued)

Table 1 (cont.)
Proportions of Area under the Standard Normal Curve: The z-Table

A	B	C	A	B	C	A	B	C
	Area	Area		Area	Area		Area	Area
	between	beyond z		between	beyond z		between	beyond z
z	Mean and z	in Tail	z	Mean and z	in Tail	z	Mean and z	in Tail
3.00	.4987	.0013	3.12	.4991	.0009	3.24	.4994	.0006
3.01	.4987	.0013	3.13	.4991	.0009	3.25	.4994	.0006
3.02	.4987	.0013	3.14	.4992	.0008	3.30	.4995	.0005
3.03	.4988	.0012	3.15	.4992	.0008	3.35	.4996	.0004
3.04	.4988	.0012	3.16	.4992	.0008	3.40	.4997	.0003
3.05	.4989	.0011	3.17	.4992	.0008	3.45	.4997	.0003
3.06	.4989	.0011	3.18	.4993	.0007	3.50	.4998	.0002
3.07	.4989	.0011	3.19	.4993	.0007	3.60	.4998	.0002
3.08	.4990	.0010	3.20	.4993	.0007	3.70	.4999	.0001
3.09	.4990	.0010	3.21	.4993	.0007	3.80	.4999	.0001
3.10	.4990	.0010	3.22	.4994	.0006	3.90	.49995	.00005
3.11	.4991	.0009	3.23	.4994	.0006	4.00	.49997	.00003

Table 2
Critical Values of *t*: The *t*-Table
(Note: Values of $-t_{crit}$ = values of $+t_{crit}$.)

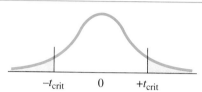

Two-Tailed Test

$-t_{crit}$ 0 $+t_{crit}$

Alpha Level

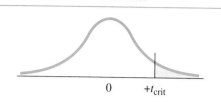

One-Tailed Test

0 $+t_{crit}$

Alpha Level

df	$\alpha = .05$	$\alpha = .01$	df	$\alpha = .05$	$\alpha = .01$
1	12.706	63.657	1	6.314	31.821
2	4.303	9.925	2	2.920	6.965
3	3.182	5.841	3	2.353	4.541
4	2.776	4.604	4	2.132	3.747
5	2.571	4.032	5	2.015	3.365
6	2.447	3.707	6	1.943	3.143
7	2.365	3.499	7	1.895	2.998
8	2.306	3.355	8	1.860	2.896
9	2.262	3.250	9	1.833	2.821
10	2.228	3.169	10	1.812	2.764
11	2.201	3.106	11	1.796	2.718
12	2.179	3.055	12	1.782	2.681
13	2.160	3.012	13	1.771	2.650
14	2.145	2.977	14	1.761	2.624
15	2.131	2.947	15	1.753	2.602
16	2.120	2.921	16	1.746	2.583
17	2.110	2.898	17	1.740	2.567
18	2.101	2.878	18	1.734	2.552
19	2.093	2.861	19	1.729	2.539
20	2.086	2.845	20	1.725	2.528
21	2.080	2.831	21	1.721	2.518
22	2.074	2.819	22	1.717	2.508
23	2.069	2.807	23	1.714	2.500
24	2.064	2.797	24	1.711	2.492
25	2.060	2.787	25	1.708	2.485
26	2.056	2.779	26	1.706	2.479
27	2.052	2.771	27	1.703	2.473
28	2.048	2.763	28	1.701	2.467
29	2.045	2.756	29	1.699	2.462
30	2.042	2.750	30	1.697	2.457
40	2.021	2.704	40	1.684	2.423
60	2.000	2.660	60	1.671	2.390
120	1.980	2.617	120	1.658	2.358
∞	1.960	2.576	∞	1.645	2.326

From Table 12 of E. Pearson and H. Hartley, *Biometrika Tables for Statisticians*, Vol. 1, 3rd ed. (Cambridge: Cambridge University Press, 1966.) Reprinted with the permission of the Biometrika Trustees.

Table 3

Critical Values of the Pearson Correlation Coefficient: The *r*-Table

Two-Tailed Test

−r_{crit} 0   +r_{crit}

Alpha Level

df (no. of pairs − 2)	α = .05	α = .01
1	.997	.9999
2	.950	.990
3	.878	.959
4	.811	.917
5	.754	.874
6	.707	.834
7	.666	.798
8	.632	.765
9	.602	.735
10	.576	.708
11	.553	.684
12	.532	.661
13	.514	.641
14	.497	.623
15	.482	.606
16	.468	.590
17	.456	.575
18	.444	.561
19	.433	.549
20	.423	.537
21	.413	.526
22	.404	.515
23	.396	.505
24	.388	.496
25	.381	.487
26	.374	.479
27	.367	.471
28	.361	.463
29	.355	.456
30	.349	.449
35	.325	.418
40	.304	.393
45	.288	.372
50	.273	.354
60	.250	.325
70	.232	.302
80	.217	.283
90	.205	.267
100	.195	.254

One-Tailed Test

0   +r_{crit}

Alpha Level

df (no. of pairs − 2)	α = .05	α = .01
1	.988	.9995
2	.900	.980
3	.805	.934
4	.729	.882
5	.669	.833
6	.622	.789
7	.582	.750
8	.549	.716
9	.521	.685
10	.497	.658
11	.476	.634
12	.458	.612
13	.441	.592
14	.426	.574
15	.412	.558
16	.400	.542
17	.389	.528
18	.378	.516
19	.369	.503
20	.360	.492
21	.352	.482
22	.344	.472
23	.337	.462
24	.330	.453
25	.323	.445
26	.317	.437
27	.311	.430
28	.306	.423
29	.301	.416
30	.296	.409
35	.275	.381
40	.257	.358
45	.243	.338
50	.231	.322
60	.211	.295
70	.195	.274
80	.183	.256
90	.173	.242
100	.164	.230

From R. A. Fisher and F. Yates, *Statistical Tables for Biological, Agricultural and Medical Research*, 6th ed. Copyright © 1963, R. A. Fisher and F. Yates. Reprinted by permission of Pearson Education Limited.

Table 4
Critical Values of *F*: The *F*-Table

Critical values for α = .05 are in **dark numbers.**
Critical values for α = .01 are in light numbers.

0   F_{crit}

Degrees of Freedom within Groups (degrees of freedom in denominator of *F*-ratio)	α	Degrees of Freedom between Groups (degrees of freedom in numerator of *F*-ratio)														
		1	2	3	4	5	6	7	8	9	10	11	12	14	16	20
1	.05	161	200	216	225	230	234	237	239	241	242	243	244	245	246	248
	.01	4,052	4,999	5,403	5,625	5,764	5,859	5,928	5,981	6,022	6,056	6,082	6,106	6,142	6,169	6,208
2	.05	18.51	19.00	19.16	19.25	19.30	19.33	19.36	19.37	19.38	19.39	19.40	19.41	19.42	19.43	19.44
	.01	98.49	99.00	99.17	99.25	99.30	99.33	99.34	99.36	99.38	99.40	99.41	99.42	99.43	99.44	99.45
3	.05	10.13	9.55	9.28	9.12	9.01	8.94	8.88	8.84	8.81	8.78	8.76	8.74	8.71	8.69	8.66
	.01	34.12	30.82	29.46	28.71	28.24	27.91	27.67	27.49	27.34	27.23	27.13	27.05	26.92	26.83	26.69
4	.05	7.71	6.94	6.59	6.39	6.26	6.16	6.09	6.04	6.00	5.96	5.93	5.91	5.87	5.84	5.80
	.01	21.20	18.00	16.69	15.98	15.52	15.21	14.98	14.80	14.66	14.54	14.45	14.37	14.24	14.15	14.02
5	.05	6.61	5.79	5.41	5.19	5.05	4.95	4.88	4.82	4.78	4.74	4.70	4.68	4.64	4.60	4.56
	.01	16.26	13.27	12.06	11.39	10.97	10.67	10.45	10.27	10.15	10.05	9.96	9.89	9.77	9.68	9.55
6	.05	5.99	5.14	4.76	4.53	4.39	4.28	4.21	4.15	4.10	4.06	4.03	4.00	3.96	3.92	3.87
	.01	13.74	10.92	9.78	9.15	8.75	8.47	8.26	8.10	7.98	7.87	7.79	7.72	7.60	7.52	7.39
7	.05	5.59	4.47	4.35	4.12	3.97	3.87	3.79	3.73	3.68	3.63	3.60	3.57	3.52	3.49	3.44
	.01	12.25	9.55	8.45	7.85	7.46	7.19	7.00	6.84	6.71	6.62	6.54	6.47	6.35	6.27	6.15
8	.05	5.32	4.46	4.07	3.84	3.69	3.58	3.50	3.44	3.39	3.34	3.31	3.28	3.23	3.20	3.15
	.01	11.26	8.65	7.59	7.01	6.63	6.37	6.19	6.03	5.91	5.82	5.74	5.67	5.56	5.48	5.36
9	.05	5.12	4.26	3.86	3.63	3.48	3.37	3.29	3.23	3.18	3.13	3.10	3.07	3.02	2.98	2.93
	.01	10.56	8.02	6.99	6.42	6.06	5.80	5.62	5.47	5.35	5.26	5.18	5.11	5.00	4.92	4.80
10	.05	4.96	4.10	3.71	3.48	3.33	3.22	3.14	3.07	3.02	2.97	2.94	2.91	2.86	2.82	2.77
	.01	10.04	7.56	6.55	5.99	5.64	5.39	5.21	5.06	4.95	4.85	4.78	4.71	4.60	4.52	4.41
11	.05	4.84	3.98	3.59	3.36	3.20	3.09	3.01	2.95	2.90	2.86	2.82	2.79	2.74	2.70	2.65
	.01	9.65	7.20	6.22	5.67	5.32	5.07	4.88	4.74	4.63	4.54	4.46	4.40	4.29	4.21	4.10
12	.05	4.75	3.88	3.49	3.26	3.11	3.00	2.92	2.85	2.80	2.76	2.72	2.69	2.64	2.60	2.54
	.01	9.33	6.93	5.95	5.41	5.06	4.82	4.65	4.50	4.39	4.30	4.22	4.16	4.05	3.98	3.86
13	.05	4.67	3.80	3.41	3.18	3.02	2.92	2.84	2.77	2.72	2.67	2.63	2.60	2.55	2.51	2.46
	.01	9.07	6.70	5.74	5.20	4.86	4.62	4.44	4.30	4.19	4.10	4.02	3.96	3.85	3.78	3.67
14	.05	4.60	3.74	3.34	3.11	2.96	2.85	2.77	2.70	2.65	2.60	2.56	2.53	2.48	2.44	2.39
	.01	8.86	6.51	5.56	5.03	4.69	4.46	4.28	4.14	4.03	3.94	3.86	3.80	3.70	3.62	3.51
15	.05	4.54	3.68	3.29	3.06	2.90	2.79	2.70	2.64	2.59	2.55	2.51	2.48	2.43	2.39	2.33
	.01	8.68	6.36	5.42	4.89	4.56	4.32	4.14	4.00	3.89	3.80	3.73	3.67	3.56	3.48	3.36
16	.05	4.49	3.63	3.24	3.01	2.85	2.74	2.66	2.59	2.54	2.49	2.45	2.42	2.37	2.33	2.28
	.01	8.53	6.23	5.29	4.77	4.44	4.20	4.03	3.89	3.78	3.69	3.61	3.55	3.45	3.37	3.25

(continued)

Table 4 (cont.)
Critical Values of *F:* The *F*-Table

Degrees of Freedom within Groups (degrees of freedom in denominator of *F*-ratio)	α	Degrees of Freedom between Groups (degrees of freedom in numerator of *F*-ratio)														
		1	2	3	4	5	6	7	8	9	10	11	12	14	16	20
17	.05	4.45	3.59	3.20	2.96	2.81	2.70	2.62	2.55	2.50	2.45	2.41	2.38	2.33	2.29	2.23
	.01	8.40	6.11	5.18	4.67	4.34	4.10	3.93	3.79	3.68	3.59	3.52	3.45	3.35	3.27	3.16
18	.05	4.41	3.55	3.16	2.93	2.77	2.66	2.58	2.51	2.46	2.41	2.37	2.34	2.29	2.25	2.19
	.01	8.28	6.01	5.09	4.58	4.25	4.01	3.85	3.71	3.60	3.51	3.44	3.37	3.27	3.19	3.07
19	.05	4.38	3.52	3.13	2.90	2.74	2.63	2.55	2.48	2.43	2.38	2.34	2.31	2.26	2.21	2.15
	.01	8.18	5.93	5.01	4.50	4.17	3.94	3.77	3.63	3.52	3.43	3.36	3.30	3.19	3.12	3.00
20	.05	4.35	3.49	3.10	2.87	2.71	2.60	2.52	2.45	2.40	2.35	2.31	2.28	2.23	2.18	2.12
	.01	8.10	5.85	4.94	4.43	4.10	3.87	3.71	3.56	3.45	3.37	3.30	3.23	3.13	3.05	2.94
21	.05	4.32	3.47	3.07	2.84	2.68	2.57	2.49	2.42	2.37	2.32	2.28	2.25	2.20	2.15	2.09
	.01	8.02	5.78	4.87	4.37	4.04	3.81	3.65	3.51	3.40	3.31	3.24	3.17	3.07	2.99	2.88
22	.05	4.30	3.44	3.05	2.82	2.66	2.55	2.47	2.40	2.35	2.30	2.26	2.23	2.18	2.13	2.07
	.01	7.94	5.72	4.82	4.31	3.99	3.76	3.59	3.45	3.35	3.26	3.18	3.12	3.02	2.94	2.83
23	.05	4.28	3.42	3.03	2.80	2.64	2.53	2.45	2.38	2.32	2.28	2.24	2.20	2.14	2.10	2.04
	.01	7.88	5.66	4.76	4.26	3.94	3.71	3.54	3.41	3.30	3.21	3.14	3.07	2.97	2.89	2.78
24	.05	4.26	3.40	3.01	2.78	2.62	2.51	2.43	2.36	2.30	2.26	2.22	2.18	2.13	2.09	2.02
	.01	7.82	5.61	4.72	4.22	3.90	3.67	3.50	3.36	3.25	3.17	3.09	3.03	2.93	2.85	2.74
25	.05	4.24	3.38	2.99	2.76	2.60	2.49	2.41	2.34	2.28	2.24	2.20	2.16	2.11	2.06	2.00
	.01	7.77	5.57	4.68	4.18	3.86	3.63	3.46	3.32	3.21	3.13	3.05	2.99	2.89	2.81	2.70
26	.05	4.22	3.37	2.98	2.74	2.59	2.47	2.39	2.32	2.27	2.22	2.18	2.15	2.10	2.05	1.99
	.01	7.72	5.53	4.64	4.14	3.82	3.59	3.42	3.29	3.17	3.09	3.02	2.96	2.86	2.77	2.66
27	.05	4.21	3.35	2.96	2.73	2.57	2.46	2.37	2.30	2.25	2.20	2.16	2.13	2.08	2.03	1.97
	.01	7.68	5.49	4.60	4.11	3.79	3.56	3.39	3.26	3.14	3.06	2.98	2.93	2.83	2.74	2.63
28	.05	4.20	3.34	2.95	2.71	2.56	2.44	2.36	2.29	2.24	2.19	2.15	2.12	2.06	2.02	1.96
	.01	7.64	5.45	4.57	4.07	3.76	3.53	3.36	3.23	3.11	3.03	2.95	2.90	2.80	2.71	2.60
29	.05	4.18	3.33	2.93	2.70	2.54	2.43	2.35	2.28	2.22	2.18	2.14	2.10	2.05	2.00	1.94
	.01	7.60	5.42	4.54	4.04	3.73	3.50	3.33	3.20	3.08	3.00	2.92	2.87	2.77	2.68	2.57
30	.05	4.17	3.32	2.92	2.69	2.53	2.42	2.34	2.27	2.21	2.16	2.12	2.09	2.04	1.99	1.93
	.01	7.56	5.39	4.51	4.02	3.70	3.47	3.30	3.17	3.06	2.98	2.90	2.84	2.74	2.66	2.55
32	.05	4.15	3.30	2.90	2.67	2.51	2.40	2.32	2.25	2.19	2.14	2.10	2.07	2.02	1.97	1.91
	.01	7.50	5.34	4.46	3.97	3.66	3.42	3.25	3.12	3.01	2.94	2.86	2.80	2.70	2.62	2.51
34	.05	4.13	3.28	2.88	2.65	2.49	2.38	2.30	2.23	2.17	2.12	2.08	2.05	2.00	1.95	1.89
	.01	7.44	5.29	4.42	3.93	3.61	3.38	3.21	3.08	2.97	2.89	2.82	2.76	2.66	2.58	2.47
36	.05	4.11	3.26	2.86	2.63	2.48	2.36	2.28	2.21	2.15	2.10	2.06	2.03	1.98	1.93	1.87
	.01	7.39	5.25	4.38	3.89	3.58	3.35	3.18	3.04	2.94	2.86	2.78	2.72	2.62	2.54	2.43
38	.05	4.10	3.25	2.85	2.62	2.46	2.35	2.26	2.19	2.14	2.09	2.05	2.02	1.96	1.92	1.85
	.01	7.35	5.21	4.34	3.86	3.54	3.32	3.15	3.02	2.91	2.82	2.75	2.69	2.59	2.51	2.40
40	.05	4.08	3.23	2.84	2.61	2.45	2.34	2.25	2.18	2.12	2.07	2.04	2.00	1.95	1.90	1.84
	.01	7.31	5.18	4.31	3.83	3.51	3.29	3.12	2.99	2.88	2.80	2.73	2.66	2.56	2.49	2.37
42	.05	4.07	3.22	2.83	2.59	2.44	2.32	2.24	2.17	2.11	2.06	2.02	1.99	1.94	1.89	1.82
	.01	7.27	5.15	4.29	3.80	3.49	3.26	3.10	2.96	2.86	2.77	2.70	2.64	2.54	2.46	2.35

Table 4 (cont.)
Critical Values of F: The F-Table

Degrees of Freedom within Groups (degrees of freedom in denominator of F-ratio)	α	\multicolumn{17}{c}{Degrees of Freedom between Groups (degrees of freedom in numerator of F-ratio)}															
		1	2	3	4	5	6	7	8	9	10	11	12	14	16	20
44	.05	4.06	3.21	2.82	2.58	2.43	2.31	2.23	2.16	2.10	2.05	2.01	1.98	1.92	1.88	1.81
	.01	7.24	5.12	4.26	3.78	3.46	3.24	3.07	2.94	2.84	2.75	2.68	2.62	2.52	2.44	2.32
46	.05	4.05	3.20	2.81	2.57	2.42	2.30	2.22	2.14	2.09	2.04	2.00	1.97	1.91	1.87	1.80
	.01	7.21	5.10	4.24	3.76	3.44	3.22	3.05	2.92	2.82	2.73	2.66	2.60	2.50	2.42	2.30
48	.05	4.04	3.19	2.80	2.56	2.41	2.30	2.21	2.14	2.08	2.03	1.99	1.96	1.90	1.86	1.79
	.01	7.19	5.08	4.22	3.74	3.42	3.20	3.04	2.90	2.80	2.71	2.64	2.58	2.48	2.40	2.28
50	.05	4.03	3.18	2.79	2.56	2.40	2.29	2.20	2.13	2.07	2.02	1.98	1.95	1.90	1.85	1.78
	.01	7.17	5.06	4.20	3.72	3.41	3.18	3.02	2.88	2.78	2.70	2.62	2.56	2.46	2.39	2.26
55	.05	4.02	3.17	2.78	2.54	2.38	2.27	2.18	2.11	2.05	2.00	1.97	1.93	1.88	1.83	1.76
	.01	7.12	5.01	4.16	3.68	3.37	3.15	2.98	2.85	2.75	2.66	2.59	2.53	2.43	2.35	2.23
60	.05	4.00	3.15	2.76	2.52	2.37	2.25	2.17	2.10	2.04	1.99	1.95	1.92	1.86	1.81	1.75
	.01	7.08	4.98	4.13	3.65	3.34	3.12	2.95	2.82	2.72	2.63	2.56	2.50	2.40	2.32	2.20
65	.05	3.99	3.14	2.75	2.51	2.36	2.24	2.15	2.08	2.02	1.98	1.94	1.90	1.85	1.80	1.73
	.01	7.04	4.95	4.10	3.62	3.31	3.09	2.93	2.79	2.70	2.61	2.54	2.47	2.37	2.30	2.18
70	.05	3.98	3.13	2.74	2.50	2.35	2.23	2.14	2.07	2.01	1.97	1.93	1.89	1.84	1.79	1.72
	.01	7.01	4.92	4.08	3.60	3.29	3.07	2.91	2.77	2.67	2.59	2.51	2.45	2.35	2.28	2.15
80	.05	3.96	3.11	2.72	2.48	2.33	2.21	2.12	2.05	1.99	1.95	1.91	1.88	1.82	1.77	1.70
	.01	6.96	4.88	4.04	3.56	3.25	3.04	2.87	2.74	2.64	2.55	2.48	2.41	2.32	2.24	2.11
100	.05	3.94	3.09	2.70	2.46	2.30	2.19	2.10	2.03	1.97	1.92	1.88	1.85	1.79	1.75	1.68
	.01	6.90	4.82	3.98	3.51	3.20	2.99	2.82	2.69	2.59	2.51	2.43	2.36	2.26	2.19	2.06
125	.05	3.92	3.07	2.68	2.44	2.29	2.17	2.08	2.01	1.95	1.90	1.86	1.83	1.77	1.72	1.65
	.01	6.84	4.78	3.94	3.47	3.17	2.95	2.79	2.65	2.56	2.47	2.40	2.33	2.23	2.15	2.03
150	.05	3.91	3.06	2.67	2.43	2.27	2.16	2.07	2.00	1.94	1.89	1.85	1.82	1.76	1.71	1.64
	.01	6.81	4.75	3.91	3.44	3.14	2.92	2.76	2.62	2.53	2.44	2.37	2.30	2.20	2.12	2.00
200	.05	3.89	3.04	2.65	2.41	2.26	2.14	2.05	1.98	1.92	1.87	1.83	1.80	1.74	1.69	1.62
	.01	6.76	4.71	3.88	3.41	3.11	2.90	2.73	2.60	2.50	2.41	2.34	2.28	2.17	2.09	1.97
400	.05	3.86	3.02	2.62	2.39	2.23	2.12	2.03	1.96	1.90	1.85	1.81	1.78	1.72	1.67	1.60
	.01	6.70	4.66	3.83	3.36	3.06	2.85	2.69	2.55	2.46	2.37	2.29	2.23	2.12	2.04	1.92
1000	.05	3.85	3.00	2.61	2.38	2.22	2.10	2.02	1.95	1.89	1.84	1.80	1.76	1.70	1.65	1.58
	.01	6.66	4.62	3.80	3.34	3.04	2.82	2.66	2.53	2.43	2.34	2.26	2.20	2.09	2.01	1.89
∞	.05	3.84	2.99	2.60	2.37	2.21	2.09	2.01	1.94	1.88	1.83	1.79	1.75	1.69	1.64	1.57
	.01	6.64	4.60	3.78	3.32	3.02	2.80	2.64	2.51	2.41	2.32	2.24	2.18	2.07	1.99	1.87

Table 5
Values of Studentized Range Statistic, q_k

For a one-way ANOVA, or a comparison of the means from a main effect, the value of k is the number of means in the factor.
To compare the means from an interaction, find the appropriate design (or number of cell means) in the table below and obtain the adjusted value of k. Then use adjusted k as k to find the value of q_k.

Values of Adjusted k

Design of Study	Number of Cell Means in Study	Adjusted Value of k
2×2	4	3
2×3	6	5
2×4	8	6
3×3	9	7
3×4	12	8
4×4	16	10
4×5	20	12

Values of q_k for $\alpha = .05$ are **dark numbers** and for $\alpha = .01$ are light numbers.

Degrees of Freedom within Groups (degrees of freedom in denominator of F-ratio)	α	\multicolumn{11}{c}{k = Number of Means Being Compared}

Degrees of Freedom within Groups	α	2	3	4	5	6	7	8	9	10	11	12
1	.05	**18.00**	**27.00**	**32.80**	**37.10**	**40.40**	**43.10**	**45.40**	**47.40**	**49.10**	**50.60**	**52.00**
	.01	90.00	135.00	164.00	186.00	202.00	216.00	227.00	237.00	246.00	253.00	260.00
2	.05	**6.09**	**8.30**	**9.80**	**10.90**	**11.70**	**12.40**	**13.00**	**13.50**	**14.00**	**14.40**	**14.70**
	.01	14.00	19.00	22.30	24.70	26.60	28.20	29.50	30.70	31.70	32.60	33.40
3	.05	**4.50**	**5.91**	**6.82**	**7.50**	**8.04**	**8.48**	**8.85**	**9.18**	**9.46**	**9.72**	**9.95**
	.01	8.26	10.60	12.20	13.30	14.20	15.00	15.60	16.20	16.70	17.10	17.50
4	.05	**3.93**	**5.04**	**5.76**	**6.29**	**6.71**	**7.05**	**7.35**	**7.60**	**7.83**	**8.03**	**8.21**
	.01	6.51	8.12	9.17	9.96	10.60	11.10	11.50	11.90	12.30	12.60	12.80
5	.05	**3.64**	**4.60**	**5.22**	**5.67**	**6.03**	**6.33**	**6.58**	**6.80**	**6.99**	**7.17**	**7.32**
	.01	5.70	6.97	7.80	8.42	8.91	9.32	9.67	9.97	10.20	10.50	10.70
6	.05	**3.46**	**4.34**	**4.90**	**5.31**	**5.63**	**5.89**	**6.12**	**6.32**	**6.49**	**6.65**	**6.79**
	.01	5.24	6.33	7.03	7.56	7.97	8.32	8.61	8.87	9.10	9.30	9.49
7	.05	**3.34**	**4.16**	**4.69**	**5.06**	**5.36**	**5.61**	**5.82**	**6.00**	**6.16**	**6.30**	**6.43**
	.01	4.95	5.92	6.54	7.01	7.37	7.68	7.94	8.17	8.37	8.55	8.71
8	.05	**3.26**	**4.04**	**4.53**	**4.89**	**5.17**	**5.40**	**5.60**	**5.77**	**5.92**	**6.05**	**6.18**
	.01	4.74	5.63	6.20	6.63	6.96	7.24	7.47	7.68	7.87	8.03	8.18
9	.05	**3.20**	**3.95**	**4.42**	**4.76**	**5.02**	**5.24**	**5.43**	**5.60**	**5.74**	**5.87**	**5.98**
	.01	4.60	5.43	5.96	6.35	6.66	6.91	7.13	7.32	7.49	7.65	7.78

Table 5 (cont.)

Values of Studentized Range Statistic, q_k

Degrees of Freedom within Groups (degrees of freedom in denominator of F-ratio)	α	\multicolumn{11}{c}{k = Number of Means Being Compared}
		2	3	4	5	6	7	8	9	10	11	12
10	.05	3.15	3.88	4.33	4.65	4.91	5.12	5.30	5.46	5.60	5.72	5.83
	.01	4.48	5.27	5.77	6.14	6.43	6.67	6.87	7.05	7.21	7.36	7.48
11	.05	3.11	3.82	4.26	4.57	4.82	5.03	5.20	5.35	5.49	5.61	5.71
	.01	4.39	5.14	5.62	5.97	6.25	6.48	6.67	6.84	6.99	7.13	7.26
12	.05	3.08	3.77	4.20	4.51	4.75	4.95	5.12	5.27	5.40	5.51	5.62
	.01	4.32	5.04	5.50	5.84	6.10	6.32	6.51	6.67	6.81	6.94	7.06
13	.05	3.06	3.73	4.15	4.45	4.69	4.88	5.05	5.19	5.32	5.43	5.53
	.01	4.26	4.96	5.40	5.73	5.98	6.19	6.37	6.53	6.67	6.79	6.90
14	.05	3.03	3.70	4.11	4.41	4.64	4.83	4.99	5.13	5.25	5.36	5.46
	.01	4.21	4.89	5.32	5.63	5.88	6.08	6.26	6.41	6.54	6.66	6.77
16	.05	3.00	3.65	4.05	4.33	4.56	4.74	4.90	5.03	5.15	5.26	5.35
	.01	4.13	4.78	5.19	5.49	5.72	5.92	6.08	6.22	6.35	6.46	6.56
18	.05	2.97	3.61	4.00	4.28	4.49	4.67	4.82	4.96	5.07	5.17	5.27
	.01	4.07	4.70	5.09	5.38	5.60	5.79	5.94	6.08	6.20	6.31	6.41
20	.05	2.95	3.58	3.96	4.23	4.45	4.62	4.77	4.90	5.01	5.11	5.20
	.01	4.02	4.64	5.02	5.29	5.51	5.69	5.84	5.97	6.09	6.19	6.29
24	.05	2.92	3.53	3.90	4.17	4.37	4.54	4.68	4.81	4.92	5.01	5.10
	.01	3.96	4.54	4.91	5.17	5.37	5.54	5.69	5.81	5.92	6.02	6.11
30	.05	2.89	3.49	3.84	4.10	4.30	4.46	4.60	4.72	4.83	4.92	5.00
	.01	3.89	4.45	4.80	5.05	5.24	5.40	5.54	5.56	5.76	5.85	5.93
40	.05	2.86	3.44	3.79	4.04	4.23	4.39	4.52	4.63	4.74	4.82	4.91
	.01	3.82	4.37	4.70	4.93	5.11	5.27	5.39	5.50	5.60	5.69	5.77
60	.05	2.83	3.40	3.74	3.98	4.16	4.31	4.44	4.55	4.65	4.73	4.81
	.01	3.76	4.28	4.60	4.82	4.99	5.13	5.25	5.36	5.45	5.53	5.60
120	.05	2.80	3.36	3.69	3.92	4.10	4.24	4.36	4.48	4.56	4.64	4.72
	.01	3.70	4.20	4.50	4.71	4.87	5.01	5.12	5.21	5.30	5.38	5.44
∞	.05	2.77	3.31	3.63	3.86	4.03	4.17	4.29	4.39	4.47	4.55	4.62
	.01	3.64	4.12	4.40	4.60	4.76	4.88	4.99	5.08	5.16	5.23	5.29

From B. J. Winer, *Statistical Principles in Experimental Design,* McGraw-Hill, Copyright © 1962. Reproduced by permission of the McGraw-Hill Companies, Inc.

Table 6
Critical Values of Chi Square: The χ^2-Table

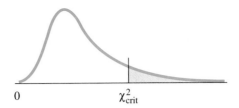

0   χ^2_{crit}

	Alpha Level	
df	$\alpha = .05$	$\alpha = .01$
1	3.84	6.64
2	5.99	9.21
3	7.81	11.34
4	9.49	13.28
5	11.07	15.09
6	12.59	16.81
7	14.07	18.48
8	15.51	20.09
9	16.92	21.67
10	18.31	23.21
11	19.68	24.72
12	21.03	26.22
13	22.36	27.69
14	23.68	29.14
15	25.00	30.58
16	26.30	32.00
17	27.59	33.41
18	28.87	34.80
19	30.14	36.19
20	31.41	37.57
21	32.67	38.93
22	33.92	40.29
23	35.17	41.64
24	36.42	42.98
25	37.65	44.31
26	38.88	45.64
27	40.11	46.96
28	41.34	48.28
29	42.56	49.59
30	43.77	50.89
40	55.76	63.69
50	67.50	76.15
60	79.08	88.38
70	90.53	100.42

From R. A. Fisher and F. Yates, *Statistical Tables for Biological, Agri-cultural and Medical Research*, 6th ed. Copyright © 1963, R. A. Fisher and F. Yates. Reprinted by permission of Pearson Education Limited.

Answers
to Odd-Numbered
Questions

Chapter 1
Using What You Know

1. To understand the laws of nature pertaining to the behaviors of living organisms.

3. In an experiment, the researcher has control over and manipulates one of the variables (the independent variable). In a correlational study, the researcher simply measures the participants' scores on two variables. Inferential statistics are used to decide whether the sample accurately represents the population.

5. A representative sample is a sample that accurately reflects the scores and relationship in the population.

7. A continuous scale allows for fractional amounts. A discrete scale can only be measured in fixed amounts, which cannot be broken into smaller amounts.

9. Sample A.

11. (a) Experiment. (b) Correlational study. (c) Correlational study. (d) Experiment. (e) Correlational study. (f) Experiment.

13. (a) Sample. (b) Population. (c) Population. (d) Sample.

15.

Variable	Type of Measurement Scale	Continuous or Discrete
Nationality	nominal	discrete
Hand pressure	ratio	continuous
Baseball team rank	ordinal	discrete
Letter grade on a test	ordinal	discrete
Pregnancy test	nominal	discrete
Checkbook balance	interval	continuous

Chapter 1
Review and Application Questions

1. So that they understand the language and procedures used in research.

3. A relationship exists when certain scores on one variable are associated with certain scores on the other variable so as the scores on one variable change, the scores on the other variable change in a consistent fashion.

5. The design of the study and the scale of measurement used.

7. The independent variable is the overall variable the researcher is interested in; the conditions are the specific amounts or categories of the independent variable under which participants are tested.

9. They are used to organize, summarize, and describe the characteristics of a sample of scores.

11. (a) A statistic describes a characteristic of a sample of scores. A parameter describes a characteristic of a population of scores. (b) Statistics use letters from the English alphabet. Parameters use letters from the Greek alphabet.

13. Jeremiah's is an experiment because he manipulated alcohol consumption; Orville's is a correlational study because he merely measured participants on both variables.

15. (a) Each score on X is matched with one and only one Y score. (b) A different batch of Y scores is paired with each X score. (c) Virtually the same batch of Y scores occurs with every X score.

17. Samples A (Y scores increase) and D (Y scores increase then decrease).

19.

Variable	Type of Measurement Scale	Continuous or Discrete
Personality type	nominal	discrete
Academic major	nominal	discrete
Number of minutes before and after an event	interval	continuous
Restaurant ratings (best, next best, etc.)	ordinal	discrete
Speed	ratio	continuous
Number of dollars in your pocket	ratio	discrete
Position in line	ordinal	discrete
Change in weight (lbs.)	interval	continuous

Chapter 2
Using What You Know

1. A frequency distribution is a set of data organized by each score's frequency.

3. When a score is in a tail of the normal distribution, it is an extreme score. It is relatively far away from the middle of the distribution and occurs with a relatively low frequency.

5. A negatively skewed distribution indicates the test was relatively easy for most students. This is because the most frequent scores are middle or high scores and seldom are there relatively low scores.

7. The proportion of the area under the curve to the left of a score reflects the proportion of all scores below that score. By multiplying this proportion by 100, we obtain the percentage of all scores below the score, which is the score's percentile.

9. It tells you only 25% of all those who took the test scored better than you did. Therefore, you should feel pretty good about your intelligence score on this test.

11. (a) It indicates many employees are paid a relatively low salary and only a few make a high salary. (b) It means you are paid more than most other people at the company.

13.

Blood Type	f
B	3
AB−	3
A−	2
0	7
A	5

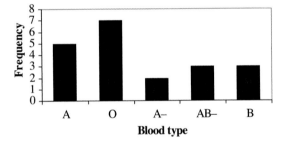

15.

Scores	f	Relative Frequency	Percent
87	6	.21	21
86	10	.34	34
85	5	.17	7
84	2	.07	7
83	4	.14	14
82	0	.00	0
81	2	.07	7

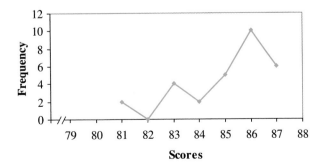

17. (a)

Score	f	Relative Frequency	Percent
25	4	.08	8
24	5	.10	10
23	6	.12	12
22	6	.12	12
21	8	.16	16
20	7	.14	14
19	5	.10	10
18	4	.08	8
17	5	.10	10

(b) The sum of all relative frequencies in the table equals 1.00. (c) The sum of all percents equals 100%.

Chapter 2
Review and Application Questions

1. N is the number of scores in a sample; f is frequency, the number of times a score occurs.
3. (a) In a bar graph adjacent bars do not touch; in a histogram they do. (b) Bar graphs are used with nominal or ordinal scores; histograms are used with interval or ratio scores. (c) A histogram has a bar above each score; a polygon has data points above the scores that are connected by straight lines. (d) Histograms are used with a few different interval or ratio scores; polygons are used with a wide range of interval/ratio scores.
5. A data point.
7. A positively skewed distribution has only one tail at the extreme high scores; a negatively skewed distribution has only one tail at the extreme low scores.
9. Scores in the left-hand tail are low-frequency *low* scores; in the right-hand tail are low-frequency *high* scores.
11. (a) The middle IQ score has the highest frequency in a symmetrical distribution; the higher and lower scores have lower frequencies, and the highest and lowest scores have a relatively very low frequency. (b) The agility scores form a symmetrical distribution containing two distinct "humps" where there are two scores that occur more frequently than the surrounding scores. (c) The memory scores form an asymmetrical distribution in which there are some very infrequent, extremely low scores, but there are not correspondingly infrequent high scores.

13. It indicates that the test was difficult for the class, because most often the scores are low or middle scores, and seldom are there high scores.
15. (a) Bar graph. (b) Polygon. (c) Bar graph. (d) Histogram.
17. (a) 35% of the sample scored below your score. (b) The score occurred 40% of the time. (c) It is one of the highest and least frequent scores. (d) It is one of the lowest and least frequent scores. (e) 60% of the area under the curve and thus 60% of the distribution is to the left of (below) your score.
19. (a) 70, 72, 60, 85, 45. (b) 10th percentile. (c) .50. (d) .50 − .10 = .40. (e) A total of .50 + .35 = .85 of the curve is to the left of 80, so it is at the 85th percentile.
21.

Score	f	Relative Frequency
53	1	.06
52	3	.17
51	2	.11
50	5	.28
49	4	.22
48	0	.00
47	3	.17

23.

Score	f	Relative Frequency
16	5	.33
15	1	.07
14	0	.00
13	2	.13
12	3	.20
11	4	.27

Chapter 3
Using What You Know

1. (a) They indicate where most of the scores in a distribution are located. (b) The mean, median, and mode. (c) The type of measurement scale and, for interval and ratio scores, the shape of the distribution.
3. Because the mean, at the mathematical center of the distribution, is not located near most of the scores.
5. The distribution is a perfect normal distribution.

7. (a) A = −7, B = −2, C = 0, D = +1, E = +5.
   (b) +5, +1, 0, −2, −7. (c) 0, +1, −2, +5, −7.
9. Because overall, overestimates and underestimates from such predictions will cancel out. Mathematically, the total error is the sum of the deviations, or $\Sigma(X - \overline{X})$, which equals 0.
11. We must assume the population is normally distributed.
13. (a) A line graph because the data are interval.
    (b) A line graph because the data are ratio.
    (c) A bar graph because the data are ordinal.
    (d) A bar graph because the data are nominal.
15. Graph A: (a) the means are different; (b) the scores tend to be different; (c) the μs differ; (d) there is a relationship.
    Graph B: (a) the means do not differ; (b) the scores do not differ; (c) the μs do not differ; (d) there is no relationship.
17. No, because the summary data points form a horizontal line.
19. Because the mean is larger than the median, which is larger than the mode, this is a positively skewed distribution.
21. This is a trick question: The mode is the most frequent score.

Chapter 3
Review and Application Questions

1. It indicates where on a variable most scores tend to be located.
3. The mode is a most frequently occurring score, used with nominal scores.
5. The mean is the average score—the mathematical center of a distribution, used with symmetrical distributions of interval or ratio scores.
7. Because here the mean is not near most of the scores.
9. μ is the symbol for a population mean, estimated from a sample mean.
11. (a) $X - \overline{X}$. (b) $\Sigma(X - \overline{X})$. (c) The total distance scores are above the mean equals the total distance scores are below the mean.
13. (a) Mean. (b) Median (these ratio scores are skewed). (c) Mode (this is a nominal variable). (d) Median (this is an ordinal variable).
15. Sample A: $\Sigma X = 8$, $N = 4$, $\overline{X} = 2.00$, Mdn = 2, Mode = 2. Sample B: $\Sigma X = 31$, $N = 6$, $\overline{X} = 5.1666 = 5.17$, Mdn = 5, Mode = 5. Sample C: $\Sigma X = 40$, $N = 5$, $\overline{X} = 8.00$, Mdn = 8, Mode = 8. Sample D: $\Sigma X = 59$, $N = 6$, $\overline{X} = 9.833 = 9.83$, Mdn = 10, Mode = 10 and 11.

17. Mean errors do not change until there has been 5 hours of sleep deprivation. Mean errors then increase as a function of increasing sleep deprivation.
19. (a) (1) income on Y axis, age on X axis; (2) line graph; (3) find median income per age group (income is skewed). (b) (1) positive votes on Y axis, presence or absence of a wildlife refuge on X axis; (2) bar graph; (3) find mean number of votes if normally distributed. (c) (1) running speed on Y axis, amount of carbohydrates consumed on X axis; (2) line graph; (3) find mean running speed if normally distributed. (d) (1) alcohol abuse on Y axis, ethnic group on X axis; (2) bar graph; (3) find mean rate of alcohol abuse per group if normally distributed.
21. (a) The means for Conditions 1, 2, and 3 are 15, 12, and 9, respectively.
    (b)

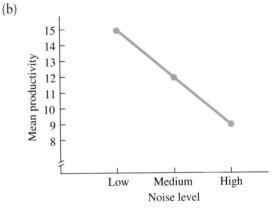

(c)

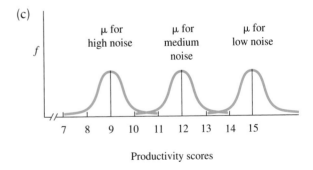

(d) In nature, as noise level increases, the typical productivity score decreases from around 15 to around 12 to around 9.

Chapter 4
Using What You Know

1. A measure of central tendency and a measure of variability.

3. (a) S_X indicates a sample statistic describing the sample; σ_X indicates a true population parameter describing the population; and s_X indicates an estimate of the population parameter computed from a sample. (b) When the squared sign (^{2}) is present, the symbol indicates a variance; when the squared sign (^{2}) is not present, the symbol indicates a standard deviation.

5. (a) They are both ways to describe the variability or spread of the scores around the mean. (b) The standard deviation. It measures in the same units as the raw scores, and so it more directly indicates the "average" deviation of the scores from the mean. It also can be applied directly to the normal curve.

7. (a) The mean and standard deviation of the scores in each condition. (b) A relationship is present if the means change as the conditions change.

9. (a) $s_X^2 = 5.46$; $s_X = 2.34$. (b) Because the final division here is by $N - 1$, which results in a larger answer. (c) Of the N scores in a sample, only $N - 1$ of them reflect the variability in the population. Thus, to find the "average" of this variability, we divide by $N - 1$.

11. (a) You computed the sample variance instead of the unbiased estimator. (b) With a final division of 29 instead of 30, $s_X^2 = 11.099$.

13. (a) You have used the formula for the unbiased estimator of the population variance to describe the sample and the formula for the sample variance to estimate the population. (b) The estimated variance is always larger than the sample variance (and s_X is larger than S_X).

15. (a) The mean, 32. (b) Class A. The scores in that class are closer to the mean.

Chapter 4
Review and Application Questions

1. (a) The distribution's shape, its central tendency, and its variability. (b) To know how spread out scores are and how accurately the mean summarizes them.

3. (a) The range is the distance between the highest and lowest scores in a distribution. (b) Because it includes only the most extreme and often least-frequent scores, so it does not summarize most of the differences in a distribution. (c) With nominal or ordinal scores or with interval/ratio scores that cannot be accurately described by other measures.

5. (a) Variance is the average of the squared deviations around the mean. (b) Variance equals the squared standard deviation, and the standard deviation equals the square root of the variance.

7. Because the unbiased estimates divide by the quantity $N - 1$, resulting in a slightly larger estimate.

9. (a) Range $= 8 - 0 = 8$, so the scores span 8 different scores. (b) $\Sigma X = 41$, $\Sigma X^2 = 231$, $N = 10$, so $S_X^2 = (231 - 168.1)/10 = 6.29$: the average squared deviation of creativity scores from the mean is 6.29. (c) $S_X = \sqrt{6.29} = 2.51$: the "average deviation" of creativity scores from the mean is 2.51.

11. About 160 people. $\overline{X} = 4.10$, so $X = 1.59$ is the score one S_X below the mean $(4.10 - 2.51 = 1.59)$. About 16% of the scores are below this X, and 16% of 1000 is 160.

13. (a) Because the sample tends to be normally distributed, the population should be normal, too. (b) Because $\overline{X} = 1297/17 = 76.29$, we would estimate the μ to be 76.29. (c) The estimated population variance is $(99,223 - 98,953.47)/16 = 16.85$. (d) The estimated standard deviation is $\sqrt{16.85} = 4.10$. (e) Between 72.19 $(76.29 - 4.10)$ and 80.39 $(76.29 + 4.10)$.

15. (a) Guchi, because his standard deviation is larger, his scores are spread out around the mean, so he tends to be a more inconsistent student. (b) Pluto, because his scores are closer to the mean so it more accurately describes all of his scores. (c) Pluto, because we predict each will score at his mean score, and Pluto's scores are closer to his mean than Guchi's are to his mean. (d) Guchi, because his scores vary more widely above and below 60.

17. (a) Compute the mean and sample standard deviation in each condition. (b) Changing conditions A, B, and C changes dependent scores from around 11.00 to 32.75 to 48.00, respectively. (c) The S_X for the three conditions are .71, 1.09, and .71, respectively. Participants scored rather consistently in each condition.

19. (a) The scores in distribution A are less spread out than in B. (b) Find the scores at $\pm 1 S_X$ from the mean: 68% of the scores are between 35 and 45 in A; between 30 and 50 in B.

21. (a) For conditions 1, 2, and 3, we'd expect μs of about 13.33, 8.33, and 5.67, respectively. (b) Somewhat inconsistently: based on s_X we'd expect a σ_X of 4.51, 2.52, and 3.06, respectively.

Chapter 5
Using What You Know

1. (a) z indicates how far a score is from the mean when measured in standard deviation units. (b) The size of the raw score's deviation and the size of S_X.

3. z-scores standardize or equate different distributions so they can be compared and graphed on the same set of axes.

5. Four uses of z-scores are (1) interpreting raw scores, (2) comparing different variables, (3) computing relative frequency, and (4) describing sample means.

7. (a) $(80-86)/12 = -.5$. (b) $(98-86)/12 = +1.0$. (c) $(-1.5)(12) + 86 = 68$. (d) $(1.0)(12) + 86 = 98$.

9. (a) -1.7. (b) $+1.0$. (c) $-.7$. (d) 0.

11. (a) $z = -2.3$. (b) $z = -.6$.

13. (a) In the z-table, you should locate $z = +.97$ and find the relative frequency (given as a proportion in the table) associated with the area under the curve to the right of $z = +.97$. This is the expected relative frequency in each distribution. (b) You should compute the $\overline{X}$ and S_X for each distribution and transform 100 to a z-score in each distribution. Then, in the z-table, you should locate each of these z-scores and find the expected relative frequency of scores between the mean and a raw score of 100 associated with each of the z-scores, as given in the table.

15. (a) $z = (76 - 100)/12 = -2.0$, relative frequency = .4772. (b) $z = (112 - 100)/12 = +1.0$, relative frequency = .1587. (c) $z = (106 - 100)/12 = +.5$, relative frequency below .1915 + .50 = .6915, so about the 69th percentile. (d) $z = (84 - 100)/12 = -1.33$, relative frequency below .0918, so about the 9th percentile.

17. (a) Because she can't determine the sample's relative standing in this way. (b) Compute the sample's z-score. (c) $\sigma_{\overline{X}} = 5/\sqrt{25} = 1$; $z = (53 - 50)/1 = +3$. (d) A z-score of $+3$ indicates that among all possible samples of 25 students, yours has one of the highest and least frequent mean self-esteem scores you could ever obtain.

Chapter 5
Review and Application Questions

1. (a) A z-score indicates the distance, measured in standard deviation units, that a score is above or below the mean. (b) z-scores can be used to interpret scores from any normal distribution of interval or ratio scores.

3. It is the distribution after transforming a distribution of raw scores into z-scores.

5. Because z-scores standardize or equate different distributions so that they can be compared.

7. (a) It is our model of the perfect normal z-distribution. (b) It is a model of any normal distribution of raw scores after being transformed to z-scores. (c) The raw scores should be approximately normally distributed interval or ratio scores and the sample should be relatively large.

9. (a) That it is normally distributed, that its μ equals the μ of the raw score population, and that its standard deviation (the standard error of the mean) equals the raw score population's standard deviation divided by the square root of N. (b) Because it indicates the characteristics of any sampling distribution, without our having to actually measure all possible sample means.

11. (a) He should consider the size of each class' standard deviation. (b) Small. This will give him a large positive z-score, placing him at the top of his class. (c) Large. Then he will have a small negative z and be close to the mean.

13. $\Sigma X = 103$, $\Sigma X^2 = 931$, and $N = 12$, so $S_X = 1.98$ and $\overline{X} = 8.58$. (a) For $X = 10$, $z = (10 - 8.58)/1.98 = +.72$. (b) For $X = 6$, $z = (6 - 8.58)/1.98 = -1.30$.

15. (a) $z = +1.0$ (b) $z = -2.8$ (c) $z = -.70$ (d) $z = -2.0$

17. (a) .4706 (b) .0107 (c) .3944 + .4970 = .8914 (d) .0250 + .0250 = .05

19. From the z-table, the 25th percentile is at approximately $z = -.67$. The cutoff score is then $X = (-.67)(10) + 75 = 68.3$.

21. For City A, her salary has a z of $(57,000 - 70,000)/15,000 = -.87$. For City B, her salary has a z of $(42,000 - 48,000)/12,000 = -.5$. City B is the better offer, because her income will be closer to the average cost of living in that city.

Chapter 6
Using What You Know

1. (a) The relative frequency of the event in the population. (b) Probability is expressed as a proportion and is always between 0 and 1.

3. It provides the relative frequency, and thus the probability, of any range of sample means when all samples have the same N and are randomly selected from a population having a certain μ and σ_X.

5. (a) We decide the corresponding sample does not represent the underlying raw score population.

(b) We decide the corresponding sample may represent the underlying raw score population.

7. (a) $p = 13/30 = .433$. (b) $p = 2/30 = .067$. (c) $p = 10/30 = .333$. (d) $p = 5/30 = .167$. (e) p = the sum of all p, or 1.0.

9. (a) .0179 (b) .3340 + .3340 = .6680 (c) .0630 + .0057 = .0687 (d) .5000 + .4147 = .9147

11. (a) $z = (64 - 46)/8 = +2.25$; $p = .0122$ (b) $z = (40 - 46)/8 = -.75$; $z = (50 - 46)/8 = +.50$; $p = .2734 + .1915 = .4649$ (c) $z = (48 - 46)/8 = +.25$; $p = .0987 + .50 = .5987$

13. (a)

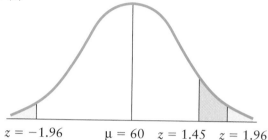

$z = -1.96$ $\mu = 60$ $z = 1.45$ $z = 1.96$

(b) No, the researcher should not reject the idea that the sample may be representative of the population with $\mu = 60$.

15. (a) Either (1) $\overline{X}$ represents μ, but differs from μ by chance, or (2) $\overline{X}$ differs from μ because $\overline{X}$ represents some other population. (b) In problem 13, the difference is likely to be due to chance, and so $\overline{X}$ is likely to represent that μ. In problem 14, the difference is unlikely to be due to chance, and so we decide that $\overline{X}$ represents a different μ.

17. (a) $\sigma_{\overline{X}} = 8/\sqrt{25} = 1.60$, $z = (69 - 72)/1.60 = -1.875$. With a critical value of -1.645, this sample is unrepresentative of the population where $\mu = 72$. (b) $\mu = 69$. (c) Comedians who are not as funny as successful late-night comedians.

19. $\overline{X} = 1{,}492/12 = 124.333$; $\sigma_{\overline{X}} = 30/\sqrt{12} = 8.660$; $z = (124.333 - 100)/8.660 = 2.81$. Reject the idea that Mr. Smart's class is representative of the population where $\mu = 100$.

Chapter 6
Review and Application Questions

1. (a) The likelihood of, or our confidence in, the event. (b) The relative frequency of the event in the population.

3. It is selecting a sample so that all elements or individuals have the same chance of selection.

5. Because of chance, the members of a sample are unrepresentative of the population, so that $\overline{X}$ is different from μ, or the sample represents some other population.

7. It indicates whether or not the sample's z-score (and the sample $\overline{X}$) lies in the region of rejection.

9. No. The probability of a boy is still .5.

11. The p of a hurricane is $160/200 = .80$. The uncle is looking at an unrepresentative sample over the past 13 years. Walter fails to realize that p is based on the long run, and so in the next few years there may not be a hurricane.

13. Terry may have obtained an unrepresentative sample, so that the population does not actually prefer sauerkraut juice.

15. No. With a $z = (24 - 18)/2.19 = +2.74$, this mean falls beyond the critical value of 1.96. It is unlikely to represent this population.

17. (a) The $\overline{X} = 321/9 = 35.67$; $\sigma_{\overline{X}} = 5/\sqrt{9}$, $= 1.67$. Then $z = (35.67 - 30)/1.67 = +3.40$. With a critical value of ± 1.96, conclude that the football players do not represent this population. (b) Football players, as represented by your sample, form a population different from non-football players, having a μ of about 35.67.

19. (a) Chance should not produce a sample that is very unlikely to come from the population, so we reject that it does. (b) These samples are likely to occur in the population, so we accept that the sample represents the population.

21. (a) For Fred's sample, $\mu = 26$, and for Ethel's, $\mu = 18$. (b) The population with $\mu = 26$ is most likely to produce a sample with $\overline{X} = 26$, and the population with $\mu = 18$ is most likely to produce a sample with $\overline{X} = 18$.

Chapter 7
Using What You Know

1. (a) To decide whether sample data represent a particular relationship in the population. (b) Parametric and nonparametric.

3. (a) Either $\overline{X}$ is different from this μ because of sampling error (the sample is unrepresentative of this population), or $\overline{X}$ represents a different value of μ (the sample represents a different population). (b) The appropriate inferential statistical procedure must be performed.

5. (a) H_0 describes the value of μ represented by the sample mean if the predicted relationship does not exist. (b) H_a describes the value of μ represented by the sample mean if the predicted relationship exists. (c) H_0 always maintains the sample data reflect sampling error and thus poorly represents the relationship is not found in the population.

7. (a) To predict whether changing the independent variable causes the dependent scores to increase

or to decrease. (b) You perform a two-tailed test when the direction of the relationship is not predicted; you perform a one-tailed test when the direction of the relationship is predicted.

9. (a) We have *evidence* the independent variable works as predicted. (b) We have shown H_0 is *unlikely to be true*. (d) We have *confidence* our sample mean represents a μ around a certain value. (e) We *can argue* the independent variable causes scores to change. (f) We have *shown* the difference between $\overline{X}$ and μ is *not likely to be* due to sampling error. (i) We have *significant* results.

11. (b) We have *failed to show* that the independent variable *may cause* scores to change as predicted. (c) We are *unconvinced* that the independent variable *does work*. (d) We *cannot* conclude that there *is* a relationship in the population. (g) We have *nonsignificant* results.

13. The probability of rejecting a false H_0.

15. (a) Andy should do a one-tailed test because he has predicted the sample $\overline{X}$ will be significantly higher than the population μ. (b) $H_0: \mu \le 57$; $H_a: \mu > 57$. (c) $z_{crit} = +1.645$. (d) $\sigma_{\overline{X}} = 7/\sqrt{25} = 1.40$; $z_{obt} = (60 - 57)/1.40 = 2.14$. (e) $z = 2.14, p < .05$. (f) The results are significant. Therefore, Andy should conclude that adolescent males enrolled in this anger-management course exhibit increased aggressive behaviors.

17. (a) Denise should do a two-tailed test because she has predicted the sample $\overline{X}$ will be significantly "different" from the population μ, but she did not predict whether the results would be higher or lower than the population. (b) $H_0: \mu = 120$; $H_a: \mu \ne 120$. (c) $z_{crit} = \pm 1.96$. (d) $\sigma_{\overline{X}} = 10/\sqrt{80} = 1.12$; $z_{obt} = (122 - 120)/1.12 = 1.79$. (e) $z = 1.78$, $p > .05$. (f) These results are not significant. There is insufficient evidence children from low-income families have coordination skills different from the population.

19. (a) She should do a two-tailed test because she is interested in either more or fewer prescriptions. (b) $H_0: \mu = 5$; $H_a: \mu \ne 5$. (c) $z_{crit} = \pm 1.96$. (d) $\sigma_{\overline{X}} = 3/\sqrt{20} = .671$; $z_{obt} = (3.45 - 5)/.671 = -2.31$. (e) $z = -2.31, p < .05$. (f) The results are significant. Therefore, Cherise should conclude older persons who are treated by gerontologists take a different number of prescription drugs than the population average.

21. (a) You should use a one-tailed test because you hypothesized the children will be more relaxed with their mothers. (b) $H_0: \mu \le 44$; $H_a: \mu > 44$.

(c) With $\alpha = .05$, $z_{crit} = +1.645$. (d) $\overline{X} = 757/16 = 47.31$; $\sigma_{\overline{X}} = 6.32/\sqrt{16} = 1.58$; $z_{obt} = (47.31 - 44)/1.58 = +2.09$. (e) $z = +2.09$, $p < .05$. (f) The results are significant. You should conclude that having the mother present significantly increases a child's relaxation score.

Chapter 7
Review and Application Questions

1. Because a sample may (1) poorly represent one population so there is not really a relationship, or (2) represent some other population and there is a relationship.

3. α stands for the criterion probability; it determines the size of the region of rejection, and it is the theoretical probability of a Type I error.

5. They describe the predicted relationship that may or may not be demonstrated in an experiment.

7. (a) Use a one-tailed test when predicting the direction the scores will change. (b) Use a two-tailed test when predicting a relationship but not the direction that scores will change.

9. (a) Power is the probability of not making a Type II error. (b) So we do not miss relationships when they exist. (c) When results are not significant, we worry if we missed a real relationship.

11. (a) Changing the independent variable from not finals week to finals week increases the dependent variable of amount of pizza consumed; the experiment will not demonstrate an increase. (b) Changing the independent variable from not performing breathing exercises to performing them changes the dependent variable of blood pressure; the experiment will not demonstrate a change. (c) Changing the independent variable by increasing hormone levels changes the dependent variable of pain sensitivity; the experiment will not demonstrate a change. (d) Changing the independent variable by increasing amount of light will decrease the dependent variable of frequency of dreams; the experiment will not demonstrate a decrease.

13. (a) A two-tailed test: we do not predict the direction that scores will change. (b) $H_0: \mu = 50$, $H_a: \mu \ne 50$. (c) $\sigma_{\overline{X}} = 12/\sqrt{49} = 1.714$; $z_{obt} = (54.63 - 50)/1.714 = +2.70$. (d) $z_{crit} = \pm 1.96$. (e) Yes, because z_{obt} is beyond z_{crit}, the results are significant: Changing from the condition of no music to the condition of music results in test scores changing from a μ of 50 to a μ around 54.63.

15. (a) The probability of a Type I error is $p < .05$. The error would be concluding that music

influences scores when really it does not. (b) By rejecting H_0, there is no chance of making a Type II error. It would be concluding that music does not influence scores when really it does.

17. (a) He is incorrect. In both studies the researchers decided the results were unlikely to reflect sampling error from the H_0 population; they merely defined "unlikely" differently. (b) The probability of a Type I error is less in Study B.

19. This study seeks to prove the null hypothesis, seeking to reject H_a and accept H_0 (which cannot be done). At best, both hypotheses will be retained.

21. (a) The researcher decided that the difference between the scores for Brand X and other brands is too large to have resulted by chance. (b) The $p < .44$ indicates the probability is .44 that the researcher made a Type I error. This p is far too large for us to accept the conclusion.

Chapter 8
Using What You Know

1. We use the z-test when the true standard deviation of the population, σ_X, is known. We use the t-test when it must be estimated based on the sample.

3. (a) It is a model of the distribution of all possible values of t and corresponding sample means when samples have equal Ns and all are randomly selected from one particular raw score population. (b) The larger the df, the more closely the t-distribution approximates the standard normal curve. (c) There is a different t-distribution with a different t_{crit} for each df value.

5. (a) Because $N - 1$ provides us with an unbiased estimator of the population's standard deviation. (b) $N - 1$ is called the degrees of freedom or df.

7. (a) Point estimation—estimating that μ equals $\overline{X}$; interval estimation—creating an interval containing a range of values, one of which is likely to equal μ. (b) Interval estimation, because it takes into account the possibility that $\overline{X}$ suffers from sampling error. (c) It provides a low and a high value, and we are 95% confident the interval contains the μ represented by $\overline{X}$.

9. (a) $H_0: \mu \leq 2.3$; $H_a: \mu > 2.3$. (b) $s_{\overline{X}} = \sqrt{1.96/12} = .40$; $t_{obt} = (3.2 - 2.3)/.40 = 2.25$. (c) $df = 12 - 1 = 11$. (d) $t_{crit} = 1.796$ (one-tailed test, $\alpha = .05$, $df = 11$). (e) Since $t_{obt} > t_{crit}$, reject H_0 and accept H_a. (f) $(.40)(-2.201) + 3.2 \leq \mu \leq (.40)(2.201) + 3.2 = 2.32 \leq \mu \leq 4.08$.

11. (a) The condition in the population is nonsmoking. The condition in the sample is smoking.

The dependent variable is birth weight. (b) $H_0: \mu \geq 116$; $H_a: \mu < 116$. (c) $s_{\overline{X}} = \sqrt{8/18} = .667$; $t_{obt} = (113.6 - 116)/.667 = -2.4/.667 = -3.60$. (d) For $df = 17$, $t_{crit} = -1.74$. (e) $t(17) = -3.60$, $p < .05$. (f) $(.67)(-2.110) + 113.6 \leq \mu \leq (.667)(+2.110) + 113.6 = 112.19 \leq \mu \leq 115.01$. (g) The sample mean for smokers is significantly less than the μ for nonsmokers. Thus, there is a relationship in the population such that for nonsmokers, the μ for birth weights is 116, and for smokers, the μ for birth weights is between 112.19 and 115.01.

13. (a) $t_{obt} = +2.26$. (Changing α does not change t_{obt}.) (b) For $\alpha = .01$ and $df = 11$, $t_{crit} = \pm 3.106$. (Changing α from .05 to .01 changes t_{crit}.) (c) $t(11) = +2.26$, $p > .01$. There is insufficient evidence of a relationship. (d) Since the t-test was not significant, a confidence interval should not be computed. (e) As α becomes smaller (changing from .05 to .01), the value of t_{crit} becomes larger (changing from ± 2.201 to ± 3.106). This makes it harder to obtain a significant result.

15. (a) $H_0: \mu = 40.5$; $H_a: \mu \neq 40.5$. (b) $t_{obt} = +2.03$. (Changing from a one-tailed test to a two-tailed test does not change t_{obt}.) $t_{crit} = \pm 2.069$. (Changing from a one-tailed test to a two-tailed test *does* change t_{crit}.) (c) $t(24) = +2.03$, $p > .05$. There is insufficient evidence of a relationship. (d) As we learned earlier, a one-tailed test is more powerful than a two-tailed test. Here, the two-tailed test failed to find a relationship when the one-tailed test did find a relationship.

17. (a) $H_0: \mu = 32$; $H_a: \mu \neq 32$. (The hypotheses do not change.) (b) $s_{\overline{X}} = .80$; $t_{obt} = -3.75$. (Changing N changes $s_{\overline{X}}$, which in turn changes t_{obt}.) (c) $t_{crit} = \pm 2.045$. (Changing N changes the degrees of freedom, which in turn changes t_{crit}.) (d) $t(30) = -3.75$; $p < .05$. There is a relationship. (e) As the sample size (N) increases, the value of t_{obt} becomes larger while the value of t_{crit} becomes smaller, thus making it easier to find evidence of a relationship.

Chapter 8
Review and Application Questions

1. (a) The t-test and the z-test. (b) Compute z if the standard deviation of the population (σ_X) is known; compute t if σ_X is estimated by s_X.

3. (a) $s_{\overline{X}}$ is the estimated standard error of the mean; $\sigma_{\overline{X}}$ is the true standard error of the mean. (b) Both are used as a standard deviation to locate a sample mean on the sampling distribution of means.

5. (a) Degrees of freedom. (b) Because the appropriate t_{crit} depends on the df. (c) $df = N - 1$.

7. To describe the relationship and interpret it.

9. It describes a range of values of μ, within which we are confident our μ falls.

11. (a) $H_0: \mu = 68.5$; $H_a: \mu \neq 68.5$. (b) $s_{\bar{X}} = \sqrt{130.42/10} = 3.611$; $t_{obt} = (78.5 - 68.5)/3.61 = +2.77$. (c) With $df = 9$, $t_{crit} = \pm2.262$. (d) This book produces a significant improvement in scores: $t_{obt}(9) = 2.77, p < .05$. (e) $(3.61)(-2.262) + 78.5 \leq \mu \leq (3.61)(+2.262) + 78.5 = 70.33 \leq \mu \leq 86.67$.

13. (a) $H_0: \mu = 50$; $H_a: \mu \neq 50$. (b) $s_{\bar{X}} = \sqrt{569.869/8} = 8.44$; $t_{obt} = (53.25 - 50)/8.44 = +.39$. (c) For $df = 7$, $t_{crit} = \pm2.365$. (d) $t(7) = +.39, p > .05$. (e) The results are not significant, so do not compute the confidence interval. (f) She has no evidence that strong arguments change people's attitudes toward this issue.

15. (a) $H_0: \mu = 12$; $H_a: \mu \neq 12$. $\bar{X} = 8.667$, $s_X^2 = 4.67$; $s_{\bar{X}} = \sqrt{4.67/6} = .882$; $t_{obt} = (8.667 - 12)/.882 = -3.78$. $df = 5$, $t_{crit} = \pm2.571$. The results are significant, so there is evidence of a relationship: Without computers $\mu = 12$; with computers μ is around 8.67. (b) The 95% confidence interval for μ is $(.882)(-2.571) + 8.67 \leq \mu \leq (.882)(+2.571) + 8.67$, so μ is between 6.40 and 10.94.

17. (a) No; the closest df is 30, with $t_{crit} = +1.697$, and the results are not beyond it. (b) Yes; the t_{obt} is beyond 1.697.

19. (a) $t(42) = +6.72, p < .01$. (b) $t(5) = -1.72$, $p > .05$.

Chapter 9
Using What You Know

1. (a) The t-test for independent samples and the t-test for related samples. (b) Whether the difference between two sample means represents the predicted difference between two μs.

3. We assume scores are measured using an interval or a ratio scale and the raw score populations are normally distributed and have homogeneous variance.

5. (a) It is the standard deviation of the sampling distribution of all differences between two sample means when the samples represent the populations described by H_0. (b) It is the standard deviation of the sampling distribution of all means of differences between scores when the sample of differences represents the population of differences described by H_0.

7. (a) Independent samples. (b) Related samples. (c) Related samples. (d) Related samples. (e) Independent samples.

9. (a) $H_0: \mu_1 - \mu_2 = 0$; $H_a: \mu_1 - \mu_2 \neq 0$. (b) $s_{\bar{X}_1 - \bar{X}_2} = .48$, $t_{obt} = (12.10 - 9.60)/.48 = 5.21$. (c) With $df = 40$, $t_{crit} = \pm2.021$, and so $t(40) = 5.21, p < .05$. (d) The difference is significant. Therefore, we have evidence there is a difference in attention span scores between children who play video games and those who do not.

11. (a) She should use a related-samples t-test because participants were measured before and after treatment (repeated measures). (b) The independent variable is mild insomnia treatment condition (use of the visualization therapy versus no use of the therapy). The dependent variable is length of time (in minutes) it took to fall asleep. (c) $H_0: \mu_D \leq 0$; $H_a: \mu_D > 0$, where $D = $ no treatment $-$ treatment, because the scores are predicted to be lower after the treatment than before the treatment. (d) $t_{crit} = 1.833$, one-tailed test, $df = 9$. (e) $s_D^2 = [118 - (20^2/10)]/9 = 8.667$; $t_{obt} = 2/\sqrt{8.667/10} = 2/\sqrt{.867} = 2.15$. (f) $t(9) = 2.15, p < .05$. (g) Based on the related-samples t-test, Martha should conclude that the therapy is effective in decreasing the amount of time it takes these individuals to fall asleep. (h) $d = 2/\sqrt{8.667} = .679$.

Chapter 9
Review and Application Questions

1. (a) The independent-samples t-test and the related-samples t-test. (b) Whether the scientist created independent samples or related samples.

3. By using a matched-samples or repeated-measures design.

5. By testing the same participants under all conditions.

7. Homogeneity of variance is when the variances in the populations are equal.

9. (a) $s_{\bar{X}_1 - \bar{X}_2}$ is the standard error of the difference—the standard deviation of the sampling distribution of differences between means from independent samples. (b) $s_{\bar{D}}$ is the standard error of the mean difference, the standard deviation of the sampling distribution of $\bar{D}$ from related samples.

11. $t(40) = +4.55, p < .05$.

13. (a) H_0: $\mu_1 - \mu_2 = 0$; H_a: $\mu_1 - \mu_2 \neq 0$. (b) $s^2_{pool} = 23.695$; $s_{\bar{X}_1 - \bar{X}_2} = 1.78$; $t_{obt} = (43 - 39)/1.78 = +2.25$. (c) With $df = (15 - 1) + (15 - 1) = 28$, $t_{crit} = +2.048$. (d) The results are significant: In the population, hot baths (with μ about 43) produce different relaxation scores from those produced by cold baths (with μ about 39). (e) $r^2_{pb} = (2.25)^2/[(2.25)^2 + 28] = .15$. (f) $d = (43 - 39)/\sqrt{23.695} = .82$. (g) Yes, both r^2_{pb} and d are relatively large, so bath temperature does have a large effect.

15. (a) She should retain H_0, because in her one-tailed test the signs of t_{obt} and t_{crit} are different. (b) She probably did not subtract her sample means in the same way that she subtracted the μs in her hypotheses.

17. (a) H_0: $\mu_D = 0$; H_a: $\mu_D \neq 0$. (b) $\bar{D} = 2.625$, $s^2_D = 4.554$, $t_{obt} = (2.625 - 0)/.754 = +3.48$. (c) Low and High sunshine exposure result in a significant difference in well-being. (d) $t(7) = +3.48$, $p < .05$. (e) For Low, predict the $\bar{X}$ of 15.5. For High, predict the $\bar{X}$ of 18.13. (f) $d = 2.625/\sqrt{4.554} = 1.23$; $r^2_{pb} = 3.48^2/[(3.48)^2 + 7] = .64$; yes.

19. (a) H_0: $\mu_D \leq 0$; H_a: $\mu_D > 0$. (b) $\bar{D} = 1.2$, $s^2_D = 1.289$, $s_{\bar{D}} = .359$; $t_{obt} = (1.2 - 0)/.359 = +3.34$. (c) With $df = 9$, $t_{crit} = +1.833$. (d) The results are significant. Children exhibit more aggressive acts after watching the show (with μ about 3.9) than they do before the show (with μ about 2.7). (e) $d = 1.2/\sqrt{1.289} = 1.06$; violence is a very important variable here.

Chapter 10
Using What You Know

1. In correlational research, we do not know which variable changed first, or whether any other variables changed might be the true cause.

3. It is used when we compute a correlation, to show the nature of the relationship in the data.

5. The type (direction) of relationship and the strength of the relationship.

7. (a) The stronger the relationship, the greater the accuracy in predicting Y scores. (b) The stronger the relationship, the more consistently one value of Y, or close to one value of Y, is associated with each value of X, so that knowing X gets us closer to the actual Y.

9. (a) As X scores increase, Y scores increase in a perfectly consistent manner; the scatterplot forms a straight line; and the linear relationship is perfectly consistent. (b) As X scores increase, there is no consistent pattern of change in Y scores; the scatterplot is circular or horizontally arranged; and a consistent linear relationship does not exist. (c) As X scores increase, Y scores tend to somewhat consistently decrease; the scatterplot is slanted and somewhat narrow; and a consistent linear relationship exists to a reasonable degree.

11. Y' is the predicted Y score for a given X.

13. (a) $r = .99$. (b) As the number of hours exercised increases, the reported life satisfaction scores increase in a nearly perfectly consistent manner. That is, there is a strong positive correlation between these two variables.

15. $r^2 = -.63^2 = .40$.

Chapter 10
Review and Application Questions

1. (a) In experiments, we manipulate one variable and measure participants on another variable; in correlational studies, we measure participants on two variables. (b) In experiments, we compute the mean of the dependent (Y) scores for each condition (each X); in correlational studies, we examine the relationship over all X-Y pairs by computing a correlation coefficient.

3. (a) A scatterplot is a graph of the individual data points from a set of X-Y pairs. (b) A regression line is the summary straight line through the center of a scatterplot.

5. (a) As the X scores increase, the Y scores tend to increase. (b) As the X scores increase, the Y scores tend to decrease. (c) As the X scores increase, the Y scores do not only increase or only decrease.

7. (a) ρ stands for the Pearson correlation coefficient in the population. (b) ρ is estimated from an r calculated on a sample.

9. Compute r_{obt}, create the two- or one-tailed H_0 and H_a, set up the sampling distribution and, using $df = N - 2$, find r_{crit}; compare r_{obt} to r_{crit}.

11. (a) It is the proportion of all of the differences in Y scores that are associated with changes in the X variable. (b) Compute r^2. (c) The larger the r^2, the more useful and important the relationship is.

13. (a) He forgot the inferential test of r. (b) H_0: $\rho = 0$; H_a: $\rho \neq 0$. (c) With $df = 18$, the two-tailed $r_{crit} = \pm.444$. (d) $r(18) = +.21$, $p > .05$; r is not significant, so there is no evidence of a relationship. (e) None.

15. (a) $\Sigma X = 38$, $\Sigma X^2 = 212$, $\Sigma Y = 68$, $\Sigma Y^2 = 552$, $\Sigma XY = 317$, $N = 9$; $r = (2853 - 2584)/\sqrt{(464)(344)} = +.67$. (b) H_0: $\rho \leq 0$; H_a: $\rho > 0$. (c) With $df = 7$, the one-tailed $r_{crit} = +.582$. (d) $r(7) = +.67$, $p < .05$; this r is significantly different from zero, so there is a strong relationship here. (e) In the population of nurses, we'd expect a ρ of around $+.67$. (f) $r^2 = (.67)^2 = .45$; this is useful, because 45% of differences in absenteeism are related to differences in burnout level.

17. (a) H_0: $\rho = 0$; H_a: $\rho \neq 0$. (b) With $df = 70$, the two-tailed $r_{crit} = \pm.232$. (c) $r(70) = +.38$, $p < .05$. (d) The r is significantly different from zero, indicating a moderate relationship; in the population we expect r is around $+.38$. (e) r^2 and linear regression.

19. (a) Because we cannot compare correlations this way. (b) Assuming they are significant, computing r^2, $+.40^2 = .16$, and $+.80^2 = .64$. The study time variable is four times as important as the speed variable in these relationships.

Chapter 11
Using What You Know

1. (a) Analysis of variance. (b) An experiment involving more than two levels of the independent variable. (c) Whether the differences between two or more sample means represent differences between μs or merely reflect sampling error.

3. We assume (1) there is only one independent variable and one dependent variable, (2) each condition contains a random independent sample, (3) the raw scores are measured using an interval or a ratio scale, and (4) the populations represented are normally distributed and have homogeneous variance.

5. The mean square between groups estimates the variability of sample means in the population; the mean square within groups estimates the variability of individual scores in the population.

7. Regardless of whether one mean is smaller or larger than another mean, the value of F_{obt} will be greater than 1.0. Therefore, the ANOVA is *always* a two-tailed test.

9. (a) When F_{obt} is significant and k is greater than 2, because the F_{obt} only indicates that two or more sample means differ significantly; therefore, post hoc tests are used to determine which specific levels produced significant differences. (b) When F_{obt} is not significant, because we are not convinced

there are any differences to be found. Also, when $k = 2$, because there is only one possible difference between means in the study. (c) Tukey's HSD test is used when F_{obt} is significant, when there are more than two levels, and when the ns in all conditions are equal.

11. (a) $k = 3$. (b) $N = 15$.
(c)

Source	Sum of Squares	df	Mean Square	F
Between groups	127.60	2	63.800	1.290
Within groups	593.45	12	49.454	
Total	721.05	14		

(d) $F_{crit} = 3.88$. (e) No, the F-test is not significant. (f) Because the F-test is not significant, no other procedures should be performed.

13. (a) The independent variable is the level of interaction with the mother. The dependent variable is the aggression score. (b) She should do a between-subjects ANOVA because the scores are independent. (c) There is one factor (i.e., the independent variable). It is the level of interaction with the mother. (d) There are four levels—No Interaction, Low Interaction, Moderate Interaction, and High Interaction. (e) H_0: $\mu_1 = \mu_2 = \mu_3 = \mu_4$; H_a: not all the μs are equal.
(f)

Source	Sum of Squares	df	Mean Square	F
Between groups	9.75	3	3.250	5.91
Within groups	8.80	16	.550	
Total	18.55	19		

$SS_{tot} = 397 - (87^2/20) = 18.55$; $SS_{bn} = [(26^2/5) + (24^2/5) + (20^2/5) + (17^2/5)] - (87^2/20) = 9.75$, $SS_{wn} = 18.55 - 9.75 = 8.80$. (g) $F_{crit} = 3.24$. (h) Yes, because $F(3,16) = 5.91$ which is greater than F_{crit}. (i) Because all ns are equal, the Tukey HSD multiple comparison test is appropriate. $k = 4$, $df_{wn} = 16$, and $\alpha = .05$, so $q_k = 4.05$; $\overline{X}_1 = 5.20$; $\overline{X}_2 = 4.80$; $\overline{X}_3 = 4.00$; $\overline{X}_4 = 3.40$; $HSD = 4.05(\sqrt{.550/5}) = 1.343$
$\overline{X}_1 - \overline{X}_2 = .40$; $\overline{X}_1 - \overline{X}_3 = 1.20$; $\overline{X}_1 - \overline{X}_4 = 1.80$; $\overline{X}_2 - \overline{X}_3 = .80$; $\overline{X}_2 - \overline{X}_4 = 1.40$; $\overline{X}_3 - \overline{X}_4 = .60$.
(j) Katie should conclude the mean of the No Interaction group ($\overline{X}_1$) is significantly different from the mean of the High Interaction group ($\overline{X}_4$) and the mean of the Low Interaction group ($\overline{X}_2$) also is significantly different from the mean

of the High Interaction group ($\overline{X}_4$). No other differences exist among the group means. (k) $\eta^2 = 9.75/18.85 = .53$.

Chapter 11
Review and Application Questions

1. (a) Analysis of variance. (b) A study that contains one independent variable. (c) An independent variable. (d) A condition of the independent variable. (e) The influence that changing the independent variable had on dependent scores. (f) All samples are independent. (g) All samples are related.

3. (a) It is the probability of making a Type I error after comparing all means in an experiment. (b) Multiple *t*-tests result in an experiment-wise error rate larger than alpha, but performing ANOVA and then post hoc tests keeps the experiment-wise error rate equal to alpha.

5. (a) When F_{obt} is significant and k is greater than 2. Because the F_{obt} indicates only that two or more level means differ; post hoc tests determine which levels differ significantly. (b) When F_{obt} is not significant or when $k = 2$.

7. It compares the MS_{bn} to the MS_{wn}.

9. It indicates the effect size as the proportion of variance in dependent scores accounted for by changing the levels of the independent variable.

11. (a) $H_0: \mu_1 = \mu_2 = \mu_3 = \mu_4$. (b) H_a: not all μs are equal. (c) H_0 is that a relationship is not represented; H_a is that one is represented.

13. (a) The MS_{bn} is less than the MS_{wn} and H_0 is assumed to be true. (b) He made a computational error—F_{obt} cannot be a negative number.

15. (a) For a relationship to be potentially important, we must first believe that it's a real relationship. (b) Significance indicates the sample relationship is unlikely to occur if there is not a real relationship in the population. (c) A significant relationship is unimportant if it accounts for little of the variance.

17. (a)

	Sum of Squares	df	Mean Source	F
Between	134.80	3	44.93	17.08
Within	42.00	16	2.63	
Total	176.80	19		

$SS_{tot} = 1332 - (152^2/20) = 1332 - 1155.2 = 176.8$; $SS_{bn} = [(22^2/5) + (54^2/5) + (47^2/5) +$

$(29^2/5)] - 1155.2 = 134.8$; $SS_{wn} = SS_{tot} - SS_{bn} = 176.8 - 134.8 = 42$; $MS_{bn} = SS_{bn}/df_{bn} = 134.8/3 = 44.93$; $MS_{wn} = SS_{wn}/df_{wn} = 42.0/16 = 2.63$; $F_{obt} = MS_{bn}/MS_{wn} = 44.93/2.63 = 17.08$. (b) With $df = 3$ and 16, $F_{crit} = 3.24$, so F_{obt} is significant, $p < .05$. (c) For $k = 4$ and $df_{wn} = 16$, $q_k = 4.05$, so $HSD = (4.05)(\sqrt{2.63/5}) = 2.94$; $\overline{X}_4 = 4.4$, $\overline{X}_6 = 10.8$, $\overline{X}_8 = 9.40$, $\overline{X}_{10} = 5.8$. Only ages 4 versus 10 and ages 6 versus 8 do not differ significantly. (d) $\eta^2 = 134.8/176.8 = .76$; this relationship accounts for 76% of the variance, so it's very important.

19. (a) This is a related-samples design, probably comparing the weights of one group before and after they dieted. (b) The one-way within-subjects ANOVA.

Chatper 12
Using What You Know

1. (a) The experiment has two independent variables, with all levels of one factor combined with all levels of the other factor, in which independent samples are used. (b) We assume the scores are measured using interval or ratio scales and the populations represented are normally distributed and have homogeneous variance.

3. (a) The effect of changing the levels of that factor on the dependent variable scores. (b) We average together all scores from all levels of factor B so we have the mean for each level of A. (c) It indicates at least two of the means are likely to represent different μs.

5. For each main effect, H_0 maintains the μs represented by the levels of the factor are all equal; H_a states not all μs are equal. For the interaction, H_0 states the μs represented by the cells do not form an interaction; H_a states the μs do form an interaction.

7. (a) Post hoc comparisons must be performed when there are more than two levels in a significant factor A, or when there are more than two levels in a significant factor B, or when the interaction is significant. (b) To determine which specific levels or cells in the factor or interaction differ significantly.

9. (a) When the data are appropriate for a parametric procedure, there are two factors, and all cells involve related samples. (b) When the data are appropriate for a parametric procedure, there are two factors, and one factor is a between-subjects

factor (with independent samples) and the other is a within-subjects factor (with related samples).

11. (a) Perform a post hoc comparison (such as Tukey's) on the interaction effect. (b) Graph the interaction, since it is significant.

13. (a)

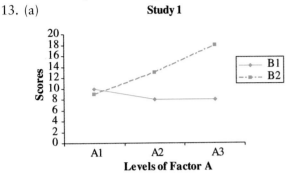

Study 1

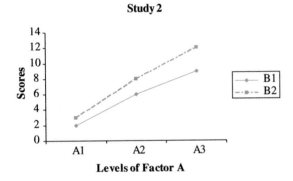

Study 2

(b) In Study 1, there does appear to be an A × B Interaction. In Study 2, there does not appear to be an A × B Interaction.

15. (a)

Source	Sum of Squares	df	Mean Square	F
Between	114.800			
Factor A	6.200	2	3.100	4.65
Factor B	90.133	1	90.133	135.13
A × B Interaction	18.467	2	9.234	13.84
Within	16.000	24	.667	
Total	130.800	29		

(b) $F_{2,24} = 3.40$, $F_{1,24} = 4.26$, $4.65 > 3.40$, so factor A is significant, $p < .05$; $135.13 > 4.26$, so factor B is significant, $p < .05$; $13.84 > 3.40$, so the A × B interaction is significant, $p < .05$. (c) Since the interaction is significant, we can conclude the level of mental acuity depends on the specific combination of fat in the diet and amount of exercise.

Chapter 12
Review and Application Questions

1. (a) She can use a *t*-test or a one-way ANOVA. (b) A two-way, between-subjects ANOVA; a two-way, within-subjects ANOVA; or a two-way, mixed-design ANOVA. (c) Whether both factors are tested using independent samples, both factors are tested using related samples, or one factor involves related samples and the other involves independent samples.

3. A main effect mean is based on scores in a level of one factor after collapsing across the other factor. A cell mean is the mean of scores from a particular combination of a level of factor A with a level of factor B.

5. (a) A confounded comparison involves two cells that differ along more than one factor. It occurs with cells diagonally positioned in a study's diagram. (b) An unconfounded comparison involves two cells that differ along only one factor. It occurs with means within the same column or within the same row of a diagram. (c) Because we cannot determine which factor produced the difference.

7. (a) H_0 is that the μs represented by the main effect means from factor A are all equal; H_a is that not all μs are equal. (b) H_0 is that the μs represented by the main effect means from factor B are all equal; H_a is that they are not all equal. (c) H_0 is that the μs represented by the cell means do not form an interaction; H_a is that they do form an interaction.

9. (a) That changing a factor produced at least two main effect means that differ significantly. (b) That the relationship between one factor and the dependent scores changes depending on which level of the other factor is present.

11. *Study 1:* For A, means are 7 and 9; for B, means are 3 and 13. Apparently there are effects for A and B but not for A × B. *Study 2:* For A, means are 7.5 and 7.5; for B, means are 7.5 and 7.5. There is no effect for A or B, but there is an effect for A × B. *Study 3:* For A, means are 8 and 8; for B, means are 11 and 5. There is no effect for A, but there are effects for B and A × B.

13. Perform Tukey's post hoc comparisons on each main effect and the interaction, graph each main effect and interaction, and compute each η².

15. Only the main effect for difficulty level is significant.

17. (a) For low reward, $\overline{X} = 8$; for medium, $\overline{X} = 10$; and for high, $\overline{X} = 12$. It appears that as reward

increases, performance increases. (b) For low practice, $\overline{X} = 7$; for medium, $\overline{X} = 8$; for high, $\overline{X} = 15$. It appears that increasing practice increases performance. (c) Yes. How the scores change with increasing reward depends on the level of practice, and vice versa. (d) By comparing the three means within each column and the three means within each row.

19. (a) $\overline{X}_{males} = 8.9$, $\overline{X}_{females} = 10.1$; that changing from males to females raises scores from around 8.9 to around 10.1. (b) $\overline{X}_{male} = 11.9$, $\overline{X}_{female} = 7.1$; that changing from a male to a female experimenter lowers scores from around 11.9 to around 7.1. (c) For males with a male, $\overline{X} = 9.0$; for females with a male, $\overline{X} = 14.8$; for males with a female, $\overline{X} = 8.8$; for females with a female, $\overline{X} = 5.4$. Changing from male to female subjects with a male experimenter increases scores from around 9.0 to around 14.8. Changing from male to female subjects with a female experimenter decreases scores from around 8.8 to around 5.4.

Chapter 13
Using What You Know

1. (a) Whether sample data are likely to represent the predicted relationship in the population. (b) When the data do not meet the assumptions of a parametric procedure.

3. (a) Observed frequency is symbolized as f_o; it is the number of participants who fall into a category or cell. (b) Expected frequency is symbolized as f_e; it is the expected number of participants who fall into a category or cell if the data perfectly represent the distribution described by H_0.

5. (a) It is a correlation coefficient used to describe the strength of the relationship in a significant two-way χ^2 involving a 2×2 design. (b) It is a correlation coefficient used to describe the strength of the relationship in a significant two-way χ^2 not involving a 2×2 design.

7. (a) The Mann-Whitney U test. (b) The Kruskal-Wallis H test for three or more independent samples. (c) The Wilcoxon T test for two related samples. (d) The one-way chi square test.

9. (a) H_0: the samples represent equal frequencies in the population; H_a: the samples represent unequal frequencies in the population. (b) $f_e = 1{,}346/2 = 673$, $\chi^2_{obt} = [(628 - 673)^2/673] + [(718 - 673)^2/673] = 6.02$. (c) For $df = 1$, $\chi^2_{crit} = 3.84$, and so results are significant. (d) $\chi^2 (1) = 6.02$, $p < .05$. There is evidence that the frequency

of males and females born in the population is not equal. Rather, based on our sample, we expect 628/1346, or 47%, are male, and 718/1346, or 53%, are female.

11. (a) She should use a one-way chi square test. (b) H_0: all frequencies in the population are equal. H_a: not all frequencies in the population are equal. (c) $f_e = 40/2 = 20$, $\chi^2_{obt} = [(23 - 20)^2/20] + [(17 - 20)^2/20] = .45 + .45 = .90$, χ^2_{crit} for $df = 2 - 1$ is 3.84. Since χ^2_{obt} is not greater than χ^2_{crit}, we fail to reject the H_0. (d) Charlene should conclude there is insufficient evidence to dispute her company's claim they hire as many women as men engineers.

13. (a) The Wilcoxon T test. (b) The related samples t-test.

Chapter 13
Review and Application Questions

1. They all test whether, due to sampling error, the data poorly represent the absence of the predicted relationship in the population.

3. (a) Either nominal or ordinal scores. (b) They may form very nonnormal distributions, or their populations may not have homogeneous variance, so they are transformed to ranks.

5. (a) When the data consist of the frequency that participants fall into each category of one or more variables. (b) When categorizing participants using one variable. (c) When simultaneously categorizing participants using two variables.

7. That the sample frequencies are unlikely to represent the distribution of frequencies in the population that are described by H_0.

9. (a) It is the correlation coefficient between the two variables used in a significant 2×2 chi square design. (b) C is the correlation coefficient between the two variables used in a significant two-way chi square that is not a 2×2 design.

11. (a) H_0: The elderly population is 30% Republican, 55% Democrat, and 15% other; H_a: Affiliations in the elderly population are not distributed this way. (b) For Republicans, $f_e = (.30)(100) = 30$; for Democrats, $f_e = (.55)(100) = 55$; and for others, $f_e = (.15)(100) = 15$. (c) $\chi^2_{obt} = 4.80 + 1.47 + .60 = 6.87$. (d) For $df = 2$, $\chi^2_{crit} = 5.99$, so the results are significant. Party membership in the population of senior citizens is different from party membership in the general population, and it is distributed as in our samples, $p < .05$.

13. (a) The frequency with which students *dislike* each professor also must be included. (b) Kerry can perform a separate one-way χ^2 on the data for each professor to test for a difference between the frequency for "like" and "dislike," or Kerry can perform a two-way χ^2 to determine if liking or disliking one professor is correlated with liking or disliking the other professor.

15. (a) H_0: Gender and political party affiliation are independent in the population. H_a: Gender and political party affiliation are dependent in the population. (b) For males, Republican $f_e = (75)(57)/155 = 27.58$, Democrat $f_e = (75)(66)/155 = 31.94$, and other $f_e = (75)(32)/155 = 15.48$. For females, Republican $f_e = (80)(57)/155 = 29.42$, Democrat $f_e = (80)(66)/155 = 34.06$, and other $f_e = (80)(32)/155 = 16.52$. (c) $\chi^2_{obt} = 3.33 + 3.83 + .14 + 3.12 + 3.59 + .133 = 14.14$. (d) With $df = 2$, $\chi^2_{crit} = 5.99$, so the results are significant. In the population, frequency of political party affiliation depends on gender, $p < .05$. (e) $C = \sqrt{14.14/(155 + 14.14)} = .29$, indicating a somewhat consistent relationship.

17. H_0 is that the relationship in the ordinal scores in the sample poorly represent that no relationship exists in the population. H_a is that the sample data accurately represent the relationship in the population.

19. (a) The design involved a within-subjects factor with two conditions. (b) The scores were ordinal scores. (c) That the rank scores in one group were significantly higher or lower than those in the other group.

statistical hypothesis testing. *See* hypothesis testing
statistics, descriptive. *See* descriptive statistics
statistics, inferential. *See* inferential statistics
statistics, logic of, 2–4
strength of a relationship, 143–146
sum of deviations around the mean, 37–38
sum of squared Xs, 50
sum of squares, 164–165
sum of squares between groups, 165
sum of squares within groups, 165
sum of the cross products, 147
sum of the squared Ys, 147
sum of X, 30–31
symmetrical distribution. *See* normal distribution or normal curve

T

t-distributions, 110
t-table, 111–112
t-test, one-sample, 106–119
  computation of t_{obt}, 108–110
  defined, 107
  estimating μ with confidence interval, 114–116
  interpreting, 112–113
  logic of, 106–108
  related-samples t-test, 129
  reporting in research literature, 116
  in SPSS, 116
  t-distribution and degrees of freedom, 110–112
t-test, two-sample, 120–137
  effect size, describing, 134–136
  independent-samples t-test, 122–128
  logic of, 120–122
  related-samples t-test, 128–133
  in research literature, 133
  in SPSS, 136
tables
  F-table, 167
  frequency tables, 18
  t-table, 111–112
  z-tables, 67–68

tail of the distribution, 22. *See also* one-tailed test; two-tailed test
tendency, 7–8. *See also* central tendency, measures of
test of independence, 190–194
theoretical probability distributions, 77
total area under the normal curve, 26
total sum of squares, 165
treatment effect, 159–160, 167–168
treatments (levels), 159–160
Tukey *HSD* multiple comparisons test, 168–169, 178, 180–181, 182
two-sample t-test. *See* t-test, two-sample
two-tailed test
  Pearson r and, 149
  setting up sampling distribution for, 94–95
  use of, 91
two-way, between-subjects ANOVA, 172–173
two-way, mixed-design ANOVA, 173
two-way, within-subjects ANOVA, 173
two-way ANOVA. *See* analysis of variance (ANOVA), two-way
two-way chi square, 190–194
two-way interaction, 176
two-way interaction effect, 177
Type I errors, 101–102, 150, 161
Type II errors, 102–103

U

U-shaped pattern, 142
unbiased estimators, 53, 106
unconfounded comparisons, 180–181
unimodal distribution, 33
unrepresentative samples, 5

V

variability, 44–57
  consistency vs., 45–46

correlation coefficient and, 144, 145
  defined, 45
  importance of, 44
  normal distributions and, 46
  population variance and population standard deviation, 52–55
  range, 47
  in research literature, 56
  sample variance and sample standard deviation, 50–52
  in SPSS, 56
  variance and standard deviation, 47–50
variables
  as continuums, 32
  defined, 6
  dependent, 11, 12, 38
  independent, 10–11
  quantitative vs. qualitative, 7
  relationships between, 6–8
  z-distribution, comparing variables with, 63–64
variance. *See also* ANOVA (analysis of variance)
  about, 47
  estimated population variance, 53–55, 109, 124–125
  estimated variance of the population of different scores, 131
  homogeneity of, 123
  pooled, 125
  population variance, 52
  proportion of variance accounted for, 135, 154, 170, 193
  sample variance, 48, 51

W

Wilcoxon T test, 195–196
within-subjects ANOVA, 159–160. *See also* ANOVA (analysis of variance)
within-subjects factor, 159–160

X

X, 30

Y

Y intercept, 153

Z

z-distribution
  comparing different variables with, 63–64
  computing relative frequency with, 64–68
  defined, 62
  interpreting scores with, 62–63
  standard normal curve, 65–67
z-scores, 58–73
  comparing critical value to, 96
  comparing different variables with z-distribution, 63–64
  computing a raw score when z is known, 62
  computing a z-score in sample or population, 61
  computing relative frequency with z-distribution, 64–68
  defined, 60–61
  describing sample means with, 68–72
  interpreting scores with z-distribution, 62–63
  logic of, 58–61
  probability of obtaining sample means and, 78–79
  and representativeness, determining, 82, 84–86
  with SPSS, 72
  standard normal curve, 65–67
  z_{obt} in two-tailed test, 95–96
z-tables, 67–68
z-test, 88–105
  assumptions, 94
  computing z_{obt}, 95
  defined, 94
  reporting in research literature, 101
  sampling distribution for two-tailed test, 94–95
  t-test compared to, 106
zero, 13–14, 38
zero and nonzero differences, 123, 126, 131
zero correlation, 146
zero relationship, 93

Learning Outcomes and Objectives

Key Terms

1.1 Why Is It Important to Learn about Sta...

1.2 The Logic of Research

**Recognize the logic of research and the purpose of statistical pr...
how to recognize a *relationship* between scores**

The gro... ...This column contains the chapter objectives with related learning outcomes and brief reviews. ...nclusion applies is the *po*... subset... ...neasured is the *sample*, and... in a sar... ...e is anything that, when measured, can produc... ...*relationship* occurs when, as the scores on one vari... ...other variable also change in a consistent fashion.

Quick Practice

Which samples show a perfect relationship? An inconsistent one? No relationship?

A		B		C		D	
X	Y	X	Y	X	Y	X	Y
2	4	80	80	33	28	40	60
2	4	80	79	33	20	40	60
3	6	85	76	43	27	45	60
3	6	85	75	43	20	45	60
4	8	90	71	53	20	50	60
4	8	90	70	53	28	50	60

Quick Practices are designed to help you make sure you understand important concepts before you move to the next section.

Answers to the Quick Practice are upside down on the back of the first Review Card to help you double check that you've grasped each concept.

1.3 Applying Descriptive and Infer...

Understand when and why *descriptive* and *inferential* statistical procedures are used

Descriptive statistics are used to describe sample data. *Inferential statistics* are for drawing inferences about the scores and relationship found in the population. A *statistic* describes a characteristic of a sample of scores... the English alphabet. A statistic is used to infer the c... describes a characteristic of a population of scores, Greek alphabet.

Here you'll find the Key Terms in the order they appear in the chapter, along with their Definitions. On the back of the first card in a similar column, you'll find the Key Formulas, grouped together for quick review.

1.4 Understanding Experiments Studies

Explain the difference between an experiment an... understand the independent variable, the conditi... variable

A *study's design* is the particular way in which the study is laid out. In an *experiment* the researcher manipulates the *independent variable* and then measures participants' scores on the *dependent variable*. Each specific amount or category of the independent variable is a *condition*. In a *correlational study* neither variable is

What's a Review Card?

To help you refresh key concepts from each chapter, we've developed a Review Card. Each card can be torn out and carried easily for quick review before quizzes and tests. Each Chapter has two cards, one with an overview of Learning Outcomes, Key Terms, and Key Formulas, and the other with exercises and technology instructions.

...complete group of scores found in any particular situation

participants The individuals who are measured in a sample

variable Anything that, when measured, can produce two or more different scores

relationship A pattern between two variables whereby a change in variable is accompanied by a ...ent change in the other

...ptive statistics Procedures ...anizing and summarizing data ...the important characteristics ...described and communicated

...ntial statistics Procedures ...ermining whether sample data represent a particular relationship in the population

statistic A number from a descriptive procedure that describes a sample of scores; symbolized by a letter from the English alphabet

parameter A number obtained using inferential procedures that describes a population of scores; symbolized by a letter from the Greek alphabet

design The way in which a study is laid out

experiment A research procedure in which one variable is actively changed or manipulated and the scores on another variable are measured to determine whether there is a relationship

SPSS Instructions

This section on your Review Cards describes how to use the computer program *SPSS* to compute the statistics described in this text. These instructions are appropriate for most recent versions of SPSS.

You'll find each chapter (except chapters 6 and 7) has step-by-step SPSS instructions for concepts covered in the chapter. For the best results, read the chapter, then practice the concepts by working through all the practice problems at the end of the chapter and on the cards. That way you will fully understand the concepts before tackling a new computer program. Then, you can use the SPSS instructions with the online problems designed for SPSS. You'll master the concepts in no time.

Some notes on terminology and layout:

- *Select* indicates you should highlight a drop-down menu or click on the button indicated, as appropriate.
- Terms in **bold** are words that appear in SPSS you will highlight or click.
- Terms in *italic* indicate your actions.
- *Move* indicates you should transfer the appropriate pieces into another box, often using an arrow button.

Brief step-by-step instructions for using SPSS appear in this column.

Review and Application Questions

1. Why is it important for stud[...] statistics?

On the second card in each chapter, you'll find an extensive list of Review and Application Questions designed to take with you for quick study sessions on the go.

2. (a) What is a population? (b) [...] make conclusions about p[...]

3. What patterns in the scores [...] two variables?

4. What is the general purpose of experiments and correlational studies?

5. What are the two aspects of a study to consider when selecting the descriptive or inferential statistics that you should employ?

6. What is the difference between an experiment and a correlational study?

7. What is the difference between the independent variable and the conditions of the independent variable?

8. In an experiment what is the dependent variable?

9. What are descriptive statistics used for?

10. What are inferential statistics used for?

11. (a) What is the difference between a statistic and a parameter? (b) What types of symbols are used for statistics and parameters?

12. (a) Define the four scales of measurement. (b) Distinguish between continuous and discrete variables. (c) Which scales are usually assumed to be discrete, and which are assumed to be continuous?

13. A student, Poindexter, gives participants various amounts of alcohol and then observes any decrease in their ability to walk. Another student, Foofy, notes the various amounts of alcohol that participants drink at a party and then observes any decrease in their ability to walk. Which study is an experiment, and which is a correlational study? Why?

14. In each of the following experiments, identify the independent variable, the conditions of the independent variable, and the dependent variable: (a) studying whether scores on a final exam are influenced by whether background music is soft, loud, or absent; (b) comparing freshmen, sophomores, juniors, and seniors with respect to how much fun they have while attending college; (c) studying whether being first-born, second-born, or third-born is related to intelligence; (d) examining whether length of daily exposure to a sun lamp (15 minutes versus 60 minutes) accounts for differences in self-reported depression; (e) studying whether being in a room with blue walls, green walls, red walls, or beige walls influences aggressive behavior in a group of adolescents.

15. What is the general pattern found in (a) a perfectly consistent relationship? (b) an inconsistent relationship? (c) no relationship?

16. Using the words *statistic* and *parameter,* how do we describe a relationship in a population?

Chapter 2

N	number of scores in the data
f	frequency

Chapter 3

ΣX	sum of X
Mdn	median
$\overline{X}$	sample mean
$X - \overline{X}$	deviation
μ	mu; population mean
$\Sigma(X - \overline{X})$	sum of deviations around the mean

Chapter 4

ΣX^2	sum of squared Xs
$(\Sigma X)^2$	squared sum of X
S_X^2	sample variance
S_X	sample standard deviation
σ_X^2	population variance
σ_X	population standard deviation
s_X^2	estimated population variance
s_X	estimated population standard deviation

Chapter 5

$\pm$	plus or minus
z	z-score
$\sigma_{\overline{X}}$	standard error of the mean

Chapter 6

p	probability

Chapter 7

$>$	greater than
$<$	less than
$\geq$	greater than or equal to
$\leq$	less than or equal to
$\neq$	not equal to
H_a	alternative hypothesis
H_0	null hypothesis
z_{obt}	obtained value of z-test
z_{crit}	critical value of z-test
α	alpha; theoretical probability of making a Type I error

Chapter 8

$s_{\overline{X}}$	estimated standard error of the mean
df	degrees of freedom
t_{obt}	obtained value in t-test
t_{crit}	critical value of t-test

Chapter 9

n	number of scores in each sample
s_{pool}^2	pooled variance
$s_{\overline{X}_1 - \overline{X}_2}$	standard error of the difference
$\overline{D}$	mean of difference scores
μ_D	mean of population of difference scores
s_D^2	estimated variance of population of difference scores
$s_{\overline{D}}$	standard error of the mean difference
r_{pb}^2	squared point-biserial correlation coefficient

Chapter 10

ΣY	sum of Ys
ΣY^2	sum of squared Ys
$(\Sigma Y)^2$	squared sum of Ys
ΣXY	sum of cross products of X and Y
r	Pearson correlation coefficient
ρ	rho; population correlation coefficient
Y'	Y prime; predicted Y score
r^2	proportion of variance accounted for

Chapter 11

ANOVA	analysis of variance
k	number of levels in a factor
MS_{bn}	mean square between groups
df_{bn}	degrees of freedom between groups
SS_{bn}	sum of squares between groups
MS_{wn}	mean square within groups
df_{wn}	degrees of freedom within groups
SS_{wn}	sum of squares within groups
SS_{tot}	total sum of squares
df_{tot}	total degrees of freedom
F_{obt}	obtained value from F-ratio
F_{crit}	critical value of F
q_k	value used in HSD test

Statistical Decision Tree

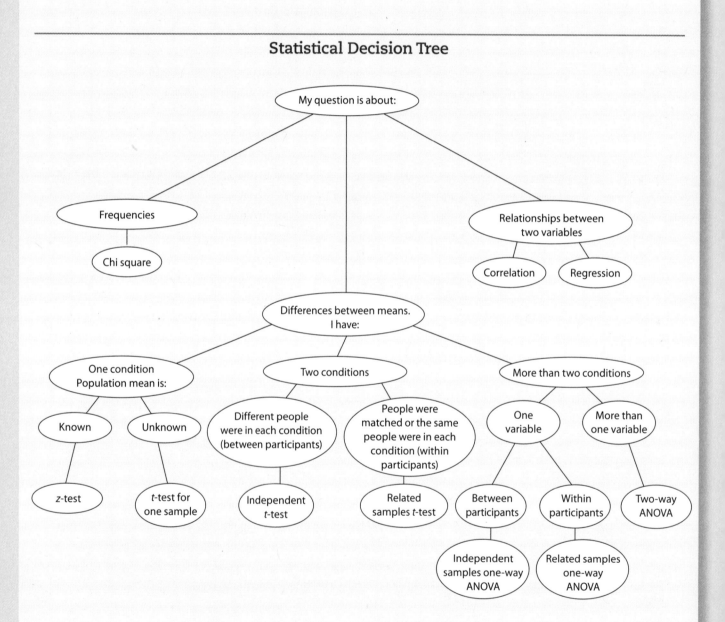

Learning Outcomes and Objectives

Key Terms

1.1 Why Is It Important to Learn about Statistics?

1.2 The Logic of Research

Recognize the logic of research and the purpose of statistical procedures * Know how to recognize a *relationship* between scores

The group of all individuals to which a conclusion applies is the *population*. The subset of the population that is actually measured is the *sample,* and the individuals in a sample are the *participants*. A *variable* is anything that, when measured, can produce two or more different scores. A *relationship* occurs when, as the scores on one variable change, the scores on the other variable also change in a consistent fashion.

Quick Practice

Which samples show a perfect relationship? An inconsistent one? No relationship?

A		B		C		D	
X	Y	X	Y	X	Y	X	Y
2	4	80	80	33	28	40	60
2	4	80	79	33	20	40	60
3	6	85	76	43	27	45	60
3	6	85	75	43	20	45	60
4	8	90	71	53	20	50	60
4	8	90	70	53	28	50	60

1.3 Applying Descriptive and Inferential Statistics

Understand when and why *descriptive* and *inferential* statistical procedures are used

Descriptive statistics are used to describe sample data. *Inferential statistics* are for drawing inferences about the scores and relationship found in the population. A *statistic* describes a characteristic of a sample of scores, symbolized by a letter from the English alphabet. A statistic is used to infer the corresponding *parameter* that describes a characteristic of a population of scores, symbolized by a letter from the Greek alphabet.

1.4 Understanding Experiments and Correlational Studies

Explain the difference between an experiment and a correlational study, and understand the independent variable, the conditions, and the dependent variable

A *study's design* is the particular way in which the study is laid out. In an *experiment* the researcher manipulates the *independent variable* and then measures participants' scores on the *dependent variable*. Each specific amount or category of the independent variable is a *condition*. In a *correlational study* neither variable is

population The large group of all possible scores that would be obtained if the behavior of every individual of interest in a particular situation could be measured

sample A relatively small subset of a population, intended to represent the population; a subset of the complete group of scores found in any particular situation

participants The individuals who are measured in a sample

variable Anything that, when measured, can produce two or more different scores

relationship A pattern between two variables whereby a change in one variable is accompanied by a consistent change in the other

descriptive statistics Procedures for organizing and summarizing data so that the important characteristics can be described and communicated

inferential statistics Procedures for determining whether sample data represent a particular relationship in the population

statistic A number from a descriptive procedure that describes a sample of scores; symbolized by a letter from the English alphabet

parameter A number obtained using inferential procedures that describes a population of scores; symbolized by a letter from the Greek alphabet

design The way in which a study is laid out

experiment A research procedure in which one variable is actively changed or manipulated and the scores on another variable are measured to determine whether there is a relationship

independent variable In an experiment, a variable that is changed or manipulated by the experimenter; a variable hypothesized to cause a change in the dependent variable

condition An amount or category of the independent variable that creates the specific situation under which participants' scores on the dependent variable are measured

dependent variable In an experiment, the behavior or attribute of participants that is expected to be influenced by the independent variable

correlational study A procedure in which participants' scores on two variables are measured, without manipulation of either variable, to determine whether they form a relationship

nominal scale A measurement scale in which each score is used simply for identification and does not indicate an amount

ordinal scale A measurement scale in which scores indicate rank order

interval scale A measurement scale in which each score indicates an actual amount and there is an equal unit of measurement between consecutive scores, but in which zero is simply another point on the scale (not zero amount)

ratio scale A measurement scale in which each score indicates an actual amount, there is an equal unit of measurement, and there is a true zero

continuous scale A measurement scale that allows for fractional amounts of the variable being measured

discrete scale A measurement scale that allows for measurement in fixed amounts, which cannot be broken into smaller amounts; usually amounts are labeled using whole numbers

actively manipulated. Scores on both variables are simply measured and then the relationship is described.

Quick Practice

1. In an experiment the _____ is changed by the researcher to see if it produces a change in participants' scores on the _____.

2. To see if drinking influences one's ability to drive, participants' coordination is measured after drinking 1, 2, or 3 ounces of alcohol. The independent variable is _____, the conditions are _____, and the dependent variable is _____.

3. In an experiment the _____ variable reflects participants' behavior or attributes.

4. We measure the age and income of fifty people to see if older people tend to make more money. What type of design is this?

1.5 The Characteristics of Scores

Understand the four scales of measurement * Explain the difference between continuous and discrete variables

The four *scales of measurement* are (1) *nominal,* in which scores identify a category; (2) *ordinal,* in which scores indicate rank order; (3) *interval,* in which scores measure a specific amount but with no true zero; and (4) *ratio,* in which scores measure a specific amount and 0 indicates truly zero amount. A *continuous* variable can be measured in fractional amounts. A *discrete* variable is measured only in fixed amounts, which cannot be broken into fractions. *Interval* and *ratio* scales are assumed to be continuous scales, which include fractional amounts; *nominal* and *ordinal* scales are assumed to be discrete scales, which do not include fractional amounts.

Quick Practice

1. Whether you are ahead or behind when gambling involves a(n) _____ scale.

2. The number of hours you slept last night involves a(n) _____ scale.

3. Your blood type involves a(n) _____ scale.

4. Whether you are a lieutenant or major in the army involves a(n) _____ scale.

5. A(n) _____ scale allows fractional scores; a(n) _____ scale allows only whole-number scores.

Review and Application Questions

1. Why is it important for students of behavioral research to understand statistics?

2. (a) What is a population? (b) What is a sample? (c) How are samples used to make conclusions about populations?

3. What patterns in the scores do you see when a relationship exists between two variables?

4. What is the general purpose of experiments and correlational studies?

5. What are the two aspects of a study to consider when selecting the descriptive or inferential statistics that you should employ?

6. What is the difference between an experiment and a correlational study?

7. What is the difference between the independent variable and the conditions of the independent variable?

8. In an experiment what is the dependent variable?

9. What are descriptive statistics used for?

10. What are inferential statistics used for?

11. (a) What is the difference between a statistic and a parameter? (b) What types of symbols are used for statistics and parameters?

12. (a) Define the four scales of measurement. (b) Distinguish between continuous and discrete variables. (c) Which scales are usually assumed to be discrete, and which are assumed to be continuous?

13. A student, Poindexter, gives participants various amounts of alcohol and then observes any decrease in their ability to walk. Another student, Foofy, notes the various amounts of alcohol that participants drink at a party and then observes any decrease in their ability to walk. Which study is an experiment, and which is a correlational study? Why?

14. In each of the following experiments, identify the independent variable, the conditions of the independent variable, and the dependent variable: (a) studying whether scores on a final exam are influenced by whether background music is soft, loud, or absent; (b) comparing freshmen, sophomores, juniors, and seniors with respect to how much fun they have while attending college; (c) studying whether being first-born, second-born, or third-born is related to intelligence; (d) examining whether length of daily exposure to a sun lamp (15 minutes versus 60 minutes) accounts for differences in self-reported depression; (e) studying whether being in a room with blue walls, green walls, red walls, or beige walls influences aggressive behavior in a group of adolescents.

15. What is the general pattern found in (a) a perfectly consistent relationship? (b) an inconsistent relationship? (c) no relationship?

16. Using the words *statistic* and *parameter,* how do we describe a relationship in a population?

SPSS Instructions

This section on your Review Cards describes how to use the computer program *SPSS* to compute the statistics described in this text. These instructions are appropriate for most recent versions of SPSS.

You'll find each chapter (except chapters 6 and 7) has step-by-step SPSS instructions for concepts covered in the chapter. For the best results, read the chapter, then practice the concepts by working through all the practice problems at the end of the chapter and on the cards. That way you will fully understand the concepts before tackling a new computer program. Then, you can use the SPSS instructions with the online problems designed for SPSS. You'll master the concepts in no time.

Some notes on terminology and layout:

- *Select* indicates you should highlight a drop-down menu or click on the button indicated, as appropriate.

- Terms in **bold** are words that appear in SPSS you will highlight or click.

- Terms in *italic* indicate your actions.

- *Move* indicates you should transfer the appropriate pieces into another box, often using an arrow button.

17. Which of the following data sets show a relationship?

Sample A		Sample B		Sample C		Sample D	
X	Y	X	Y	X	Y	X	Y
1	10	20	40	13	20	92	71
1	10	20	42	13	19	93	77
1	10	22	40	13	18	93	77
2	20	22	41	13	17	95	79
2	20	23	40	13	15	96	74
3	30	24	40	13	14	97	71
3	30	24	42	13	13	98	69

18. Which sample in problem 17 shows the most consistent relationship? How do you know?

19. In the chart below, identify the characteristics of each variable.

Variable	Type of Measurement Scale	Continuous or Discrete
Personality type		
Academic major		
Number of minutes before and after an event		
Restaurant ratings (best, next best, etc.)		
Speed (miles per hour)		
Number of dollars in your pocket		
Position when standing in line		
Change in weight (lbs.)		

20. Using the terms *sample, population, variable, statistic,* and *parameter,* summarize the steps a researcher follows, starting with a hypothesis and ending with a conclusion about a nature.

Learning Outcomes and Objectives

Key Terms

2.1 Some New Symbols and Terminology

Know what *N* and *f* mean

The number of scores in the data is symbolized by *N*. The number of times each score occurs within a set of data is symbolized by *f*.

2.2 Understanding Frequency Distributions

Understand what *frequency* is and how a *frequency distribution* is created * Know when to graph frequency distributions using a *bar graph*, *histogram*, or *polygon*

A *frequency distribution* shows the frequency of each score. The symbol for simple frequency is *f*. Graph a frequency distribution using a *bar graph* for nominal or ordinal scores, a *histogram* for a few different interval or ratio scores, and a *polygon* for many different interval or ratio scores. A dot plotted on a graph is called a *data point*.

Quick Practice

1. What is the difference between *f* and *N*?

2. Create a frequency table for these scores: 7, 9, 6, 6, 9, 7, 7, 6, 6.

3. What is the *N* here?

4. What is the frequency of 6 and 7 together?

5. A _____ has a separate, discrete bar above each score, a _____ contains bars that touch, and a _____ has dots connected with straight lines.

6. A "dot" plotted on a graph is called a _____.

7. To show the number of freshmen, sophomores, and juniors in a sample, plot a _____.

8. To show the frequency of people who are above average weight by either 0, 5, 10, or 15, pounds plot a _____.

9. To show the number of people preferring chocolate or vanilla ice cream, plot a _____.

10. To show the number of people who are above average weight by each amount between 0 and 100 pounds, plot a _____.

2.3 Types of Frequency Distributions

Explain what *normal, skewed,* and *bimodal* distributions are

In a *normal distribution* forming a *normal curve,* the *tails* contain infrequent, extreme high and low scores, while scores closer to the middle score occur more frequently. A *negatively skewed distribution* has a pronounced tail over the extreme low scores; a *positively skewed distribution* has a pronounced tail over the extreme high scores. A *bimodal distribution* contains two areas with high-frequency scores.

Quick Practice

1. Arrange the scores below from most frequent to least frequent.

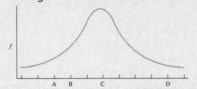

raw scores The scores initially measured in a study

frequency (*f*) The number of times each score occurs within a set of data; also called simple frequency

frequency distribution A distribution of scores, organized to show the number of times each score occurs in a set of data

bar graph A graph in which a free-standing vertical bar is centered over each score on the *X* axis; used with nominal or ordinal scores

histogram A graph similar to a bar graph but with adjacent bars touching; used to plot the frequency distribution of a small range of interval or ratio scores

frequency polygon A graph that shows interval or ratio scores (*X* axis) and their frequencies (*Y* axis), using data points connected by straight lines

data point A dot plotted on a graph to represent a pair of *X* and *Y* scores

grouped distribution A distribution created by combining individual scores into small groups and then reporting the total frequency (or other description) of each group

normal curve The symmetrical, bell-shaped curve produced by graphing a normal distribution

normal distribution A set of scores in which the middle score has the highest frequency and, proceeding toward higher or lower scores, the frequencies at first decrease slightly but then decrease drastically, with the highest and lowest scores having very low frequency

Key Terms

tail of the distribution The far-left or far-right portion of a frequency polygon, containing the relatively low-frequency, extreme scores

negatively skewed distribution A frequency polygon with low-frequency, extreme low scores but without corresponding low-frequency, extreme high ones, so that its only pronounced tail is in the direction of the lower scores

positively skewed distribution A frequency polygon with low-frequency, extreme high scores but without corresponding low-frequency, extreme low ones, so that its only pronounced tail is in the direction of the higher scores

bimodal distribution A symmetrical frequency polygon with two distinct humps where there are relatively high-frequency scores and with center scores that have the same frequency

relative frequency The proportion of time a score occurs in a distribution, which is equal to the proportion of the total number of scores that the score's simple frequency represents

proportion of the area under the curve The proportion of the total area beneath the normal curve at certain scores, which represents the relative frequency of those scores

percentile The percentage of all scores in the sample that are below a particular score

Key Formulas

Relative frequency $= \dfrac{f}{N}$

Percent $=$ (Relative frequency)(100)

Relative frequency $= \dfrac{\text{Percent}}{100}$

Learning Outcomes and Objectives

2. What label should be given to each of the following?

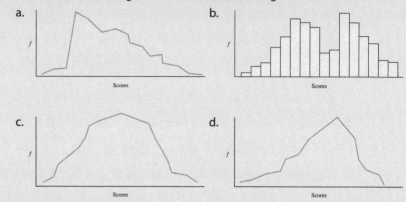

2.4 Relative Frequency and the Normal Curve

Explain what *relative frequency* is and how we use the area *under the normal curve* to compute it

Relative frequency is the proportion of time that a score occurs. The area under the normal curve corresponds to 100% of a sample, so a *proportion of area under the curve* will contain that proportion of the scores, which is their relative frequency.

Quick Practice

1. If a score occurs 23% of the time its relative frequency is _____?

2. If a score's relative frequency is .34, it occurs _____ percent of the time.

3. Scores that occupy .20 of the area under the curve have a relative frequency of _____.

4. In a sample, the scores between 15 and 20 have a relative frequency of .40, so they make up _____ of the area under the curve.

2.5 Understanding Percentile

Explain what a *percentile* is and how we use the area *under the normal curve* to compute it

Percentile is the percent of all scores in the data that are *below* a particular score. On the normal curve the percentile of a score equals the percent of the curve to the left of the score.

QUICK PRACTICE ANSWERS

2.2: 1. f is the number of times a score occurs, N is the total number of scores in the data;
2.

Scores	f
6	2
8	0
7	3
9	4

3. $N = 9$; 4. $f = 3 + 4 = 7$; 5. bar graph, histogram, polygon; 6. data point; 7. bar graph; 8. histogram; 9. bar graph; 10. polygon
2.3: 1. C, B, A, D; 2. a. positively skewed, b. bimodal, c. normal, d. negatively skewed
2.4: 1. 23%/100 = .23; 2. (.34)(100) = 34%; 3. .20, 4. .40

Review and Application Questions

1. What do the symbols N and f mean?

2. Why is it important to understand the frequency distribution produced by a set of data?

3. (a) What is the difference between a bar graph and a histogram? (b) With what kind of data is each used? (c) What is the difference between a histogram and a polygon? (d) With what kind of data is each used?

4. (a) What is the difference between a score's simple frequency and its relative frequency? (b) What is a score's percentile?

5. What is a dot plotted on a graph called?

6. (a) What is the difference between a skewed distribution and a normal distribution? (b) What is the difference between a bimodal distribution and a normal distribution?

7. What is the difference between a positively skewed distribution and a negatively skewed distribution?

8. What does it mean when a score is in a tail of a normal distribution?

9. What is the difference between a score in the left-hand tail of a normal curve and a score in the right-hand tail?

10. (a) What is a proportion? (b) How do you transform a proportion to a percentage? (c) How do you transform a percentage to a proportion?

11. In reading psychological research you encounter the following statements. Interpret each one. (a) "The IQ scores were approximately normally distributed." (b) "A bimodal distribution of physical agility scores was observed." (c) "The distribution of the patients' memory scores was severely negatively skewed."

12. From the data 1, 4, 5, 3, 2, 5, 7, 3, 4, 5, Uriah created the following frequency table. What three things did he do wrong?

Score	f
1	1
2	1
3	2
4	2
5	3
7	1
	$N = 6$

13. The distribution of scores on your next statistics test is positively skewed What does this indicate about the difficulty of the test?

14. (a) On a normally distributed set of exam scores, Gideon scored at the 10th percentile, so he claims that he outperformed 90% of his class. Why is he correct or incorrect? (b) Because Rowena's score had a relative frequency of .02, she claims she had one of the highest scores on the exam. Why is she correct or incorrect?

SPSS Instructions

Creating and Using Frequency Distributions

Frequency Distribution Table

Main Menu: *Analyze*

Select: **Descriptive Statistics→ Frequencies**

Move: **Desired variable(s) into Variable(s) box**

Select: ✓ **Display frequency table→OK**

Notes: (1) Basic statistics (e.g., mean, standard deviation, may be chosen through Statistics... option.
(2) Frequency histograms and bar charts can be chosen through Charts... option.

Frequency Bar Graphs

Main Menu: *Analyze*

Select: **Descriptive Statistics→ Frequencies**

Move: **Desired variable(s) into Variable(s) box**

Select: **Charts... →**
    ✓ **Bar Charts →**
    ✓ **Frequencies →**
    Continue → OK

Notes: May be performed simultaneously with Frequency Distribution Table

Frequency Bar Histograms

Main Menu: *Analyze*

Select: **Descriptive Statistics → Frequencies**

Move: **Desired variable(s) into Variable(s) box**

Select: **Charts... →**
    ✓ **Histograms →**
    Continue → OK

Notes: May be performed simultaneously with Frequency Distribution Table

15. What type of frequency graph is appropriate when counting the number of (a) blondes, brunettes, redheads, or "others" attending a college; (b) people having each body weight reported in a state-wide survey; (c) children in each grade at an elementary school; and (d) car owners reporting above average, average, or below average problems with their car.

16. Why are adjacent bars in a bar graph separate but the adjacent bars in a histogram and the data points in a polygon connected?

17. Interpret each of the following: (a) You scored at the 35th percentile. (b) Your score has a relative frequency of .40. (c) Your score is in the upper tail of the normal curve. (d) Your score is in the left-hand tail of the normal curve. (e) From the normal curve your score is at the 60th percentile.

18. The grades on a test form a bimodal distribution. What does this indicate about performance on the test?

19. The following shows the distribution of final exam scores in a class. The proportion of the total area under the curve is given for three segments. (a) Order the scores 45, 60, 70, 72, and 85 from most frequent to least frequent. (b) What is the percentile of a score of 55? (c) What proportion of the sample scored below 70? (d) What is the relative frequency of scores between 55 and 70? (e) What is the percentile of the score of 80?

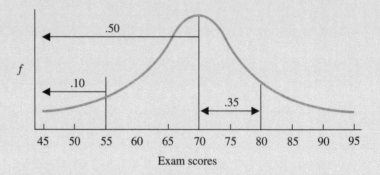

20. On a normal curve identify the approximate location of the following scores: (a) You have the most frequent score. (b) You have a low-frequency score, but the score is higher than most. (c) You have one of the lower scores, but it has a relatively high frequency. (d) Your score seldom occurred.

21. Organize the ratio scores below in a table and show their simple frequency and relative frequency.

49	52	47	52	52	47	49	47	50
51	50	49	50	50	50	53	51	49

22. Draw a frequency polygon using the data in problem 21.

23. Organize the nominal scores below in a table and show their simple frequency and relative frequency.

16	11	13	12	11	16	12	16
16	11	15	13	11	16	12	

24. Using the data in problem 23, draw the appropriate graph showing simple frequency.

Learning Outcomes and Objectives

Key Terms

3.1 Some New Symbols and Terminology

ΣX, Mdn, $\overline{X}$, $\Sigma(X - \overline{X})$, and μ

3.2 What Is Central Tendency?

Understand how measures of *central tendency* describe data

Measures of central tendency summarize the location of a distribution of scores on a variable, indicating where the center of the distribution tends to be.

3.3 Computing the Mean, Median, and Mode

Know what the *mean, median,* and *mode* indicate and when each is appropriate

The *mode* is the most frequent score in the data. The *median* (Mdn) is the score at the 50th percentile. It is used primarily with ordinal data and with skewed interval or ratio data. If *N* is an odd number, the median is the middle score. If *N* is even, the median is the average of the two middle scores. The mean is the average score, located at the mathematical center of the distribution. It is used with interval or ratio data that form a symmetrical, unimodal distribution such as a normal distribution. The symbol for a sample mean is $\overline{X}$, and the symbol for a population mean is μ.

The Mean as the Balance Point of any Distribution

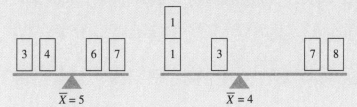

$\overline{X} = 5$ $\overline{X} = 4$

Measures of Central Tendency for Skewed Distributions
The vertical lines show the relative positions of the mean, median, and mode.

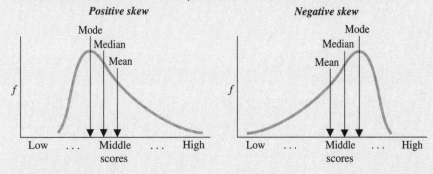

Quick Practice

1. What is the mode in 4, 6, 8, 6, 3, 6, 8, 7, 9, 8?
2. What is the median in the above scores?
3. With what types of scores is the mode preferred?
4. With what types of scores is the median preferred?
5. What is the symbol for the sample mean?

sum of *X* (ΣX) The sum of the scores in a sample

measure of central tendency Statistics that summarize the location of a distribution on a variable by indicating where the center of the distribution tends to be located

mode The most frequently occurring score in a sample

unimodal A distribution whose frequency polygon has only one hump and thus has only one score qualifying as the mode

bimodal distribution A frequency polygon with two distinct humps where there are relatively high-frequency scores and with center scores that have the same frequency

median (Mdn) The score located at the 50th percentile

mean The score located at the mathematical center of a distribution

$\overline{X}$ The symbol used to represent the sample mean

deviation The distance that separates a score from the mean and thus indicates how much the score differs from the mean

sum of the deviations around the mean [$\Sigma(X - \overline{X})$] The sum of all differences between the scores and the mean

line graph A graph of an experiment when the independent variable is an interval or ratio variable; plotted by connecting the data points with straight lines

bar graph A graph in which a free-standing vertical bar is centered over each score on the *X* axis; used with nominal or ordinal scores

μ The symbol used to represent the population mean

Key Formulas

COMPUTING A SAMPLE MEAN

$$\overline{X} = \frac{\Sigma X}{N}$$

MEDIAN

To estimate the median, order the scores from smallest to largest. If N is an odd number, the score in the middle position is the median. If N is an even number, the average of the two scores in the middle positions is the median.

6. What is the mean of 7, 6, 1, 4, 5, 2?
7. With what data is the $\overline{X}$ appropriate?
8. How is a mean interpreted?

3.4 Applying the Mean to Research

Explain what is meant by "deviations around the mean" * Understand how a sample mean is used to describe individual scores and the population

The amount a score *deviates* from the mean is computed as $X - \overline{X}$. The *sum of the deviations around the mean* is $\Sigma(X - \overline{X})$ and always equals zero. The mean is the typical score *around* which a normal distribution is located and is therefore the best estimate of any individual's score. Over the long run, errors when using the mean will cancel out to equal zero. A relationship is present in an experiment if different means occur as the conditions change. A relationship is present in the population when different distributions, having different μs, occur as the conditions change. When graphing the results of an experiment, the independent variable is placed on the X axis and the dependent variable is on the Y axis. A line graph is created for a ratio or interval independent variable. A bar graph is created for a nominal or ordinal independent variable. If the data points on a graph form a line that is not horizontal, then a relationship is present. If the data points form a horizontal line, then a relationship is not present. Envision the results of an experiment in the population as a normal distribution for each condition, with the μ equal to the $\overline{X}$ of that condition.

Quick Practice

1. The independent variable is plotted on the _____ axis, and the dependent variable is plotted on the _____ axis.

2. A _____ shows a data point above each X, with adjacent data points connected with straight lines.

3. A _____ shows a discrete bar above each X.

4. The characteristics of the _____ variable determine whether to plot a line or bar graph.

5. Create a bar graph with _____ or _____ variables.

6. Create a line graph with _____ or _____ variables.

Review and Application Questions

1. What does a measure of central tendency indicate?
2. To how many decimal places do you round off a calculation?
3. What is the mode, and with what type of data is it most appropriate?
4. What is the median, and with what type of data is it most appropriate?
5. What is the mean, and with what type of data is it most appropriate?
6. Why is it best to use the mean with a normal distribution?
7. Why is it inappropriate to use the mean with a skewed distribution?
8. Which measure of central tendency is used most often in behavioral research? Why?
9. What is μ and how do we usually determine it?
10. Why do we use the mean of a sample to predict anyone's score?
11. (a) What is the symbol for a score's deviation from the mean? (b) What is the symbol for the sum of the deviations? (c) What does it mean to say "the sum of the deviations around the mean equals zero"?
12. (a) When graphing an experiment, which variable is on the *X* axis and which is on *Y*? (b) When are bar graphs or line graphs appropriate?
13. A researcher collected the following sets of data. For each, indicate the measure of central tendency she should compute: (a) the following IQ scores: 60, 72, 63, 83, 68, 74, 90, 86, 74, 80; (b) the following error scores: 10, 15, 18, 15, 14, 13, 42, 15, 12, 14, 42; (c) the following blood types: A−, A−, O, A+, AB−, A+, O, O, O, AB+; (d) the following grades: B, D, C, A, B, F, C, B, C, D, D.
14. (a) When do you round off an answer to two decimal places? (b) When do you round off an answer to 3 decimal places? (c) An answer is 4.00. What do the zeros indicate?
15. For each sample below, compute the mean, the median, and the mode.

Sample A	Sample B	Sample C	Sample D
2	5	8	9
1	4	8	11
3	6	8	10
2	5	8	10
	5	8	11
	6	8	8

16. In experiments, how do we envision a relationship in the population?
17. For the following experimental results, interpret specifically the relationship between the independent and dependent variables:

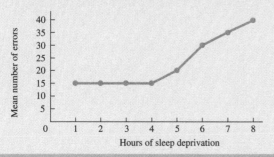

SPSS Instructions

Summarizing Scores with Measures of Central Tendency

Mode

Main Menu: *Analyze*
Select: **Descriptive Statistics → Frequencies**
Move: **Desired variable(s) into Variable(s) box**
Select: **Statistics... →**
✓ **Mode → Continue → OK**

Median

Main Menu: *Analyze*
Select: **Descriptive Statistics → Frequencies**
Move: **Desired variable(s) into Variable(s) box**
Select: **Statistics... →**
✓ **Median → Continue → OK**

Mean (Option 1)

Main Menu: *Analyze*
Select: **Descriptive Statistics → Frequencies**
Move: **Desired variable(s) into Variable(s) box**
Select: **Statistics... →** ✓ **Mean → Continue → OK**

Mean (Option 2)

Main Menu: *Analyze*
Select: **Descriptive Statistics → Descriptives**
Move: **Desired variable(s) into Variable(s) box**
Select: **Options... →** ✓ **Mean → Continue → OK**

18. (a) If you participated in the study in problem 17 and had been deprived of 5 hours of sleep, how many errors do you think you would make? (b) If we tested all people in the world after 5 hours of sleep deprivation, how many errors do you think each would make? (c) What symbol stands for your prediction in part (b)?

19. For each of the experiments listed below, determine (1) which variable should be plotted on the Y axis and which on the X axis, (2) whether to use a line graph or a bar graph, and (3) how to summarize the dependent scores: (a) a study of income as a function of the independent variable of age; (b) a study of the dependent variable of politicians' positive votes on environmental issues depending on the presence or absence of a wildlife refuge in their political district; (c) a study of running speed produced by different conditions of carbohydrates consumed; (d) a study of rates of alcohol abuse found for different ethnic groups.

20. You hear that a line graph of scores from the Grumpy Emotionality Test slants downward with increases in the amount of sunlight present on the day subjects were tested. (a) What does this tell you about the mean scores for the conditions? (b) What does this tell you about the raw scores for each condition? (c) Assuming that the samples are representative, what does this tell you about the μs? (d) What do you conclude about whether a relationship between emotionality and sunlight in nature exists?

21. You conduct a study to determine the impact that varying the amount of noise in an office has on worker productivity. You obtain the following productivity scores.

Condition 1 Low Noise	Condition 2 Medium Noise	Condition 3 Loud Noise
15	13	12
19	11	9
13	14	7
13	10	8

(a) Productivity scores are normally distributed ratio scores. Summarize the results of this experiment. (b) Draw the appropriate graph for these data. (c) Assuming the data are representative, draw how we would envision the populations produced by this experiment. (d) What conclusions should you draw from this experiment?

22. Assume that the data in problem 21 reflect a highly skewed interval variable. (a) Summarize these scores. (b) What conclusion would you draw from the sample data? (c) What conclusion would you draw about the populations produced by this experiment?

Learning Outcomes and Objectives

Key Terms

4.1 Understanding Variability

Know what is meant by *variability*

Measures of variability indicate how much the scores differ from each other, how much the distribution is spread out, how inconsistent the scores are, and how accurately the mean summarizes the scores.

4.2 The Range

Define the *range*

The *range* is the difference between the highest and the lowest score.

4.3 The Variance and Standard Deviation

Understand what the *standard deviation* and *variance* indicate about a set of scores

The *sample variance* (S_X^2) and the *sample standard deviation* (S_X) are the two statistics to use with the mean to describe variability. The standard deviation is interpreted as the average amount scores deviate from the mean, and it is the square root of the variance. There are three versions of the variance. S_X^2 describes how far the sample is spread out around $\overline{X}$, σ_X^2 describes how far the population is spread out around μ, and s_X^2 is the unbiased estimate of how far the population is spread out around μ. There are three versions of the standard deviation. S_X describes how far the sample is spread out around $\overline{X}$, σ_X describes how far the population is spread out around μ, and s_X is the unbiased estimate of how far the population is spread out around μ.

Quick Practice

1. The symbol for the sample variance is _____.

2. The symbol for the sample standard deviation is _____.

3. What is the difference between computing the standard deviation and the variance?

4. In Sample A, $S_X = 6.82$; in Sample B, $S_X = 11.41$. Sample A is _____ (more/less) variable and most scores tend to be _____ (closer to/farther from) the mean.

5. If $\overline{X} = 10$ and $S_X = 2$, then 68% of the scores fall between _____ and _____.

4.4 Computing the Sample Variance and Sample Standard Deviation

Be able to compute the *standard deviation* and *variance* in a sample

ΣX^2 indicates to find the *sum of the squared X*s. $(\Sigma X)^2$ indicates to find the *squared sum of* X. Approximately 68% of the scores in a normal distribution lie between the scores that are at one standard deviation below the mean and one standard deviation above the mean.

measures of variability Statistics that summarize the extent to which scores in a distribution differ from one another

range The distance between the highest and lowest scores in a set of data

sample variance (S_X^2) The average of the squared deviations of a sample of scores around the sample mean

sample standard deviation (S_X) The square root of the sample variance. That is, the square root of the average squared deviation of sample scores around the sample mean

sum of squared Xs (ΣX^2) Calculated by squaring each score in a sample and adding the squared scores

squared sum of X [$(\Sigma X)^2$] Calculated by adding all scores and then squaring their sum

population standard deviation (σ_X) The square root of the population variance. That is, the square root of the average squared deviation of scores around the population mean

population variance (σ_X^2) The average squared deviation of scores around the population mean

biased estimators The formula for the variance or standard deviation involving a final division by N, used to describe a sample, but which tends to underestimate the population variability

unbiased estimators The formula for the variance or standard deviation involving a final division by $N - 1$; calculated using sample data to estimate the population variability

estimated population variance (s_X^2) The unbiased estimate of the population variance calculated from sample data using $N - 1$

Key Terms

estimated population standard deviation (s_X) The unbiased estimate of the population standard deviation calculated from sample data using $N - 1$

Key Formulas

RANGE

Range = Highest score – Lowest score

SAMPLE VARIANCE

$$S_X^2 = \frac{\Sigma X^2 - \frac{(\Sigma X)^2}{N}}{N}$$

SAMPLE STANDARD DEVIATION

$$S_X = \sqrt{\frac{\Sigma X^2 - \frac{(\Sigma X)^2}{N}}{N}}$$

ESTIMATED POPULATION VARIANCE

$$s_X^2 = \frac{\Sigma X^2 - \frac{(\Sigma X)^2}{N}}{N - 1}$$

ESTIMATED POPULATION STANDARD DEVIATION

$$s_X = \sqrt{\frac{\Sigma X^2 - \frac{(\Sigma X)^2}{N}}{N - 1}}$$

Learning Outcomes and Objectives

Quick Practice

For the scores 2, 4, 5, 6, 6, 7:

1. What is $(\Sigma X)^2$?
2. What is ΣX^2?
3. What is the variance?
4. What is the standard deviation?

4.5 The Population Variance and the Population Standard Deviation

Know how to find the *estimated population variance* and the *estimated population standard deviation*

The symbols s_X and s_X^2 refer to the *estimated population standard deviation* and *variance*, respectively. When computing them, the final division involves $N - 1$.

Quick Practice

1. The symbols for the biased population estimators of the variance and standard deviation are _____ and _____.
2. The symbols for the unbiased population estimators of the variance and standard deviation are _____ and _____.
3. When do you compute the unbiased estimators?
4. When do you compute the biased estimators?
5. Compute the estimated population variance and standard deviation for the scores 1, 2, 3, 4, 4, 5.

4.6 Summary of the Variance and Standard Deviation

The usual way to summarize data is to envision a normal distribution; then compute the mean to describe its location and the standard deviation to describe its variability or spread.

4.7 Statistics in the Research Literature: Reporting Means and Variability

QUICK PRACTICE ANSWERS

4.3: 1. S_X^2; 2. S_X; 3. The standard deviation is the square root of the variance; 4. less, closer; 5. 8, 12

4.4: 1. $(30)^2 = 900$; 2. $2^2 + 4^2 + 5^2 + 6^2 + 6^2 + 7^2 = 166$;

$$S_X^2 = \frac{166 - \frac{900}{6}}{6} = 2.667; 4.\ S_X = \sqrt{2.667} = 1.63$$

3.

4.5: 1. S_X^2, S_X; 2. s_X^2, s_X; 3. To estimate the population standard deviation and variance; 4. To describe the sample standard deviation and variance;

$$s_X^2 = \frac{75 - \frac{(21)^2}{7}}{6} = 2.00; s_X = \sqrt{2.00} = 1.41$$

5.

Review and Application Questions

1. (a) In any research, what three characteristics of a distribution must the researcher describe? (b) Why is it important to describe the variability in any set of scores?

2. What do measures of variability communicate about (a) the size of differences among the scores in a distribution and (b) how consistently the participants behaved?

3. (a) What is the range? (b) Why is it not the most accurate measure of variability? (c) When is it used as the sole measure of variability?

4. (a) What do both the variance and the standard deviation tell you about a distribution? (b) Which measure will you usually want to compute? Why?

5. (a) What is the mathematical definition of the variance? (b) Mathematically, how is a sample's variance related to its standard deviation, and vice versa?

6. (a) What do S_X, s_X, and σ_X have in common in terms of what they communicate? (b) How do they differ in terms of their use?

7. Why are your estimates of the population variance and standard deviation always larger than the corresponding statistics that describe a sample from that population?

8. In an experiment what does the size of S_X in each condition indicate?

9. A researcher obtained the following creativity scores.

3	2	1	0	7	4	8	6	6	4

In terms of creativity, interpret the variability in this sample using the following: (a) the range; (b) the variance; (c) the standard deviation.

10. If you could test the entire population in question 9, what would you expect each of the following to be? (a) the typical, most common creativity score; (b) the variance; (c) the standard deviation; (d) the two scores between which about 68% of all creativity scores occur in this situation

11. Say the sample in question 9 had an N of 1000. About how many people would you expect to score below 1.59? Why?

12. As part of studying the relationship between mental and physical health, you obtain the following heart rates.

73	72	67	74	78	84	79	71	76	76	79	81	75	80	78	76	78

In terms of differences in heart rates, interpret these data using the following: (a) the range; (b) the variance; (c) the standard deviation.

13. If you could test the population in question 12, what would you expect each of the following to be? (a) the shape of the distribution; (b) the typical, most common heart rate; (c) the variance; (d) the standard deviation; (e) the two scores between which about 68% of all heart rates fall

14. Alfred has a normal distribution of scores ranging from 2 to 9. (a) He computed the variance to be −.06. What should you conclude about this answer, and why? (b) He recomputes the standard deviation to be 18. What should you conclude, and why? (c) He recomputes the standard deviation to be 1.50. What should you conclude and why?

SPSS Instructions

Summarizing Scores with Measures of Variability

Range

Main Menu:	Analyze
Select:	Descriptive Statistics → Descriptives
Move:	Desired variable(s) into Variable(s) box
Select:	Options... → ✓ Range (under Dispersion) → Continue → OK

Variance

Main Menu:	Analyze
Select:	Descriptive Statistics → Descriptives
Move:	Desired variable(s) into Variable(s) box
Select:	Options... → ✓ Variance (under Dispersion) → Continue → OK

Notes: Calculates the estimated population variance

Standard Deviation

Main Menu:	Analyze
Select:	Descriptive Statistics → Descriptives
Move:	Desired variable(s) into Variable(s) box
Select:	Options... → ✓ Std. deviation (under Dispersion) → Continue → OK

Notes: Calculates the estimated population standard deviation

15. From his statistics grades, Guchi has a $\overline{X}$ of 60 and $S_x = 20$. Pluto has a $\overline{X}$ of 60 and $S_x = 5$. (a) Who is the more inconsistent student, and why? (b) Who is more accurately described as a 60 student, and why? (c) For which student can you more accurately predict the next test score, and why? (d) Who is more likely to do either extremely well or extremely poorly on the next exam?

16. You correctly compute the variance of a distribution to be $S_X^2 = 0$. What should you conclude about this distribution?

17. Consider the results of this experiment.

Condition A	Condition B	Condition C
12	33	47
11	33	48
11	34	49
10	31	48

(a) What types of measures should you compute to summarize these data? (b) These are ratio scores. Compute the appropriate descriptive statistics and summarize the relationship in the sample data. (c) Describe how consistent participants were in each condition.

18. Say that you conducted the previous experiment on the entire population. (a) Summarize the relationship you'd expect to observe. (b) Describe how consistently you'd expect participants to behave in each condition.

19. In two studies the mean is 40, but in Study A the S_x is 5, and in Study B the S_x is 10. (a) What is the difference in the distributions from these studies? (b) Where do you expect the majority of scores to fall in each study?

20. Consider these ratio scores from an experiment.

Condition 1	Condition 2	Condition 3
18	8	3
13	11	9
9	6	5

(a) What should you do to summarize the experiment? (b) Summarize the relationship in the sample data. (c) Describe how consistent the scores were in each condition.

21. Say that you conducted the experiment in question 20 on the entire population. (a) Summarize the relationship you'd expect to observe. (b) Describe how consistently you'd expect participants to behave in each condition.

Learning Outcomes and Objectives

Key Terms

5.1 Understanding z-Scores

Explain what a z-score is and what it tells you about a raw score's relative standing

The *relative standing* of a score reflects a systematic evaluation of the score relative to a sample or population. A *z-score* indicates a score's relative standing by indicating the distance the score is from the mean when measured in standard deviations. A positive z-score indicates the raw score is above the mean; a negative z-score indicates the raw score is below the mean. The larger the absolute value of z, the farther the raw score is from the mean, so the less frequently the z-score and raw score occur. z-scores are used to describe the relative standing of raw scores, to compare raw scores from different variables, and to determine the relative frequency of raw scores.

Quick Practice

With $\bar{X} = 50$, $S_X = 10$,

1. What is z for $X = 44$?

2. What X produces $z = -1.30$?

With $\bar{X} = 100$, $S_X = 16$,

3. What is the z for a score of 132?

4. What X produces $z = +1.4$?

5.2 Using the z-Distribution to Interpret Scores

A *z-distribution* is produced by transforming all raw scores in a distribution into z-scores.

5.3 Using the z-Distribution to Compare Different Variables

z-scores can be used to compare scores across two or more variables.

5.4 Using the z-Distribution to Compute Relative Frequency

Use the *standard normal curve* with z-scores to determine relative frequency, simple frequency, and percentile

The *standard normal curve* is a perfect normal z-distribution that is our model of any approximately normal z-score distribution. Any proportion of the area under a part of the curve (found in the z-table) equals the expected relative frequency of the z-scores—and their corresponding raw scores—in that part of the curve.

To find the relative frequency of scores above or below a raw score, transform it into a *z-score*. From the z-table, find the proportion of the area under the curve above or below that z. To find the raw score at a specified relative frequency, find the proportion in the z-table and transform the corresponding z into its raw score.

relative standing A description of a particular score derived from a systematic evaluation of the score using the characteristics of the sample or population in which it occurs

z-score The statistic that indicates the distance a score is from its mean when measured in standard deviation units

z-distribution The distribution of z-scores produced by transforming all raw scores in a distribution into z-scores

standard normal curve A theoretical perfect normal curve, which serves as a model of any approximately normal z-distribution

sampling distribution of means A frequency distribution showing all possible sample means that occur when samples of a particular size are drawn from a population

central limit theorem A statistical principle that defines the mean, standard deviation, and shape of a theoretical sampling distribution

standard error of the mean $(\sigma_{\bar{X}})$ The standard deviation of the sampling distribution of means

Learning Outcomes and Objectives

TRANSFORMING A RAW SCORE IN A SAMPLE INTO A z-SCORE

$$z = \frac{X - \overline{X}}{S_X}$$

TRANSFORMING A RAW SCORE IN A POPULATION INTO A z-SCORE

$$z = \frac{X - \mu}{\sigma_X}$$

TRANSFORMING A z-SCORE IN A SAMPLE INTO A RAW SCORE

$$X = (z)(S_X) + \overline{X}$$

TRANSFORMING A z-SCORE IN A POPULATION INTO A RAW SCORE

$$X = (z)(\sigma_X) + \mu$$

THE TRUE STANDARD ERROR OF THE MEAN

$$\sigma_{\overline{X}} = \frac{\sigma_X}{\sqrt{N}}$$

TRANSFORMING A SAMPLE MEAN INTO A z-SCORE

$$z = \frac{\overline{X} - \mu}{\sigma_{\overline{X}}}$$

Quick Practice

For a sample: $\overline{X} = 65$, $S_X = 12$, and $N = 1000$.

1. What is the relative frequency of scores below 59?
2. What is the relative frequency of scores above 75?
3. How many scores are between the mean and 70?
4. What raw score delineates the top 3%?

5.5 Using z-Scores to Describe Sample Means

Explain the characteristics of a *sampling distribution of means* and what the *standard error of the mean* is * Use a sampling distribution of means with z-scores to determine the relative frequency of sample means

A *sampling distribution of means* is the distribution of all possible sample means that occur when samples of a particular size N are selected from a particular raw score population. The *central limit theorem* shows that in a sampling distribution of means (a) the distribution will be approximately normal, (b) the mean of the sampling distribution will equal the mean of the underlying raw score population, and (c) the variability of the sample means is related to the variability of the raw scores. The true *standard error of the mean* ($\sigma_{\overline{X}}$) is the standard deviation of the sampling distribution of means. The location of a sample mean on the sampling distribution of means can be described by calculating a z-score. Then the standard normal curve model can be applied to the sampling distribution to determine the expected relative frequency of the sample means above or below that z-score.

Quick Practice

In a population of raw scores, $\mu = 75$, $\sigma_X = 22$, $N = 100$, and $\overline{X} = 80$.

1. The μ of the sampling distribution here equals _____.
2. The symbol for the standard error of the mean is _____ and here it equals _____.
3. The z-score for a sample mean of 80 is _____.
4. How often will sample means between 75 and 80 occur in this situation?

QUICK PRACTICE ANSWERS

5.1: 1. $z = (44 - 50)/10 = -.60$; 2. $X = (-1.30)(10) + 50 = 37$;

3. $z = (132 - 100)/16 = +2.00$; 4. $X = (+1.40)(16) + 100 = 122.4$

5.4: 1. $z = (59 - 65) / 12 = -.50$; "below" is the lower tail, so from column C is .3085; 2. $z = (75 - 65) / 12 = +.83$; "above" is the upper tail, so from column C is .2033; 3. $z = (70 - 65) / 12 = +.42$; from column B is .1628; (.1628)(1000) gives about 163 scores; 4. The "top" is the upper tail, so from column C the proportion closest to .03 is .0301, with $z = +1.88$; so $X = (+1.88)(12) + 65 = 87.56$.

5.5: 1. 75; 2. $\sigma_{\overline{X}} = \dfrac{22}{\sqrt{100}} = 2.20$; 3. $z = \dfrac{(80 - 75)}{2.20} = +2.27$;

4. From column B: .4884 of the time.

Review and Application Questions

1. (a) What does a z-score indicate? (b) Why are z-scores important?

2. On what factors does the size of a z-score depend?

3. What is a z-distribution?

4. What are the four uses of z-scores?

5. Why are z-scores referred to as standard scores?

6. Why is using z-scores and the standard normal curve model so useful?

7. (a) What is the standard normal curve? (b) How is it applied to a set of data? (c) What three criteria should be met for the model to give an accurate description of a sample?

8. (a) What is a sampling distribution of means? (b) When is it used? (c) Why is it useful?

9. (a) What three things does the central limit theorem tell us about the sampling distribution of means? (b) Why is this useful?

10. What does the standard error of the mean indicate?

11. Levi received a grade of 55 on a biology test ($\overline{X} = 50$) and a grade of 45 on a philosophy test ($\overline{X} = 50$). He is considering whether to ask his two professors to curve the grades using z-scores. (a) What other information should he consider before making his request? (b) Does he want the S_x to be large or small in biology? Why? (c) Does he want the S_x to be large or small in philosophy? Why?

12. In freshman English, Pat earned a 76 ($\overline{X} = 85$, $S_x = 10$). In a different class, Sam earned a 60 ($\overline{X} = 50$, $S_x = 4$). Should Pat be bragging about doing better than Sam? Why?

13. For the data 9, 5, 10, 7, 9, 10, 11, 8, 12, 7, 6, 9: (a) compute the z-score for the raw score of 10. (b) Compute the z-score for the raw score of 6.

14. For the data in question 13 find the raw scores that correspond to the following:
    (a) $z = +1.22$;
    (b) $z = -.48$

15. Which z-score in each of the following pairs corresponds to the smaller raw score?
    (a) $z = +1.0$ or $z = +2.3$;
    (b) $z = -2.8$ or $z = -1.7$;
    (c) $z = -.70$ or $z = +.20$;
    (d) $z = .0$ or $z = -2.0$

16. For each pair in question 15, which z-score has the higher frequency?

17. In a normal distribution, what proportion of scores are expected to fall into each of the following areas? (a) between the mean and $z = +1.89$; (b) below $z = -2.30$; (c) between $z = -1.25$ and $z = +2.75$; (d) above $z = +1.96$ and below -1.96

18. For a distribution in which $\overline{X} = 100$, $S_x = 16$, and $N = 500$: (a) What is the relative frequency of scores between 76 and the mean? (b) How many participants are expected to score between 76 and the mean? (c) What is the percentile of someone scoring 76?

SPSS Instructions

Describing Data with z-Scores and the Normal Curve

z-Scores

Main Menu: *Analyze*

Select:	**Descriptive Statistics → Descriptives**
Move:	**Desired variable(s) into Variable(s) box**
Select:	**→ Save standardized values as variables → OK**

Notes: At least one descriptive statistic must be selected under Options... (see Chapter 4). This procedure does not generate output as usual, but generates a new variable in the data set containing the standardized scores for each value of the designated variable.

Normal Curves

Main Menu: *Analyze*

Select:	**Descriptive Statistics → Frequencies**
Move:	**Desired variable(s) into Variable(s) box**
Select:	**Charts... → ✓ Histograms ✓ Show normal curve on histogram → Continue → OK**

Notes: Allows you to compare your distribution to that of a normal curve.

19. Homer may be classified as having a math dysfunction—and not have to take statistics—if he scores below the 25th percentile on a diagnostic test. The μ of the test is 75 ($\sigma_x = 10$). Approximately what raw score is the cutoff score for avoiding statistics?

20. A researcher obtained a sample mean of 68.4 with a sample of 49 participants. In the population the mean is 65 (and $\sigma_x = 10$). How often can he expect to obtain a sample mean that is higher than 68.4?

21. A recent graduate has two job offers and must decide which to accept. The job in City A pays $57,000. The average cost of living there is $70,000, with a standard deviation of $15,000. The job in City B pays $42,000. The average cost of living there is $48,000, with a standard deviation of $12,000. Which job offer should she accept? Why?

22. You own shares of a company's stock and, over the past ten trading days, its mean selling price is $14.89. Over the years the mean price has been $10.43 ($\sigma_x = 5.60). Can the mean selling price over the next ten days be expected to go higher? Should you wait to sell, or should you sell now?

Learning Outcomes and Objectives

Key Terms

6.1 Understanding Probability

Explain what *probability* communicates

Probability (p) indicates the likelihood of an event when random chance is operating. The probability of an event is equal to its relative frequency in the population.

6.2 Probability Distributions

Understand how probability is an event's relative frequency in the population

A *probability distribution* is a model of the relative frequencies of all events in a population when random chance is operating.

Quick Practice

1. The probability of any event equals its _____ in the _____.

2. The *p* of an event decreases as the event's relative frequency _____, so our confidence that the event will occur _____.

3. A _____ shows the probabilities for all events in a population.

6.3 Obtaining Probability from the Standard Normal Curve

Compute the probability of sample means using *z*-scores and the standard normal curve

To find the probability of particular sample means we envision the sampling distribution, compute, and apply the *z*-table. The farther into the tail of the sampling distribution a sample mean falls, the less likely it is to occur when sampling from the underlying raw score population.

Quick Practice

1. What is the probability of selecting a sample mean having a positive *z*-score?

2. Approximately, what is the probability of selecting an SAT sample mean having a *z*-score between ± 1?

3. If, for some raw scores with $\mu = 100$, are we more likely to obtain a sample mean that is close to 100 or a mean that is very different from 100?

4. Which has the lower probability, a sample mean with $z = +1.75$ or a sample mean with $z = -1.82$?

6.4 Random Sampling and Sampling Error

Explain what sampling error is and how random sampling can produce representative or unrepresentative samples

In a *representative sample* the individuals and scores in the sample accurately reflect those found in the population. *Sampling error* results when chance produces a sample statistic, such as $\overline{X}$, that is different from the population parameter, μ, that it represents.

random sampling A method of selecting samples so that all members of the population have the same chance of being selected for a sample

probability (*p*) The likelihood of an event when a particular population is randomly sampled; equal to the event's relative frequency in the population

probability distribution The probability of every possible event in a population, derived from the relative frequency of every possible event in that population

representative sample A sample whose characteristics accurately reflect those of the population

sampling error The difference, due to random chance, between a sample statistic and the population parameter it represents

region of rejection That portion of a sampling distribution containing values considered too unlikely to occur by chance, found in the tail or tails of the distribution

criterion The probability that defines whether a sample is unlikely to have occurred by chance and thus is unrepresentative of a particular population

critical value The score that marks the inner edge of the region of rejection in a sampling distribution; values that fall beyond it lie in the region of rejection

Key Formulas

TRANSFORMING A SAMPLE MEAN INTO A z-SCORE

$$z = \frac{\overline{X} - \mu}{\sigma_{\overline{X}}}$$

where

$$\sigma_{\overline{X}} = \frac{\sigma_X}{\sqrt{N}}$$

When the criterion is .05, the critical value of z is 1.96 when using two tails, and it is 1.645 when using one tail.

Learning Outcomes and Objectives

6.5 Deciding Whether a Sample Represents a Population

Explain how to use a sampling distribution of means to determine whether a sample is likely to represent a particular population

If the z-score shows that a sample mean is *unlikely* in the sampling distribution, *reject* that the sample is merely poorly representing the underlying raw score population. If the z-score shows that a sample mean is *likely* in the sampling distribution, *retain* that the sample is representing the underlying raw score population, albeit somewhat poorly. To decide if a sample represents a particular raw score population, compute the sample mean's z-score and compare it to the critical value on the sampling distribution.

Quick Practice

1. _____ communicates that a sample mean is different from the μ it represents.

2. Sampling error occurs because of _____.

3. A sample mean has a $z = +1$ on the sampling distribution created from the population of psychology majors. Is this likely to be a sample of psychology majors?

4. A sample mean has a $z = -4.0$ on the previous sampling distribution. Is this likely to be a sample of psychology majors?

5. The _____ contains those samples considered to be **likely/unlikely** to represent the underlying population.

6. The _____ defines the z-score that is required for a sample to be in the region of rejection.

7. For a sample to be in the region of rejection, its z-score must be **smaller/ larger** than the critical value.

8. On a test, $\mu = 60$ and $\sigma_X = 18$. A sample ($N = 100$) produces $\overline{X} = 65$. Using the .05 criterion, does this sample represent this population?

reviewcard

CHAPTER 6
USING PROBABILITY TO MAKE DECISIONS ABOUT DATA

Review and Application Questions

1. (a) What does a probability convey about the occurrence of a random event? (b) What is the probability of a random event based on?

2. Why is the proportion of the total area under the normal curve equal to probability?

3. What is a random sampling?

4. What does the term *sampling error* convey about a sample mean?

5. A sample produces a mean that is different from the μ of the population that we think the sample may represent. What are the two possible reasons for this difference?

6. What is (a) the criterion? (b) the region of rejection? (c) the critical value?

7. What does comparing the critical value to a sample's z-score indicate?

8. What is the difference between using both tails versus one tail of the sampling distribution in terms of (a) the location of the region of rejection and (b) the critical value?

9. A couple with eight daughters decides to have one more baby, because they think the next one is bound to be a boy! Is this reasoning accurate?

10. Eleanor read in the newspaper that there is a .05% chance of swallowing a spider while you sleep. She subsequently developed insomnia. (a) What is the probability of swallowing a spider? (b) Why isn't her insomnia justified? (c) Why is her insomnia justified?

11. Walter's uncle is building a house on land that has been devastated by hurricanes 160 times in the past 200 years. Because there hasn't been a major storm there in 13 years, his uncle says this is a safe investment. His nephew argues that he is wrong, because a hurricane must be due soon. What are the fallacies in the reasoning of both men?

12. Four airplanes from different airlines have crashed in the past two weeks. This terrifies Norma. Her travel agent claims that the probability of a plane crash is minuscule. Who is correctly interpreting the situation? Why?

13. Based on a survey of a random sample of college students, Terry concludes that the favorite beverage in the population of college students is sauerkraut juice! What is the statistical argument for not accepting this conclusion?

14. (a) Why does random sampling produce representative samples? (b) Why does random sampling produce unrepresentative samples?

15. The mean of a population of raw scores is 18 ($\sigma_x = 12$). With $N = 30$, you obtained a sample mean of 24. Using the .05 criterion with the region of rejection in both tails of the sampling distribution, should you consider the sample to be representative of the population with $\mu = 18$? Why?

16. Madge computes the $\overline{X}$ from data that her professor says is a random sample from population Q. This sample mean has a z-score of $+41$ on the sampling distribution for population Q. Madge claims she has *proven* that this could not be a random sample from population Q. Do you agree or disagree? Why?

SPSS Instructions

There are no SPSS instructions for Chapters 6 or 7. Your next set of instructions is on the Chapter 8 review cards.

17. While conducting a study, you obtain the following data representing the aggressive tendencies of some football players:

40	30	39	40	41	39	31	28	33

(a) Researchers have found that μ is 30 in the population of non–football players ($\sigma_x = 5$). Using both tails of the sampling distribution, determine whether your football players represent a different population. (b) Using your sample, what do you conclude about the population of football players and its μ?

18. You obtain a sample mean of 45 ($N = 40$). Using the .05 criterion with the region of rejection in only the lower tail of the distribution, is the sample representative of the population in which $\mu = 50$ ($\sigma_x = 18$)? Why?

19. (a) Why do we conclude that a low-probability sample does not represent a particular population? (b) Why do we conclude that a high-probability sample does represent a particular population?

20. On a test of motor coordination, the population of average bowlers has a mean score of 24, with a standard deviation of 6. A random sample of 30 bowlers at Fred's Bowling Alley has a sample mean of 26. A second random sample of 30 bowlers at Ethel's Bowling Alley has a mean of 18. Using the criterion of $p = .05$ and both tails of the sampling distribution, what should we conclude about each sample's representativeness of the population of average bowlers?

21. (a) In question 20, if a particular sample does not represent the population of average bowlers, what is your best estimate of the μ of the population it does represent? (b) Explain the logic behind this conclusion.

Learning Outcomes and Objectives

7.1 The Role of Inferential Statistics in Research

Understand why the possibility of sampling error leads us to perform inferential statistical procedures

Inferential statistics are procedures for deciding whether sample data represent a particular relationship in the population. *Parametric statistics* are inferential procedures that require assumptions about the raw score populations the data represent. *Nonparametric statistics* are inferential procedures that do not require such assumptions.

7.2 Setting Up Inferential Procedures

Know when *experimental hypotheses* lead to either a *one-tailed* or a *two-tailed* test * Understand how to create the *null* and *alternative* hypotheses

The *alternative hypothesis* (H_a) describes the population μ being represented by our sample mean if the predicted relationship exists. The *null hypothesis* (H_0) describes the population μ being represented by our sample mean if the predicted relationship does not exist. A *two-tailed test* is used when we do not predict the direction in which the dependent scores will change. A *one-tailed test* is used when the direction of the relationship is predicted.

Quick Practice

1. A _____ test is used when we do *not* predict the direction that scores will change; a _____ test is used when we *do* predict the direction that scores will change.

2. The _____ hypothesis is that the sample data represent a population where the predicted relationship exists. The _____ hypothesis is that the sample data represent a population where the predicted relationship does not exist.

3. The μ for adults on a personality test is 140. We test a sample of children to see if they are different from adults. What are H_a and H_0?

4. The μ for days absent among workers is 15.6. We train a sample of new workers and ask whether the training changes worker absenteeism. What are H_a and H_0?

7.3 Performing the z-Test

Know when and how to perform the z-test

The z-*test* is the parametric procedure used in a one-sample experiment if (a) the population contains normally distributed, interval or ratio dependent scores, and (b) the standard deviation of the population (σ_X) is *known*.

7.4 Interpreting Significant and Nonsignificant Results

Know how to interpret *significant* results * Know how to interpret *nonsignificant* results

If z_{obt} lies beyond z_{crit}, reject H_0, say the results are significant, and conclude there is evidence for the predicted relationship. Otherwise, the results are not significant, and you should make no conclusion about the relationship.

Key Terms

inferential statistics Procedures for determining whether sample data represent a particular relationship in the population

parametric statistics Inferential procedures that require certain assumptions about the raw score population represented by the sample data; used when we compute the mean of the scores

nonparametric statistics Inferential procedures that do not require stringent assumptions about the raw score population represented by the sample data

experimental hypotheses Two statements made before a study is begun, describing the predicted relationship that may or may not be demonstrated by the study

two-tailed test The test used to evaluate a statistical hypothesis that predicts a relationship but not whether scores will increase or decrease

one-tailed test The test used to evaluate a statistical hypothesis that predicts that scores will only increase or only decrease

statistical hypotheses Two statements that describe the population parameters the sample statistics will represent if the predicted relationship exists or does not exist

alternative hypothesis (H_a) The hypothesis describing the population parameters that the sample data represent if the predicted relationship does exist

null hypothesis (H_0) The hypothesis describing the population parameters that the sample data represent if the predicted relationship does not exist

z-test The parametric procedure used to test the null hypothesis for a single-sample experiment when the true standard deviation of the raw score population is known

alpha (α) The Greek letter that symbolizes the criterion probability

significant Describes results that are too unlikely to accept as resulting from sampling error when the predicted relationship does not exist; it indicates rejection of the null hypothesis

nonsignificant Describes results that are considered likely to result from sampling error when the predicted relationship does not exist; it indicates failure to reject the null hypothesis

Type I error Deciding to reject the null hypothesis when the null hypothesis is true (that is, when the predicted relationship does not exist)

Type II error Deciding to retain the null hypothesis when the null hypothesis is false (that is, when the predicted relationship does exist)

power The probability that we will detect a true relationship and correctly reject a false null hypothesis; the probability of avoiding a Type II error

Key Formula

z-TEST FORMULA

$$z_{obt} = \frac{\overline{X} - \mu}{\sigma_{\overline{X}}} \quad \text{where} \quad \sigma_{\overline{X}} = \frac{\sigma_X}{\sqrt{N}}$$

When $\alpha = .05$, the z_{crit} is ±1.96 when using two tails or the z_{crit} is 1.645 when using one tail.

7.5 Summary of the z-Test

Quick Practice

We test whether a sample of 36 successful dieters are more or less satisfied with their appearance than in the population of nondieters, where $\mu = 40$ ($\sigma_X = 12$).

1. What are H_0 and H_a?
2. The $\overline{X}$ for dieters is 44. Compute z_{obt}.
3. Set up the sampling distribution.
4. What should we conclude?

7.6 The One-Tailed Test

Perform a one-tailed test when predicting the specific direction the scores will change. When predicting the $\overline{X}$ will be higher than μ, the region of rejection is in the upper tail of the sampling distribution. When predicting the $\overline{X}$ will be lower than μ, the region of rejection is in the lower tail.

Quick Practice

You test the effectiveness of a new weight-loss diet.

1. Why is this a one-tailed test?
2. For the population of nondieters, $\mu = 155$. What are H_a and H_0?
3. In which tail is the region of rejection?
4. With $\alpha = .05$, the z_{obt} for the sample of dieters is -1.86. What do you conclude?

7.7 Errors in Statistical Decision Making

Explain *Type I errors, Type II errors,* **and** *power*

A *Type I error* occurs when a true H_0 is rejected. Its theoretical probability equals α. The probability of avoiding a Type I error by retaining a true H_0 is $1 - \alpha$. A *Type II error* occurs when a false H_0 is retained. Avoiding a Type II error is rejecting a false H_0. *Power* is the probability of rejecting a false H_0 (avoiding a Type II error).

QUICK PRACTICE ANSWERS

7.2: 1. two-tailed; 2. alternative, null; 3. $H_a: \mu \neq 140$, $H_0: \mu = 140$;
4. $H_a: \mu \neq 15.6$, $H_0: \mu = 15.6$

7.5: 1. $H_0: \mu = 40$, $H_a: \mu \neq 40$; 2. $\sigma_{\overline{X}} = 12/\sqrt{36} = 2$,
$z_{obt} = (44 - 40)/2 = +2.00$; 3. With $\alpha = .05$, the sampling distribution has a region of rejection in each tail, with $z_{crit} = \pm 1.96$ (as in Figure 7.6);
4. The z_{obt} of $+2.00$ is beyond z_{crit} of ± 1.96, so the results are significant: The population of dieters are more satisfied (at a μ around 44) than the population of nondieters (at $\mu = 40$).

7.6: 1. Because a successful diet *lowers* weight scores; 2. $H_a: \mu < 155$, and $H_0: \mu \geq 155$; 3. The left-hand tail; 4. The z_{obt} is beyond z_{crit} of -1.645, so it is significant: The μ for dieters will be less than the μ of 155 for nondieters.

Review and Application Questions

1. Why does the possibility of sampling error present a problem to researchers when inferring a relationship in the population?

2. What are inferential statistics used for?

3. What does α stand for, and what two things does it determine?

4. (a) What does H_0 communicate? (b) What does H_a communicate?

5. What are experimental hypotheses?

6. What does the term *significant* convey about the results of an experiment?

7. (a) When do you use a one-tailed test? (b) When do you use a two-tailed test?

8. (a) What are the advantage and disadvantage of two-tailed tests? (b) What are the advantage and disadvantage of one-tailed tests?

9. (a) What is power? (b) Why do researchers want to maximize power? (c) What situation makes us worry whether we have sufficient power?

10. (a) Why is obtaining a significant result a goal of research? (b) Why is declaring the results significant not the final step in conducting research?

11. Describe the experimental hypotheses and identify the independent and dependent variables when we study: (a) whether the amount of pizza consumed by college students during finals week increases relative to the rest of the semester, (b) whether breathing exercises alter blood pressure, (c) whether sensitivity to pain is affected by increased levels of hormones, and (d) whether frequency of dreaming while sleeping decreases with more light in the room.

12. For each study in question 11, indicate whether a one- or a two-tailed test should be used, and state the H_0 and H_a. Assume that $\mu = 50$ when the amount of the independent variable is zero.

13. Listening to music while taking a test may be relaxing or distracting. To determine which, 49 participants are tested while listening to music and they produce a $\overline{X} = 54.63$. In the population without music, the μ is 50 ($\sigma_X = 12$). (a) Is this a one-tailed or two-tailed test? Why? (b) What are H_0 and H_a? (c) Compute z_{obt}. (d) With $\alpha = .05$, what is z_{crit}? (e) Do we have evidence of a relationship in the population? If so, describe the relationship.

14. A researcher asks whether attending a private school leads to higher or lower performance on a test of social skills. A sample of 100 students from a private school produces a mean score of 71.30. The μ for students from public schools is 75.62 ($\sigma_X = 28$). (a) Should she use a one-tailed or a two-tailed test? Why? (b) What are H_0 and H_a? (c) Compute z_{obt}. (d) With $\alpha = .05$, what is z_{crit}? (e) What should the researcher conclude about this relationship in the population?

15. (a) What is the probability that in question 13 we made a Type I error? What would the error be in terms of the independent and dependent variables? (b) What is the probability that we made a Type II error? What would the error be in terms of the independent and dependent variables?

16. (a) What is the probability that the researcher in question 14 made a Type I error? What would the error be in terms of the independent and dependent variables? (b) What would a Type II error be in terms of the independent and dependent variables?

SPSS Instructions

There are no SPSS instructions for Chapter 7. Your next set of instructions is on the Chapter 8 review cards.

Review and Application Questions

17. Carlo reads that Study A found $z_{obt} = +1.97$ and $p < .05$. He also reads that in Study B, $z_{obt} = +14.21$ and $p < .0001$. (a) Carlo concludes that the results of Study B are more significant than those of Study A. Why is he correct or incorrect? (b) In terms of their conclusions, what is the difference between the two studies?

18. A researcher measures the self-esteem scores of a sample of statistics students, reasoning that their frustration with this course may lower their self-esteem relative to that of the typical college student where $\mu = 55$ and $\sigma_X = 11.35$. He obtains the following scores: 44, 55, 39, 17, 27, 38, 36, 24, and 36. (a) Should he use a one-tailed or two-tailed test? Why? (b) What are H_0 and H_a for this study? (c) Compute z_{obt}. (d) With $\alpha = .05$, what is z_{crit}? (e) What should the researcher conclude about the relationship between the self-esteem of statistics students and that of other students?

19. A researcher suggests that males and females are the same when it comes to intelligence. Why is this hypothesis impossible to test?

20. Researcher A finds a significant relationship between increasing stress level and ability to concentrate. Researcher B replicates this study, but finds a nonsignificant relationship. Identify the statistical error that each researcher may have made.

21. A report indicates that Brand X toothpaste significantly reduced tooth decay relative to other brands, with $p < .44$. (a) What does "significant" indicate about the researcher's decision about Brand X? (b) What makes you suspicious of the claim that Brand X works better than other brands?

Learning Outcomes and Objectives

Key Terms

8.1 Understanding the One-Sample t-Test

Explain the difference between the z-test and the t-test * Know when to use the t-test

The one-sample t-test is for testing a one-sample experiment when the dependent scores are normally distributed interval or ratio scores and the variability of the population must be estimated.

8.2 Performing the One-Sample t-Test

Know how to perform the t-test * Know the uses of the t-distribution and degrees of freedom

Perform the one-sample t-test in a one-sample experiment when you do not know the population standard deviation. The *estimated standard error of the mean* ($s_{\bar{X}}$) is an estimate of the standard deviation of the sampling distribution. A t-*distribution* is the sampling distribution of all possible values of t when a raw score population is infinitely sampled using a particular N. The appropriate t-distribution and thus the correct t_{crit} to use for a particular study is the one identified by our *degrees of freedom* (*df*). In the one-sample t-test, $df = N - 1$.

Quick Practice

In a study, H_0 is that $\mu = 9$. The data are 6, 7, 9, 8, 8.

1. To compute the t_{obt}, what two statistics are computed first?

2. What do you compute next?

3. Compute the t_{obt}.

8.3 Interpreting the t-Test

8.4 Summary of the One-Sample t-Test

In a one-sample experiment when σ_X is unknown, perform the one-sample t-test

Quick Practice

We test if artificial sunlight during the winter *lowers* one's depression. Without artificial sunlight, a depression test has $\mu = 8$. With the light, our sample scored 4, 5, 6, 7, 8.

1. What are the hypotheses?

2. Compute t_{obt}.

3. What is t_{crit}?

4. What is the conclusion?

8.5 Estimating μ by Computing a Confidence Interval

Explain the confidence interval for μ * Compute the confidence interval for μ

In *point estimation* μ is assumed to equal $\bar{X}$. In *interval estimation* a μ is assumed to lie within a specified interval. The *confidence interval for μ* describes a range of μs, within which we are confident our μ falls. Our confidence level equals $(1 - \alpha)(100)$.

one-sample t-test The parametric procedure used to test the null hypothesis for a one-sample experiment when the standard deviation of the raw score population must be estimated

estimated standard error of the mean ($s_{\bar{X}}$) An estimate of the standard deviation of the sampling distribution of means, used in calculating the one-sample t-test

t-distribution The sampling distribution of all possible values of t that occur when samples of a particular size are selected from the raw score population described by the null hypothesis

degrees of freedom (*df*) The number of scores in a sample that reflect the variability in the population; used when estimating the population variability

point estimation A way to estimate a population parameter by describing a point on the variable at which the population parameter is expected to fall

interval estimation A way to estimate a population parameter by describing an interval within which the population parameter is expected to fall

confidence interval for μ A range of values of μ, within which we are confident that the actual μ is found

Key Formulas

THE ESTIMATED STANDARD ERROR OF THE MEAN

$$s_{\bar{X}} = \sqrt{\frac{s_X^2}{N}}$$

THE ONE-SAMPLE t-TEST

$$t_{obt} = \frac{\bar{X} - \mu}{s_{\bar{X}}}$$

DEGREES OF FREEDOM IN A ONE-SAMPLE t-TEST

$df = N - 1$, where N is the number of scores in the sample

THE CONFIDENCE INTERVAL FOR μ

$(s_{\bar{X}})(-t_{crit}) + \bar{X} \leq \mu \leq (s_{\bar{X}})(+t_{crit}) + \bar{X}$
where t_{crit} is the t-value for two tails with $df = N - 1$

Learning Outcomes and Objectives

Quick Practice

1. What does this 95% confidence indicate: $15 \leq \mu \leq 20$?

2. With $N = 22$ you perform a one-tailed t-test ($\alpha = .05$). What is the t_{crit} for then computing the confidence interval?

3. Compute the 95% confidence interval when $\bar{X} = 35$, $s_{\bar{X}} = 3.33$, and $N = 22$.

8.6 Statistics in the Research Literature: Reporting t

Know what information is included when reporting the results of the t-test

---

QUICK PRACTICE ANSWERS

8.2: 1. $\underline{X}$ and s_X^2; 2. $s_{\bar{X}}$; 3. $\bar{X} = 7.6$, $s_X^2 = 1.30$, $N = 5$; $s_{\bar{X}} = \sqrt{1.3/5}$; $t_{obt} = (7.6 - 9)/.51 = -2.75$

8.4: 1. This is a one-tailed test: $H_a: \mu < 8$; $H_0: \mu \geq 8$; 2. $\bar{X} = 6$, $s_X^2 = 6$, $s_{\bar{X}} = 2.5$, $s_{\bar{X}} = \sqrt{2.5/5} = .707$, $t_{obt} = (6 - 8)/.707 = -2.83$; 3. With $\alpha = .05$ and $df = 4$, $t_{crit} = -2.132$; 4. t_{obt} is beyond t_{crit}: Artificial sunlight significantly lowers depression scores from a μ of 8 to a μ around 6.

8.5: 1. We are 95% confident that our $\bar{X}$ represents a μ between 15 and 20. 2. With $df = 21$, the two-tailed $t_{crit} = \pm 2.080$; 3. (3.33)(–2.080) + 35 = 28.07 $\leq \mu \leq$ (3.33)(+2.080) + 35 = 41.93.

Review and Application Questions

1. A scientist has conducted a one-sample experiment. (a) What two parametric procedures are available to her? (b) What is the deciding factor for selecting between them?

2. What are the assumptions of the t-test?

3. (a) What is the difference between $s_{\bar{X}}$ and $\sigma_{\bar{X}}$? (b) How is their use the same?

4. Why are there different values of t_{crit} when samples have different dfs?

5. (a) What does df symbolize? (b) Why must you compute df when performing the t-test? (c) What does df equal in the one-sample t-test?

6. Summarize the steps involved in analyzing the results of a one-sample experiment.

7. What is the final step in examining the data in any study?

8. Say you have a sample mean of 44 in a study. (a) Estimate the corresponding μ using point estimation. (b) Why is computing a confidence interval better than using a point estimate?

9. What does a confidence interval for μ indicate?

10. (a) Is the one-tailed or two-tailed t_{crit} used to compute a confidence interval? (b) Why?

11. You wish to determine whether using this textbook is beneficial or detrimental to students learning statistics. On a national statistics exam, $\mu = 68.5$ for those not using this book. A random sample of students who used this book scored as follows:

64	69	92	77	71	99	82	74	69	88

(a) What are H_0 and H_a here? (b) Compute t_{obt}. (c) With $\alpha = .05$, what is t_{crit}? (d) What do you conclude about the use of this book? (e) Compute the confidence interval for μ.

12. A researcher predicts that smoking cigarettes decreases a person's sense of smell. On a test of olfactory sensitivity, the μ for nonsmokers is 18.4. People who smoke a pack a day produced the following scores:

16	14	19	17	16	18	17	15	18	9	12	14

(a) What are H_0 and H_a? (b) Compute t_{obt}. (c) With $\alpha = .05$, what is t_{crit}? (d) What should the researcher conclude about this relationship? (e) Compute the confidence interval for μ.

13. Hattie conducts a study to determine if hearing an argument in favor of an issue alters participants' attitudes toward the issue one way or the other. She presents a 30-second speech in favor of an issue to 8 people. In a national survey, the mean attitude score toward this issue was $\mu = 50$. She obtains the following scores:

10	33	86	55	67	60	44	71

(a) What are H_0 and H_a? (b) Compute t_{obt}. (c) With $\alpha = .05$, what is t_{crit}? (d) What are the statistical results? (e) If appropriate, compute the confidence interval for μ. (f) What conclusions should Hattie draw about the relationship?

SPSS Instructions

One-Sample t-Test

Main Menu: Analyze

Select:	**Compare Means → One-Sample t-Test**
Move:	**Desired dependent variable into Test Variable(s) box**
Enter:	**The population mean value in the Test Value box**
Optional:	**Options... → Confidence Interval Percentage (default is a 95% confidence interval) → Continue**
Select:	**OK**

Notes: (1) You must enter the value for the population mean against which you wish to test into the Test Value box.

(2) Significance level given is for a two-tailed test. If a one-tailed test is desired, you will need to compare the calculated t-test against a one-tailed critical t-value from the t-table using the correct degrees of freedom.

14. For the study in question 13, (a) What statistical principle should Hattie be concerned with? (b) Why?

15. We ask whether a computer word-processing program leads to more or fewer grammatical errors. On a typing test performed without a computer word-processing program, $\mu = 12$. A sample using a computer word-processing program scores 8, 12, 10, 9, 6, 7. (a) Perform the t-test and draw the appropriate conclusion. (b) Compute the confidence interval.

16. While reading a published research report, you encounter the following statement. Identify the N, the procedure performed and its outcome, the relationship, and the type of error possibly being made. "When we examined the perceptual skills data, the mean of 55 for the sample of adolescents differed significantly from the population mean of 70 for adults, $t(45) = 3.76, p < .01$."

17. In a one-tailed test ($\alpha = .05$), our $N = 33$. Is t_{obt} significant if it equals: (a) $+1.61$? Why? (b) $+1.785$? Why?

18. Buford performed a two-tailed experiment in which $N = 20$. He couldn't find his t-table, but somehow he remembered the t_{crit} at $df = 10$. He decided to compare his t_{obt} to this t_{crit}. Why is this a correct or incorrect approach? (*Hint:* Consider whether t_{obt} turns out to be significant or nonsignificant at this t_{crit}.)

19. (a) How would you report your results if $\alpha = .01$, $N = 43$, and the $t_{obt} = +6.72$ is significant? (b) How would you report your results if $\alpha = .05$, $N = 6$, and the $t_{obt} = -1.72$ is not significant?

20. A research report indicates that for a t-test, $p = .03$. What does this indicate?

Learning Outcomes and Objectives

9.1 Understanding the Two-Sample Experiment

Explain the logic of a two-sample experiment

9.2 The Independent-Samples t-Test

Know when and how to perform the independent-samples t-test

Two samples are *independent* when participants are randomly selected for a sample without regard to who is selected for the other sample. The independent-samples t-test requires (a) independent samples, (b) normally distributed interval or ratio scores, and (c) homogeneous variance. *Homogeneity of variance* means that the σ_X^2 in the populations being represented are equal. A significant t_{obt} from the independent-samples t-test indicates that the difference between $\overline{X}_1$ and $\overline{X}_2$ is unlikely to represent the difference between μ_1 and μ_2 described by H_0. To perform the test, compute $\overline{X}_1$, s_1^2, and n_1; $\overline{X}_2$, s_2^2, and n_2. Then compute the pooled variance (s_{pool}^2). Then compute the standard error of the difference ($s_{\overline{X}_1 - \overline{X}_2}$). Then compute t_{obt}.

Quick Practice

After testing two conditions, $\overline{X}_2 = 33$, $s_1^2 = 16$, $n_1 = 21$, $\overline{X}_2 = 27$, $s_2^2 = 13$, $n_2 = 21$.

1. Compute the pooled variance (s_{pool}^2).

2. Compute the standard error of the difference ($s_{\overline{X}_1 - \overline{X}_2}$).

3. Compute t_{obt}.

9.3 Summary of the Independent-Samples t-Test

Perform the independent-samples t-test in experiments that test two independent conditions

Quick Practice

To test whether "cramming" for an exam is beneficial or harmful to grades, Condition 1 crams for a pretend exam but Condition 2 does not. Each $n = 31$, the cramming $\overline{X}$ is 43 ($s_X^2 = 64$), and the no-cramming $\overline{X}$ is 48 ($s_X^2 = 83.6$).

1. What are H_0 and H_a?

2. Compute t_{obt}.

3. What do you conclude about this relationship?

4. If we predicted cramming would lower grades, what are H_0 and H_a if we subtract cramming from no cramming?

5. Would the t_{crit} be positive or negative?

9.4 The Related-Samples t-Test

Know when and how to perform the related-samples t-test * Explain the difference between independent samples and related samples

Two samples are *related* when each participant in one condition is *matched* with someone in the other condition, or when we have *repeated measures* of the same participants under both conditions. The *related-samples* t-test is a one-sample t-test performed on the difference scores. A significant t_{obt} indicates that the mean of the difference scores ($\overline{D}$) is significantly different from the μ_D described by H_0. Therefore, the means of the raw scores in the conditions also differ significantly.

Key Terms

independent-samples t-test The parametric procedure used for significance testing of sample means from two independent samples

independent samples Samples created by selecting each participant for one condition, without regard to the participants selected for any other condition

homogeneity of variance A characteristic of data describing populations represented by samples in a study that have the same variance

sampling distribution of differences between means A frequency distribution showing all possible differences between two means that occur when two independent samples of a particular size are drawn from the population of scores described by the null hypothesis

pooled variance (s_{pool}^2) The weighted average of the sample variances in a two-sample experiment

standard error of the difference ($s_{\overline{X}_1 - \overline{X}_2}$) The estimated standard deviation of the sampling distribution of differences between the means of independent samples in a two-sample experiment

related-samples t-test The parametric procedure used for significance testing of sample means from two related samples

related samples Samples created by matching each participant in one condition with a participant in the other condition or by repeatedly measuring the same participant under all conditions

matched-samples design An experiment in which each participant in one condition is matched on an extraneous variable with a participant in the other conditions

repeated-measures design
A related-samples design in which the same subjects are measured repeatedly under all conditions of an independent variable

standard error of the mean difference ($s_{\bar{D}}$) The standard deviation of the sampling distribution of mean differences between related samples in a two-sample experiment

effect size An indicator of the amount of influence that changing the conditions of the independent variable had on dependent scores

Cohen's *d* A measure of effect size in a two-sample experiment that reflects the magnitude of the differences between the means of the conditions, relative to the variability of the scores

proportion of variance accounted for In experiments, the proportion of all differences in dependent scores that is associated with changing the independent variable

squared point-biserial correlation coefficient (r^2_{pb}) Indicates the proportion of variance in dependent scores that is accounted for by the independent variable in a two-sample experiment

9.5 Summary of the Related-Samples *t*-Test

With a matched-groups or repeated-measures design, perform the related-samples *t*-test. Find the difference between each pair of raw scores, and then perform the one-sample *t*-test on the difference scores.

Quick Practice

In a two-tailed study, we test the same participants in both Conditions A and B, with the data to the right:

A	B
8	7
10	5
9	6
8	5
11	6

1. This way of producing related samples is called _____.
2. What are H_0 and H_a?
3. Subtracting A − B, perform the *t*-test.
4. What μs do you expect in the raw score populations?
5. Subtracting A − B, what are H_0 and H_a if we predicted that B would lower scores?

9.6 Statistics in the Research Literature: Reporting a Two-Sample Study

Know what information is included when reporting the results of the *t*-test

9.7 Describing Effect Size

Explain what effect size is and how it is measured using Cohen's *d* and r^2_{pb}

Effect size indicates the amount of influence that changing conditions of the independent variable had on dependent scores. In a two-sample experiment, *Cohen's* d measures effect size as the magnitude of the difference between the conditions. The *proportion of variance accounted for* measures effect size as the proportion of all differences in dependent scores that can be attributed to changing the conditions. It is computed in a two-sample experiment using the *squared point-biserial correlation coefficient*, r^2_{pb}.

Key Formulas

CALCULATING THE VARIANCE IN A CONDITION

$$s^2_X = \frac{\Sigma X^2 - \frac{(\Sigma X)^2}{n}}{n - 1}$$

THE POOLED VARIANCE

$$s^2_{pool} = \frac{(n_1 - 1)s^2_1 + (n_2 - 1)s^2_2}{(n_1 - 1) + (n_2 - 1)}$$

THE STANDARD ERROR OF THE DIFFERENCE

$$s_{\bar{X}_1 - \bar{X}_2} = \sqrt{(s^2_{pool})\left(\frac{1}{n_1} + \frac{1}{n_2}\right)}$$

THE INDEPENDENT-SAMPLES *t*-TEST

$$t_{obt} = \frac{(\bar{X}_1 - \bar{X}_2) - (\mu_1 - \mu_2)}{s_{\bar{X}_1 - \bar{X}_2}}$$

DEGREES OF FREEDOM IN THE INDEPENDENT-SAMPLES *t*-TEST

$df = (n_1 - 1) + (n_2 - 1)$, where each *n* is the number of scores in a condition.

THE FORMULA FOR s^2_D

$$s^2_D = \frac{\Sigma D^2 - \frac{(\Sigma D)^2}{N}}{N - 1}$$

STANDARD ERROR OF THE MEAN DIFFERENCE

$$s_{\bar{D}} = \sqrt{\frac{s^2_D}{N}}$$

THE RELATED-SAMPLES *t*-TEST

$$t_{obt} = \frac{\bar{D} - \mu_D}{s_{\bar{D}}}$$

THE DEGREES OF FREEDOM IN THE RELATED-SAMPLES *t*-TEST

$df = N - 1$, where *N* is the number of difference scores.

COHEN'S *d* INDEPENDENT-SAMPLES *t*-TEST

$$d = \frac{\bar{X}_1 - \bar{X}_2}{\sqrt{s^2_{pool}}}$$

COHEN'S *d* RELATED-SAMPLES *t*-TEST

$$d = \frac{\bar{D}}{\sqrt{s^2_D}}$$

COMPUTING EFFECT SIZE IN A TWO-SAMPLE EXPERIMENT

$$r^2_{pb} = \frac{(t_{obt})^2}{(t_{obt})^2 + df}$$

Review and Application Questions

1. A scientist has conducted a two-sample experiment. (a) What two parametric procedures are available to him? (b) What is the deciding factor for selecting between them?

2. How do you create independent samples?

3. What are the two ways to create related samples?

4. How are matched samples created?

5. How is a repeated-measures study conducted?

6. In addition to independent or related samples, the two-sample t-test has what other assumptions?

7. What is homogeneity of variance?

8. What is the difference between n and N?

9. (a) What is $s_{\bar{X}_1 - \bar{X}_2}$? (b) What is $s_{\bar{D}}$?

10. (a) What does effect size indicate? (b) What does d indicate? (c) What does r^2_{pb} indicate?

11. With $\alpha = .05$ and $df = 40$, a significant independent-samples t_{obt} was $+4.55$. How would this be reported in the literature?

12. For each of the following, which type of t-test is required? (a) An investigation of the effects of a new memory-enhancing drug on the memory of Alzheimer's patients, testing a group of patients before and after administration of the drug. (b) An investigation of the effects of alcohol on motor coordination, comparing one group of participants given a moderate dose of alcohol to the population μ for people given no alcohol. (c) An investigation of whether males and females rate differently the persuasiveness of an argument delivered by a female speaker. (d) The study described in part c, but with the added requirement that for each male of a particular age, there is a female of the same age.

13. In an experiment, a researcher seeks to demonstrate a relationship between hot or cold baths and the amount of relaxation they produce. He obtains the following relaxation scores from two independent samples.

    Sample 1 (hot): $\bar{X} = 43$, $s^2_X = 22.79$, $n = 15$

    Sample 2 (cold): $\bar{X} = 39$, $s^2_X = 24.6$, $n = 15$

    (a) What are H_0 and H_a? (b) Compute t_{obt}. (c) With $\alpha = .05$, what is t_{crit}? (d) What should the researcher conclude about this relationship? (e) What is the proportion of variance accounted for by changing the bath temperature? (f) How large is the effect of temperature in terms of the difference it produces in relaxation scores? (g) Does this variable have a large influence on scores?

14. A researcher investigates whether a period of time feels longer or shorter when people are bored compared to when they are not bored. Using independent samples, the researcher obtains the following estimates of the time period (in minutes):

    Sample 1 (bored): $\bar{X} = 14.5$, $s^2_X = 10.22$, $n = 28$

    Sample 2 (not bored): $\bar{X} = 9.0$, $s^2_X = 14.6$, $n = 34$

    (a) What are H_0 and H_a? (b) Compute t_{obt}. (c) With $\alpha = .05$, what is t_{crit}? (d) What should the researcher conclude about this relationship? (e) What is the proportion of variance accounted for by boredom levels? (f) Using our two approaches, how important is this variable in determining someone's time estimate?

SPSS Instructions

Hypothesis Testing Using the Two-Sample t-Test

Two-Sample t-Test for Independent Samples

Main Menu: Analyze

Select: Compare Means → Independent-Samples t-Test

Move: Desired dependent variable into Test Variable(s) box Independent variable into Grouping Variable box

Select: Define Groups

Enter: Value used for independent variable group 1 (e.g., "a" or 1) Value used for independent variable group 2 (e.g., "b" or 2)

Optional: Options... → Confidence Interval Percentage (default is a 95% confidence interval) → Continue

Select: OK

Notes: (1) Significance level given is for a two-tailed test. If a one-tailed test is desired, you will need to compare the calculated t-test against a one-tailed critical t-value from the tables using the correct degrees of freedom. (2) Although the SPSS t-test does not provide information about effect size, both the r^2_{pb} and the Cohen's d can be calculated easily using the information provided on the printout and the formulas from the textbook.

Two-Sample t-Test for Related Samples

Main Menu: *Analyze*

Select: **Compare Means →**
Paired-Samples *t*-Test

Move: **Variable containing data for first condition of the related samples into Variable 1 of Pair 1 in the Paired Variable(s) box**
Variable containing data for second condition of the related samples into Variable 2 of Pair 1 in the Paired Variable(s) box

Optional: **Options... →**
Confidence Interval Percentage (default is a 95% confidence interval)
→ Continue

Select: **OK**

Notes: (1) Significance level given is for a two-tailed test. If a one-tailed test is desired, you will need to compare the calculated *t*-test against a one-tailed critical *t*-value from the tables using the correct degrees of freedom.

(2) This test works for either the matched-samples or the repeated-measures *t*-test.

(3) Although the SPSS *t*-test does not provide information about effect size, both the r_{pb}^2 and the Cohen's *d* can be calculated easily using the information provided on the printout and the formulas from the textbook.

15. Genevieve predicts that students who use a computer program that corrects spelling errors will receive higher grades on a term paper. She uses an independent-samples design in which Group A uses a spelling checker and Group B does not. She tests $H_0: \mu_A - \mu_B \leq 0$ and $H_a: \mu_A - \mu_B > 0$. She obtains a negative value of t_{obt}. (a) What should she conclude about this study? (b) Say that her sample means actually support her predictions. What miscalculation is she likely to have made?

16. A rather dim student proposes testing the conditions of "male" and "female" using a repeated-measures design. What's wrong with this idea?

17. We ask whether people will score higher or lower on a questionnaire measuring their well-being when they are exposed to sunshine compared to when they're not exposed to sunshine. A sample of 8 people is first measured after low levels of sunshine exposure and then again after high levels of exposure. We get the scores at right:

Low:	High:
14	18
13	12
17	20
15	19
18	22
17	19
14	19
16	16

(a) Subtracting low from high, what are H_0 and H_a? (b) Compute the appropriate *t*-test. (c) What should we conclude about this study? (d) Report your results using the correct format. (e) What is the predicted well-being score for someone tested under low sunshine or under high sunshine? (f) Compute the two statistics that reflect the amount of influence that changing sunshine has on well-being. Is it large?

18. A researcher investigates whether classical music is more or less soothing than modern music. She plays a classical selection to one group and a modern selection to another (each $n = 6$). She gives each person an irritability questionnaire and, after computing the independent-samples *t*-test, finds $t_{obt} = +1.38$. (a) Are the results significant? (b) Report the results using the correct format. (c) What should she conclude about the relationship in nature between listening to music and irritability? (d) What error and statistical principle (from previous chapters) should concern her?

19. A researcher investigates whether children exhibit a higher number of aggressive acts after watching a violent television show. The aggressive acts for the same 10 participants before and after watching the show are shown in the table to the right. (a) Subtracting before scores from after scores, what are H_0 and H_a? (b) Compute t_{obt}. (c) With $\alpha = .05$, what is t_{crit}? (d) What should the researcher conclude about this relationship? (e) How large is the effect of violence in terms of the *difference* it produces in aggression scores?

Sample 1 (After)	5	6	4	4	7	3	2	1	4	3
Sample 2 (Before)	4	6	3	2	4	1	0	0	5	2

Learning Outcomes and Objectives

10.1 Understanding Correlations

Create and interpret a *scatterplot* * Explain what a *regression line* is

A *scatterplot* is a graph that shows the location of each pair of *X-Y* scores in the data. As *X* scores increase: in a positive linear relationship the *Y* scores tend to increase, and in a negative linear relationship the *Y* scores tend to decrease. In a *nonlinear relationship,* as the *X* scores increase, the *Y* scores do not only increase or only decrease. The *strength* of a relationship is the extent to which one value of *Y* is consistently paired with only one value of *X*. A smaller correlation coefficient indicates a weaker relationship, with greater variability in *Y* scores at each *X*, greater vertical spread in the scatterplot, and the less accuracy there is in predicting *Y* scores based on correlated *X* scores.

Quick Practice

1. In a _____ relationship, the *Y* scores increase or decrease only. This is not true in a _____ relationship.

2. The more that you smoke cigarettes, the lower is your healthiness. This is a _____ linear relationship, producing a scatterplot that slants _____ as *X* increases.

3. The more that you exercise, the better is your muscle tone. This is a _____ linear relationship, producing a scatterplot that slants _____ as *X* increases.

4. In a stronger relationship the variability among the *Y* scores at each *X* is _____, producing a scatterplot that forms a ellipse.

5. The _____ line summarizes the scatterplot.

10.2 The Pearson Correlation Coefficient

Know when and how to compute the *Pearson* r * Perform the significance test of the *Pearson* r

A *Pearson* r describes the strength and type of linear relationship between two interval and/or ratio variables: *r* in the sample, ρ in the population. We always perform hypothesis testing on the Pearson *r*. A *restricted range* occurs when the range of scores on a variable is limited. The *r* is then smaller than it would be if the range were not restricted.

Quick Practice

1. Compute *r* for the following:

X	Y
1	1
1	3
2	2
2	4
3	4

For the following, we predict a negative relationship and obtain $r_{obt} = -.44$.

Key Terms

correlation coefficient A number that describes the type and the strength of the relationship present in a set of data

scatterplot A graph of the individual data points from a set of *X-Y* pairs

linear relationship A relationship in which the *Y* scores tend to change in only one direction as the *X* scores increase, forming a slanted straight regression line on a scatter plot

linear regression line The straight line that summarizes the scatter plot of a linear relationship by passing through the center of the scatterplot

positive linear relationship A linear relationship in which the *Y* scores tend to increase as the *X* scores increase

negative linear relationship A linear relationship in which the *Y* scores tend to decrease as the *X* scores increase

nonlinear (curvilinear) relationship A relationship in which the *Y* scores change their direction of change as the *X* scores change

strength of a relationship The extent to which one value of *Y* within a relationship is consistently associated with one and only one value of *X;* also called the *degree of association*

Pearson correlation coefficient (r) The correlation coefficient that describes the linear relationship between two interval or ratio variables

restricted range Occurs when the range of scores on the *X* or *Y* variable is limited, producing an *r* that is smaller than it would be if the range were not restricted

Key Terms

sampling distribution of r A frequency distribution showing all possible values of r that occur when samples are drawn from a population in which ρ is zero

linear regression The procedure used to predict Y scores based on correlated X scores

predictor variable The X variable that has the known scores from which unknown Y scores are predicted when using the linear regression equation

criterion variable The Y variable that has the unknown scores that are predicted based on a correlated X score when using the linear regression equation

predicted Y score (Y′) In linear regression, the best prediction of the Y scores at a particular X, based on the linear relationship summarized by the regression line

linear regression equation The equation that produces the value of Y' at each X and defines the straight line that summarizes a relationship

proportion of variance accounted for The proportion of the differences in Y scores that is associated with changes in the X variable

Key Formulas

THE PEARSON CORRELATION COEFFICIENT

$$r = \frac{N(\Sigma XY) - (\Sigma X)(\Sigma Y)}{\sqrt{[N(\Sigma X^2) - (\Sigma X)^2][N(\Sigma Y^2) - (\Sigma Y)^2]}}$$

THE DEGREES OF FREEDOM IN THE PEARSON CORRELATION COEFFICIENT

$df = N - 2$, where N is the number of X-Y pairs in the data

THE PROPORTION OF VARIANCE ACCOUNTED FOR

Proportion of variance accounted for $= r^2$

Learning Outcomes and Objectives

2. What are H_0 and H_a?
3. With $\alpha = .05$ and $N = 10$, what is r_{crit}?
4. What do you conclude about r_{obt}?
5. What do you conclude about the relationship in the population?

10.3 Statistics in the Research Literature: Reporting r

Know what information is included when reporting r

10.4 Understanding Linear Regression

Explain the logic of *linear regression*

Linear regression is used to predict Y scores (the *criterion variable*) by using the relationship with X (the *predictor variable*). For each X, the regression equation produces Y', which is the *predicted Y score*.

10.5 The Proportion of Variance Accounted For: r^2

Explain the logic of r^2

The *proportion of variance accounted for* equals r^2. It indicates the proportion of all differences in Y that occur with changes in X, and the proportional improvement in predicting Y that the relationship provides. It is used to judge the scientific importance of a relationship.

QUICK PRACTICE ANSWERS

10.1: 1. linear; nonlinear; 2. negative; down; 3. positive; up; 4. smaller/narrower; 5. regression

10.2: 1. $r = \dfrac{5(28) - (9)(14)}{\sqrt{[5(19) - 81][5(46) - 196]}}$

$= \dfrac{+14}{\sqrt{(14)(34)}} = \dfrac{+14}{21.817} = +.64$;

2. $H_0: \rho \geq 0$; $H_a: \rho < 0$; 3. $df = 8$, $r_{crit} = -.549$; 4. Not Significant; 5. Make no conclusion about the relationship in the population.

Review and Application Questions

1. What is the difference between an experiment and a correlational study in terms of how the researcher (a) collects the data? (b) examines the relationship?

2. (a) You have collected data that you think show a relationship. What do you do next? (b) What is the advantage of computing a correlation coefficient? (c) What two characteristics of a linear relationship are described by a correlation coefficient?

3. (a) What is a scatterplot? (b) What is a regression line?

4. When do you compute a Pearson r?

5. (a) Define a positive linear relationship. (b) Define a negative linear relationship. (c) Define a curvilinear relationship.

6. As the value of r approaches ± 1, what does it indicate about the following: (a) the shape of the scatterplot; (b) the variability of the Y scores at each X; (c) the accuracy with which we can predict Y if X is known?

7. (a) What does ρ stand for? (b) How is the value of ρ determined?

8. Why can't you obtain a correlation coefficient greater than ± 1?

9. Summarize the steps involved in hypothesis testing of the Pearson correlation coefficient.

10. (a) How do researchers use linear regression to predict unknown Y scores? (b) What is the symbol for a predicted Y score?

11. (a) What is the proportion of variance accounted for? (b) How is it measured in a correlational study? (c) How do we use the size of r^2 to judge a relationship?

12. For each of the following, indicate whether it is a positive linear, negative linear, or nonlinear relationship: (a) Quality of performance (Y) increases with increased arousal (X) up to an optimal level; then quality of performance decreases with increased arousal. (b) Heavier jockeys (X) tend to win fewer horse races (Y). (c) As number of minutes of exercise increases each week (X), dieting individuals lose more pounds (Y). (d) The number of bears in an area (Y) decreases as the area becomes increasingly populated by humans (X).

13. Marcellus wondered if there is a relationship between the quality of sneakers worn by 20 volleyball players and their average number of points scored per game. He computed $r = +.21$ and immediately claimed this was evidence that better-quality sneakers are related to better performance. (a) What did he forget to do? (b) What are his H_0 and H_a? (c) With $\alpha = .05$, what is r_{crit}? (d) Report the results, and what should he conclude? (e) What other computations should he perform to describe this relationship?

14. A researcher has just completed a correlational study measuring the number of boxes of tissue purchased per week and the number of vitamin tablets consumed per week for each participant. (a) Which is the independent and which is the dependent variable? (b) Which variable is X? Which is Y?

15. You want to know if a nurse's absences from work in one month (Y) are positively correlated with her score on a test of psychological "burnout" (X). (a) Compute the correlation coefficient using the sample data on the back of this card. (b) What are H_0 and H_a? (c) With $\alpha = .05$, what is r_{crit}? (d) Report the results, and what do you conclude about the relationship? (e) What do you conclude about the relationship that occurs in the population? (f) How scientifically useful is this relationship?

SPSS Instructions

Describing Relationships Using Correlation and Regression

Scatterplots

Main Menu: *Graphs*

Select: **Legacy Designs → Scatter/Dots → Simple Scatter → Define**

Move: **Desired X variable into X-axis box**
Y variable into Y-axis box

Select: **OK**

Pearson r and Significance Test

Main Menu: *Analyze*

Select: **Correlate → Bivariate**

Move: **Desired X and Y variables into Variable(s) box**

Select: **✓ Pearson (under Correlation Coefficients) → ✓ Two-tailed or One-tailed significance test → OK**

Notes: (1) It does not matter which variable is the X and which is the Y for this procedure, as both go into the Variable(s) box.
(2) Two-tailed significance test is the default.

Regression

Main Menu: *Analyze*

Select: **Regression → Linear**

Move: **Desired X variable into Independent(s) box**
Y variable into Dependent(s) box

Select: **OK**

Notes: Values given in Coefficients box, Unstandardized Coefficients B column are the values for the regression line. The value in row (Constant) is the Y-intercept value (a) and the value in the second row with the variable name for the Dependent variable is the slope (b).

Participant	Burnout X	Absences Y
1	2	4
2	1	7
3	2	6
4	3	9
5	4	6
6	4	8
7	7	7
8	7	10
9	8	11

16. A researcher asks if there is a relationship between the X variable of number of errors on a math test and the Y variable of the person's level of satisfaction with his/her performance. (a) Summarize the relationship in the data below. (b) What are H_0 and H_a? (c) With $\alpha = .05$, what is r_{crit}? (d) Report the results, and what do you conclude about the relationship? (e) What do you conclude about the relationship in the population? (f) What proportion of the differences in satisfaction scores occurs with different test scores?

Participant	Errors X	Satisfaction Y
1	9	3
2	8	2
3	4	8
4	6	5
5	7	4
6	10	2
7	5	7

17. A scientist suspects that as a person's stress level changes, so does the amount of his or her impulse buying. He collects data from 72 people and obtains an $r = +.38$. (a) What are H_0 and H_a? (b) With $\alpha = .05$, what is r_{crit}? (c) Report these results using the correct format. (d) What conclusions should he draw? (e) What other calculations should be performed to describe the relationship in these data?

18. Esther computes the correlation between participants' physical strength and their college grade average, obtaining a significant $r = +.08$. She claims she has uncovered a useful tool for predicting which college applicants are likely to succeed academically. Do you agree or disagree? Why?

19. Rufus finds $r = +.80$ for the variables of number of hours studied and number of errors on a statistics test, and $r = +.40$ for the variables of speed of taking the test and the number of errors on the test. He concludes that study time forms twice as strong a relationship and is therefore twice as important as the speed of taking the test. (a) Why is he incorrect? (b) Compare these relationships and draw the correct conclusion.

20. (a) Why must a relationship be significant in order to be important? (b) Why can a relationship be significant and still be unimportant?

Learning Outcomes and Objectives

11.1 An Overview of the Analysis of Variance

Understand the terminology and logic of *analysis of variance*

The general terms used previously and their corresponding ANOVA terms are shown in this table:

General Term	=	ANOVA Term
independent variable	=	factor
one independent variable	=	one-way design
independent samples	=	between-subjects
related samples	=	within-subjects
condition	=	level (treatment)
sum of squared deviations	=	sum of squares (SS)
variance (s_X^2)	=	mean square (MS)
effect of independent variable	=	treatment effect

The one-way ANOVA is performed when testing two or more conditions from one independent variable. The *experiment-wise error rate* is the probability that a Type I error will occur in an experiment. ANOVA keeps the experiment-wise error rate equal to α. The ANOVA tests the H_0 that all μs being represented are equal; H_a is that not all μs are equal. The one-way, between-subjects ANOVA assumes (1) normally distributed, interval or ratio dependent scores, (2) one independent variable is tested using independent samples, and (3) the populations being represented have homogeneous variance.

Quick Practice

1. A study involving one independent variable is a _____ design.

2. Perform the _____ when a study involves independent samples; perform the _____ when it involves related samples.

3. An independent variable is also called a _____ and a condition is also called a _____ or _____.

4. The _____ will indicate whether any of the conditions differ, and then the _____ will indicate which specific conditions differ.

5. The probability of a Type I error in the study is called the _____.

11.2 Components of the ANOVA

Compute F_{obt}

The MS_{wn} estimates the variability of individual scores in the population. The MS_{bn} estimates the variability of sample means in the population. The F-ratio produces F_{obt} and equals MS_{bn}/MS_{wn}. The F_{obt} may be greater than 1 because either (a) there is no treatment effect, but the data are unrepresentative of this, or (b) the means from two or more levels represent different μs. If F_{obt} is significant, the data are so unlikely to represent one population that we conclude that a relationship is present.

Quick Practice

1. MS_{wn} is the symbol for the _____, and MS_{bn} is the symbol for the _____.

Key Terms

factor In ANOVA, an independent variable

levels (treatments) The conditions of the independent variable

treatment effect The result of changing the conditions of an independent variable so that different populations of scores having different μs are produced

one-way ANOVA The analysis of variance performed when an experiment has only one independent variable

between-subjects factor An independent variable that is studied using independent samples in all conditions

between-subjects ANOVA The type of ANOVA that is performed when a study involves between-subjects factors

within-subjects factor The type of factor created when an independent variable is studied using related samples in all conditions because participants are either matched or repeatedly measured

within-subjects ANOVA The type of ANOVA performed when a study involves within-subjects factors

analysis of variance (ANOVA) The parametric procedure for determining whether significant differences exist in an experiment containing two or more sample means

experiment-wise error rate The probability of making a Type I error when comparing all means in an experiment

post hoc comparisons In ANOVA, statistical procedures used to compare all possible pairs of sample means in a significant effect to determine which means differ significantly from each other

Key Terms

mean square (MS) In ANOVA, an estimated population variance

mean square within groups (MS_{wn}) In ANOVA, the variability in scores that occurs in the conditions

mean square between groups (MS_{bn}) In ANOVA, the variability that occurs between the levels in a factor

F-ratio In ANOVA, the ratio of the mean square between groups to the mean square within groups

sum of squares (SS) The sum of the squared deviations of a set of scores around the mean of those scores

F-distribution The sampling distribution of all possible values of F that occur when the null hypothesis is true and all conditions represent one population μ

Tukey's *HSD* multiple comparisons test The post hoc procedure performed with ANOVA to compare means from a factor in which all levels have equal n

eta squared (η^2) The proportion of variance in the dependent scores that is accounted for by changing the levels of a factor, and thus a measurement of effect size

Key Formulas

THE TOTAL SUM OF SQUARES

$$SS_{tot} = \Sigma X_{tot}^2 - \frac{(\Sigma X_{tot})^2}{N}$$

SUM OF SQUARES BETWEEN GROUPS

$$SS_{bn} = \Sigma\left(\frac{(\Sigma X \text{ in column})^2}{n \text{ in column}}\right) - \frac{(\Sigma X_{tot})^2}{N}$$

SUM OF SQUARES WITHIN GROUPS

$$SS_{wn} = SS_{tot} - SS_{bn}$$

THE MEAN SQUARE BETWEEN GROUPS

$$MS_{bn} = \frac{SS_{bn}}{df_{bn}}$$

Learning Outcomes and Objectives

2. Differences between the individual scores in the population are estimated by _____.

3. Differences between sample means in the population are estimated by _____.

4. When H_0 is true, the F_{obt} should equal _____.

5. The larger the F_{obt}, the _____ likely that H_0 is false.

11.3 Performing the ANOVA

Compute Tukey's *HSD*

To compute F_{obt}, compute the SS_{tot}, SS_{bn}, and SS_{wn}, and df_{tot}, df_{bn}, and df_{wn}. Dividing SS_{bn} by df_{bn} gives the MS_{bn}; dividing SS_{wn} by df_{wn} gives MS_{wn}. Dividing MS_{bn} by MS_{wn} gives F_{obt}. Compare F_{obt} to F_{crit}. Post hoc comparisons indicate which specific levels differ significantly. Perform the Tukey *HSD* test when F_{obt} is significant, there are more than two levels, and all *ns* are equal. The F-*distribution* is the sampling distribution of all possible values of F_{obt} when H_0 is true. If F_{obt} is significant with more than two levels, perform *post hoc comparisons* to determine which means differ significantly. If all *ns* are equal, perform *Tukey's* HSD *test*.

Quick Practice

We have level means of $\overline{X}_1 = 16.50$, $\overline{X}_2 = 11.50$, and $\overline{X}_3 = 8.92$ (with $n = 21$, $MS_{wn} = 63.44$, and $df_{wn} = 60$).

1. What is q_k here?
2. What is the *HSD*?
3. Which means differ significantly?

11.4 Summary of the One-Way ANOVA

11.5 Statistics in the Research Literature: Reporting ANOVA

Know what information is included when reporting the ANOVA

11.6 Effect Size and Eta²

Understand effect size in ANOVA * Know how to compute eta²

Eta squared (η^2) describes the *effect size*—the *proportion of variance* in dependent scores accounted for by changing the levels of the independent variable.

For Chapter 11, find Quick Practice answers after the Review and Application questions.

THE MEAN SQUARE WITHIN GROUPS

$$MS_{wn} = \frac{SS_{wn}}{df_{wn}}$$

THE *F*-RATIO

$$F_{obt} = \frac{MS_{bn}}{MS_{wn}}$$

TUKEY'S *HSD*

$$HSD = (q_k)\left(\sqrt{\frac{MS_{wn}}{n}}\right)$$

ETA SQUARED

$$\eta^2 = \frac{SS_{bn}}{SS_{tot}}$$

Review and Application Questions

1. What does each of the following terms mean? (a) ANOVA (b) one-way design (c) factor (d) level (e) treatment effect (f) between-subjects factor (g) within-subjects factor

2. (a) What is the difference between n and N? (b) What does k stand for?

3. (a) What is the experiment-wise error rate? (b) Why does ANOVA solve the problem with the experiment-wise error rate created by multiple t-tests?

4. Summarize the major steps involved in analyzing an experiment when $k > 2$.

5. (a) When is it necessary to perform post hoc comparisons? Why? (b) When is it unnecessary to perform post hoc comparisons?

6. What are the two types of mean squares, and what does each one estimate?

7. What does the F-ratio compare?

8. (a) Why should F_{obt} equal 1 if the data represent the H_0 situation? (b) Why is F_{obt} greater than 1 when the data represent the H_a situation? (c) What does a significant F_{obt} indicate about differences between the levels of a factor?

9. What does η^2 indicate?

10. With $\alpha = .05$, $F_{obt} = 6.31$ is significant, with $df_{bn} = 4$ and $df_{wn} = 30$. How is this result reported in the literature?

11. (a) In a study comparing four conditions, what is H_0? (b) What is H_a? (c) In words, what do H_0 and H_a say for this study?

12. In an experiment, scores are measured under two conditions. (a) How will the researcher know whether to perform a parametric or nonparametric procedure? (b) Which two types of parametric procedures are available? (c) If the experiment involves three levels, which two versions of a parametric procedure are available? (d) How does the researcher choose between them?

13. (a) Julius computes an F_{obt} of .63. How should this be interpreted? (b) He computes another F_{obt} of -1.7. How should this be interpreted?

14. Bess obtained a significant F_{obt} from an experiment with five levels. She says she's demonstrated a relationship in which changing each condition of the independent variable results in a significant change in the dependent variable. (a) Is she correct? (b) What must she do?

15. (a) Why must the relationship in a one-way ANOVA be significant in order to be potentially important? (b) What does "significant" tell you about the relationship? (c) Why can the relationship be significant yet scientifically unimportant?

16. In a research report, the between-subjects factor of participants' salaries produced significant differences in their self-esteem. (a) What does this tell you about the design? (b) What does it tell you about the results?

17. Here are data from an experiment studying the effect of age on creativity scores:

Age 4	Age 6	Age 8	Age 10
3	9	9	7
5	11	12	7
7	14	9	6
4	10	8	4
3	10	9	5

SPSS Instructions

Hypothesis Testing Using One-Way Analysis of Variance

One-Way ANOVA Independent Samples (Option 1)

Main Menu: *Analyze*

Select: **Compare Means → One-way ANOVA**

Move: **Desired independent variable into Factor box Dependent variable into Dependent List box**

Optional: **Post hoc... → ✓ Tukey (or other desired post hoc test) → Continue**

Select: **OK**

Notes: This procedure does not provide information regarding effect size.

One-Way ANOVA Independent Samples (Option 2)

Main Menu: *Analyze*

Select: **General Linear Model → Univariate**

Move: **Desired independent variable into Fixed Factor(s) box Dependent variable into Dependent Variable box**

Optional: **Post hoc... Move independent variable from Factor box into Post Hoc Test For box→ ✓ Tukey (or other desired post hoc test) → Continue Options... Move independent variable from Factor(s) and Factor Interactions box into Display Means For box→ ✓ Descriptive Statistics → ✓ Estimates of Effect Size → Continue**

Select: **OK**

One-Way ANOVA Related Samples

Main Menu: *Analyze*

Select: **General Linear Model → Repeated Measures**

Enter: **Number of levels for the within-subjects factor (e.g., 3).** *Scores for each level must have been entered in a separate variable in the data. You may indicate an overarching factor name.*

Select: **Define**

Move: **The variable for level 1 of the within-subjects factor into the Within-Subjects Variable (Factor 1) in the ___?___(1) slot The variable for level 2 of the within-subjects factor into the Within-Subjects Variable (Factor 1) in the ___?___(2) slot**

Optional: **Options… Move factor 1 into the Display Means For box → ✓ Descriptive Statistics → ✓ Estimates of Effect Size → Continue**

Optional: **Options… Move independent variable from Factor(s) and Factor Interactions box into Display Means For box→ ✓ Compare main effects → Confidence interval adjustment (Least Significant Difference is the default) → ✓ Descriptive Statistics → ✓ Estimates of Effect Size → Continue**

Select: **OK**

Notes: Not all versions of SPSS contain this procedure.

(a) Compute F_{obt} and create an ANOVA summary table. (b) With $\alpha = .05$ what do you conclude about F_{obt}? (c) Perform the post hoc comparisons. What should you conclude about the treatments? (d) How important is the relationship in this study?

18. A researcher investigated the number of illnesses people contract as a function of the amount of stress they experienced during a six-month period. She obtained the following data:

Amount of Stress			
Negligible Stress	Minimal Stress	Moderate Stress	Severe Stress
2	4	6	5
1	3	5	7
4	2	7	8
1	3	5	4

(a) What are H_0 and H_a? (b) Compute F_{obt} and complete the ANOVA summary table. (c) With $\alpha = .05$, what is F_{crit}? (d) Perform the post hoc comparisons. (e) What do you conclude about this study? (f) Describe the effect size and interpret it.

19. An article reports that, using a within-subjects design, a diet led to a significant decrease in weight for a group of participants. (a) What does this tell you about the design? (b) Which ANOVA was computed here?

20. In a study in which $k = 3$, $n = 21$, $\overline{X}_1 = 45.3$, $\overline{X}_2 = 16.9$, and $\overline{X}_3 = 8.2$, you compute the following sums of squares:

Source	Sum of Squares	df	Mean Square	F
Between	147.32			
Within	862.99			
Total	1010.31			

(a) Complete the ANOVA. (b) With $\alpha = .05$ what do you conclude about F_{obt}? (c) Perform the post hoc comparisons. (d) What do you conclude about this relationship? (e) What is the effect size, and what does it tell you about the influence of this independent variable?

QUICK PRACTICE ANSWERS

11.1: 1. one-way; 2. between-subjects ANOVA, within-subjects ANOVA; 3. factor, level, treatment; 4. F_{obt}; 5. post hoc comparisons; 5. experiment-wise error rate

11.2: 1. mean square within groups, mean square between groups; 2. MS_{wn}; 3. MS_{bn}; 4. 1.0; 5. more

11.3: 1. For $k = 3$ and $df_{wn} = 60$, $q_k = 3.40$; 2. $HSD = 3.40(\sqrt{63.44/21}) = 5.91$; 3. Only the difference for $\overline{X}_1 - \overline{X}_3$ of 7.58 is larger than 5.91, so only these means differ significantly.

Learning Outcomes and Objectives

12.1 Understanding the Two-Way ANOVA

Understand what a *two-way factorial ANOVA* is * Understand how to *collapse* across a factor * Explain *cell means* and *main effect means*

A *two-way, between-subjects ANOVA* involves two independent variables, and all conditions of both factors contain independent samples. A *two-way, within-subjects ANOVA* is used when both factors involve related samples. A *two-way, mixed-design ANOVA* is used when one factor is tested using independent samples and one factor is tested using related samples. In a *complete factorial design,* each level of one factor is combined with all levels of the other factor. Each *cell* is formed by the combining of a level from each factor. A *main effect* is the influence of manipulating an independent variable alone. The *main effect means* for a factor are obtained by *collapsing* across (combining the scores from) the levels of the other factor. Collapsing across factor B produces the main effect means for factor A: the overall mean for participants tested under A_1, the mean for those tested under A_2, etc. A significant F_{obt} for this factor indicates significant differences among the means that result from changing the levels of A. Collapsing across factor A produces the main effect means for factor B: the overall mean for participants tested under B_1, the mean for those tested under B_2, etc. A significant F_{obt} for this factor indicates significant differences among these means that result from changing the levels of B. A *two-way interaction* is the influence on scores created by combining each level of Factor A with each level of Factor B. An *interaction effect* occurs when the influence of changing one factor is not the same for each level of the other factor.

Quick Practice

In this study:

	A_1	A_2
B_1	2	5
	2	4
	2	3
B_2	11	7
	10	6
	9	5

1. The means produced by collapsing across factor B are _____ and _____. They are called the _____ means for factor _____.
2. What is the main effect of A?
3. The means produced by collapsing across factor A are _____ and _____. They are called the _____ means for factor _____.
4. What is the main effect of B?

(Continues)

Key Terms

two-way ANOVA The parametric inferential procedure performed when an experiment contains two independent variables

two-way, between-subjects ANOVA The parametric inferential procedure performed when both factors are between-subjects factors

two-way, within-subjects ANOVA The parametric inferential procedure performed when both factors are within-subjects factors

two-way, mixed-design ANOVA The parametric inferential procedure performed when the design involves one within-subjects factor and one between-subjects factor

cell In a two-way ANOVA, the combination of one level of one factor with one level of the other factor

complete factorial design A two-way ANOVA design in which all levels of one factor are combined with all levels of the other factor

incomplete factorial design A two-way ANOVA design in which not all levels of the two factors are combined

main effect In a two-way ANOVA, the effect on the dependent scores of changing the levels of one factor; found by collapsing over the other factor

main effect means The mean of the level of one factor after collapsing across the levels of the other factor

collapsing In a two-way ANOVA, averaging together all scores from all levels of one factor in order to calculate the main effect means for the other factor

cell mean The mean of the scores from one cell in a two-way design

two-way interaction effect In a two-way ANOVA, the effect in which the relationship between one factor and the dependent scores depends on the level of the other factor that is present

confounded comparison In a two-way ANOVA, a comparison of two cells that differ along more than one factor

unconfounded comparison In a two-way ANOVA, a comparison of two cells that differ along only one factor

Key Formulas

There are no new formulas in this chapter.

A Study Produces these cell means:

	A_1	A_2
B_1	2	5
	2	4
	2	3
	$\bar{X} = 2$	$\bar{X} = 4$
B_2	11	7
	10	6
	9	5
	$\bar{X} = 10$	$\bar{X} = 6$

5. The means to examine in the interaction are the _____ means.

6. What is the influence of changing the levels of A at B_1 and then at B_2?

7. Why is this an interaction effect?

12.2 Interpreting the Two-Way Experiment

Explain what a significant *main effect* indicates * Explain what a significant *interaction* indicates * Know how to interpret the results of a two-way experiment

A *main effect* indicates differences *somewhere* among the means, and therefore, if the *n*s are equal and we have more than two levels in the factor, we determine which specific means differ by performing Tukey's *HSD* test. An *interaction effect* indicates that the *cell means* differ significantly so that the relationship between one factor and the dependent scores depends on the level of the other factor that is present. Post hoc comparisons are performed on each significant effect having more than two levels. In the interaction perform only *unconfounded comparisons,* in which the cells differ along only one factor. Two means are *confounded* if the cells differ along more than one factor. Main effects are graphed by plotting the main effect means on *Y* and the levels of the factor on *X*. The interaction is graphed by plotting cell means on *Y* and the levels of one factor on *X*. Then a separate line connects the data points for the cell means from each level of the other factor. When graphed, an interaction effect produces nonparallel lines. Conclusions in a two-way ANOVA are usually based on the interaction effect when it is significant, because overall conclusions about main effects are contradicted by the interaction.

which level of B is present.

7. Because the influence of changing from A_1 to A_2 depends on

For B_2, changing from A_1 to A_2 decreases the means from 10 to 6;

6. For B_1, changing from A_1 to A_2 increases the means from 2 to 4.

4. Changing from B_1 to B_2 produces an increase in scores; 5. cell;

produces a decrease in scores; 3. $\bar{X}_{B_1} = 3$, $\bar{X}_{B_2} = 8$, main effect, B;

12.1: 1. $\bar{X}_{A_1} = 6$, $\bar{X}_{A_2} = 5$, main effect, A; 2. Changing from A_1 to A_2

QUICK PRACTICE ANSWERS

Review and Application Questions

1. (a) A researcher conducts a study involving one independent variable. What are the two types of parametric procedures available to her? (b) She next conducts a study involving two independent variables. What are the three versions of the parametric procedure available to her? (c) In part b what aspect of her design determines which version she should perform?

2. Identify the following terms: (a) two-way design, (b) complete factorial, (c) cell, and (d) two-way, between-subjects design.

3. What is the difference between a main effect mean and a cell mean?

4. Which type of ANOVA is used in a two-way design when: (a) both factors are tested using independent samples? (b) one factor involves independent samples and one factor involves related samples? (c) both factors involve related samples?

5. (a) What is a confounded comparison, and when does it occur in a study's diagram? (b) What is an unconfounded comparison, and when does it occur? (c) Why don't we perform post hoc tests on confounded comparisons?

6. Why do we usually base the interpretation of a two-way design on the interaction effect when it is significant?

7. For a 2×2 ANOVA, describe the following in words: (a) the statistical hypotheses for factor A, (b) the statistical hypotheses for factor B, and (c) the statistical hypotheses for A $\times$ B.

8. What does it mean to collapse across a factor?

9. (a) A significant main effect indicates what about a study? (b) A significant interaction effect indicates what about a study?

10. A student hears that a 2×3 design was conducted and concludes that six factors were examined. Is this conclusion correct? Why or why not?

11. Below are the cell means of three experiments. For each experiment compute the main effect means and indicate whether there appears to be a main effect of A or B and/or an interaction of A $\times$ B.

Study 1	A_1	A_2
B_1	2	4
B_2	12	14

Study 2	A_1	A_2
B_1	10	5
B_2	5	10

Study 3	A_1	A_2
B_1	8	14
B_2	8	2

12. In question 11, if we label the X axis with factor A and graph the cell means, what pattern will we see for each interaction?

13. After performing a 3×4 ANOVA with equal ns, you find that all Fs are significant. What other procedures should you perform?

14. A 2×2 design studies participants' frustration levels when solving problems depending on the difficulty of the problem and whether they are math or logic problems. The results are that logic problems produce significantly more frustration than math problems, that greater difficulty leads to significantly greater frustration, and that, when difficult, the math problems produce significantly greater frustration than the logic problems, but the reverse is true for easy problems. In the ANOVA for this study, what effects are significant?

SPSS Instructions

Understanding the Two-Way Analysis of Variance

Two-Way ANOVA Independent Samples

Main Menu: *Analyze*

Select: **General Linear Model → Univariate**

Move: **Desired independent variable 1 into Fixed Factor(s) box**

Independent variable 2 into Fixed Factor(s) box

Dependent variable into Dependent Variable box

Optional: **Post hoc... Move independent variable 1 and/or independent variable 2 from Factor box into Post Hoc Test For box→ ✓ Tukey (or other desired post hoc test) → Continue**

Options... Move independent variables from Factor(s) and Factor Interactions box into Display Means For box→ ✓ Descriptive Statistics → Estimates of Effect Size → Continue

Select: **OK**

15. In question 14, say instead that the researcher found no difference between math and logic problems, that frustration significantly increases with greater difficulty, and that this is true for both math and logic problems. In the ANOVA performed for this study, what effects are significant?

16. In an experiment, you measure the popularity of two brands of soft drinks (factor A), and for each brand you test males and females (factor B). The following table shows the cell means from the study:

(a) Describe the graph of the interaction when factor A is on the X axis. (b) Does there appear to be an interaction effect? Why? (c) What is the main effect (and means) of changing brands? (d) What is the main effect (and means) of changing gender? (e) Why will a significant interaction prohibit you from making conclusions based on the main effects?

		Factor A	
		Level A$_1$: Brand X	Level A$_2$: Brand Y
Factor B	Level B$_1$: Males	14	23
	Level B$_2$: Females	25	12

17. We examine performance on an eye–hand coordination task as a function of three levels of reward and three levels of practice, obtaining the following cell means.

(a) What are the main effect means for reward, and what do they indicate about its influence? (b) What are the main effect means for practice, and what do they indicate about its influence? (c) Does there appear to be an interaction effect? (d) How would you perform unconfounded post hoc comparisons of the cell means?

		Reward		
		Low	Medium	High
Practice	Low	4	10	7
	Medium	5	5	14
	High	15	15	15

18. (a) In question 17, why does the interaction contradict your conclusions about the effect of reward? (b) Why does the interaction contradict your conclusions about practice?

19. A study compared the performance of males and females tested by either a male or a female experimenter. The data are in the table at right.

(a) What are the main effect means for participants' gender? Interpret the apparent main effect. (b) What are the main effect means for experimenter's gender? Interpret the apparent main effect. (c) What are the means in the interaction? Interpret what appears to be the effect.

		Factor A: Participants	
		Level A$_1$: Males	Level A$_2$: Females
Factor B: Experimenter	Level B$_1$: Male Experimenter	6	8
		11	14
		9	17
		10	16
		9	19
	Level B$_2$: Female Experimenter	8	4
		10	6
		9	5
		7	5
		10	7

Learning Outcomes and Objectives

13.1 Parametric versus Nonparametric Statistics

Know when to use *nonparametric* statistics

Nonparametric procedures are used when data do not meet the assumptions of parametric procedures. Nonparametric procedures are less powerful than parametric procedures.

13.2 Chi Square Procedures

Chi square (χ^2) is used with one or more nominal (categorical) variables, and the data are the frequencies with which participants fall into each category.

13.3 One-Way Chi Square: The Goodness of Fit Test

Understand the logic of and how to use the *one-way chi square*

The *one-way chi square* compares the frequency of category membership along one variable. A significant χ^2_{obt} indicates that the observed frequencies are unlikely to represent the distribution of frequencies in the population described by H_0. When H_0 is that the frequencies are equal in the population, then $f_e = N/k$ for each category.

Quick Practice

1. The one-way chi square is used when we count the _____ that participants fall into different _____.

2. We find 21 participants in category A and 39 participants in category B. H_0 is that the frequencies are equal. The f_e for category A is _____, and f_e for category B is _____.

3. Compute χ^2_{obt} for question 2.

4. The *df* is _____ so, at $\alpha = .05$, χ^2_{crit} is _____.

5. The χ^2_{obt} is **significant/nonsignificant.**

6. We conclude that _____ % is in A and _____ % is in B.

13.4 The Two-Way Chi Square: The Test of Independence

Understand the logic of and how to use the *two-way chi square*

The *two-way chi square* tests whether the frequency of category membership on one variable is independent of the frequency of category membership on the other variable. A significant χ^2_{obt} indicates that the two variables are dependent, or correlated. In a significant 2×2 chi square, the strength of the relationship is described by the *phi correlation coefficient* (ϕ). In a significant two-way chi square that is not a 2×2, the strength of the relationship is described by the *contingency coefficient* (C). The larger these coefficients are, the closer the variables are to being perfectly dependent, or correlated. Squaring ϕ or C gives the proportion of variance accounted for, the proportion of the differences in category membership on one variable that is associated with category membership on the other variable.

Quick Practice

1. The two-way χ^2 is used when we count the _____ with which participants fall into the _____ of two variables.

2. The H_0 in the two-way χ^2 is that the frequencies in the categories of one variable are _____ of those of the other variable.

Key Terms

nonparametric statistics Inferential procedures that do not require stringent assumptions about the raw score population represented by the sample data

chi square procedure (χ^2) The nonparametric inferential procedure for testing whether the frequencies of category membership in the sample represent the predicted frequencies in the population

one-way chi square The chi square procedure for testing whether the sample frequencies of category membership on one variable represent the predicted distribution of frequencies in the population

observed frequency (f_o) The frequency with which participants fall into a category of a variable

expected frequency (f_e) The frequency expected in a category if the sample data perfectly represent the distribution of frequencies in the population described by the null hypothesis

χ^2-distribution The sampling distribution of all possible values of χ^2 that occur when the samples represent the distribution of frequencies described by the null hypothesis

goodness of fit test A name for the one-way chi square, because it tests how "good" the "fit" is between the data and H_0

two-way chi square The chi square procedure for testing whether, in the population, frequency of category membership on one variable is independent of frequency of category membership on the other variable

test of independence A name for the two-way chi square, because it tests whether the frequency of category membership on one variable is independent of frequency of category membership on the other variable

phi coefficient (ϕ) The statistic that describes the strength of the relationship in a two-way chi square when there are only two categories for each variable

contingency coefficient (C) The statistic that describes the strength of the relationship in a two-way chi square when there are more than two categories for either variable

Spearman rank-order correlation coefficient (r_s) The correlation coefficient that describes the linear relationship between pairs of ranked scores

Mann–Whitney U test The nonparametric version of the independent-samples t-test for ranked scores when there are two independent samples of ranked scores

Wilcoxon T test The nonparametric version of the related-samples t-test for ranked scores

Kruskal–Wallis H test The nonparametric version of the one-way, between-subjects ANOVA for ranked scores

Friedman χ^2 test The nonparametric version of the one-way, within-subjects ANOVA for ranked scores

Key Formulas

COMPUTING EXPECTED FREQUENCY IN A ONE-WAY CHI SQUARE

f_e in each category $= \dfrac{N}{k}$

CHI SQUARE

$\chi^2_{\text{obt}} = \Sigma\left(\dfrac{(f_o - f_e)^2}{f_e}\right)$

---

3. Below are the frequencies for individuals who are satisfied/dissatisfied with their job and who must/must not work overtime. What is the f_e in each cell?

	Overtime	No Overtime
Satisfied	$f_o = 11$	$f_o = 3$
Dissatisfied	$f_o = 8$	$f_o = 12$

4. Compute χ^2_{obt}.

5. The $df =$ _____, and χ^2_{crit} is _____.

6. What do you conclude about these variables?

13.5 Statistics in the Research Literature: Reporting χ^2

Know what is different about reporting the results of a χ^2

13.6 A Word about Nonparametric Procedures for Ordinal Scores

Know the names of the common nonparametric procedures for ranked data

The correlation coefficient that describes the strength and type of linear relationship that is present between two ordinal variables is the *Spearman rank-order correlation coefficient* (r_s). The *Mann–Whitney* U *test* is the nonparametric, independent-samples t-test for ordinal scores. The *Wilcoxon* T *test* is the nonparametric, related-samples t-test for ranks. The *Kruskal–Wallis* H *test* is the nonparametric, one-way, between-subjects ANOVA for ranks. The *Friedman χ^2 test* is the nonparametric, one-way, within-subjects ANOVA for ranks.

QUICK PRACTICE ANSWERS

13.3: 1. frequency, categories; 2. $f_e = 60/2 = 30$, $f_e = 60/2 = 30$; 3. $\chi^2_{\text{obt}} = \dfrac{(30-21)^2}{30} + \dfrac{(30-39)^2}{30} = 5.40$; 4. 1, 3.84; 5. significant; 6. 35, 65

13.4: 1. frequency, categories; 2. independent; 3. For satisfied–overtime, $f_e = (14)(19)/34 = 7.824$, for satisfied–no overtime, $f_e = (14)(15)/34 = 6.176$, for dissatisfied–overtime, $f_e = (20)(19)/34 = 11.176$, for dissatisfied–no overtime, $f_e = (20)(15)/34 = 8.824$;

4. $\chi^2_{\text{obt}} = \dfrac{(11 - 7.824)^2}{7.824} + \dfrac{(3 - 6.176)^2}{6.176} + \dfrac{(8 - 11.176)^2}{11.176} + \dfrac{(12 - 8.824)^2}{8.824} = 4.968$;

5. 1, 3.84; 6. χ^2_{obt} is significant: The frequency of job satisfaction/dissatisfaction depends on the frequency of overtime/no overtime.

DEGREES OF FREEDOM IN A ONE-WAY CHI SQUARE

$df = k - 1$

COMPUTING THE EXPECTED FREQUENCY IN EACH CELL OF A TWO-WAY CHI SQUARE

$f_e = \dfrac{(\text{Cell's row total } f_o)(\text{Cell's column total } f_o)}{N}$

DEGREES OF FREEDOM IN A TWO-WAY CHI SQUARE

$df = (\text{Number of rows} - 1)(\text{Number of columns} - 1)$

PHI COEFFICIENT

$\phi = \sqrt{\dfrac{\chi^2_{\text{obt}}}{N}}$

CONTINGENCY COEFFICIENT

$C = \sqrt{\dfrac{\chi^2_{\text{obt}}}{N + \chi^2_{\text{obt}}}}$

Review and Application Questions

1. What do all nonparametric procedures have in common with all parametric procedures?

2. Which variable in an experiment determines whether to use parametric or nonparametric procedures?

3. (a) With which two scales of measurement do you use nonparametric procedures? (b) What two things can be "wrong" with interval/ratio scores that lead you to use nonparametric procedures for ranked data?

4. (a) Why, if possible, should a researcher design a study so that the data meet the assumptions of a parametric procedure? (b) Why shouldn't you use parametric procedures for data that clearly violate their assumptions?

5. (a) When do you use the chi square? (b) When do you use the one-way chi square? (c) When do you use the two-way chi square?

6. (a) What is the symbol for observed frequency? What does it mean? (b) What is the symbol for expected frequency? What does it mean?

7. What does a significant one-way chi square indicate?

8. What does a significant two-way chi square indicate?

9. (a) What is the phi coefficient, and when is it used? (b) What is the contingency coefficient, and when is it used?

10. What is the nonparametric version of each of the following: (a) the one-way, between-subjects ANOVA, (b) the independent-samples t-test, (c) the related samples t-test, (d) the Pearson correlation coefficient, and (e) the one-way, within subjects ANOVA?

11. In the general population, the distribution of political party affiliation is 30% Republican, 55% Democratic, and 15% other. To determine whether this distribution is also found among the elderly, in a sample of 100 senior citizens, we find 18 Republicans, 64 Democrats, and 18 others. (a) What are H_0 and H_a? (b) What is f_e for each group? (c) Compute χ^2_{obt}. (d) With $\alpha = .05$, what do you conclude about party affiliation in the population of senior citizens?

12. A survey finds that, given the choice, 34 females prefer males much taller than themselves, and 55 females prefer males only slightly taller than themselves. (a) What are H_0 and H_a? (b) With $\alpha = .05$, what would you conclude about the preference of females in the population? (c) Describe how you would graph these results.

13. Kerry counts the students who like Professor Demented and those who like Professor Randomsampler. Kerry then performs a one-way χ^2 to determine if there is a significant difference between the frequency with which students like each professor. (a) Why is this approach incorrect? (Hint: Check the assumptions of χ^2.) (b) How should Kerry analyze the data?

14. The data to the right reflect the frequency with which people voted in the last election and were satisfied with the officials elected.

    (a) What are H_0 and H_a? (b) What is f_e in each cell? (c) Compute χ^2_{obt}. (d) With $\alpha = .05$, what do you conclude about these variables? (e) How consistent is this relationship?

		Satisfied	
		Yes	No
Voted	Yes	48	35
	No	33	52

SPSS Instructions

Chi Square and Nonparametric Statistics

One-Way Chi Square

Main Menu: Analyze

Select: **Nonparametric Tests → Legacy Dialogs → Chi-square**

Move: **Desired variable into Test Variable List box**

Select: **OK**

Notes: (1) You can either allow the computer to determine the range of values based on the observations in the data or you can specify the range of values by providing a lower and upper limit. (2) You can either choose the expected frequencies to be equal or you may specify expected frequencies.

Two-Way Chi Square

Main Menu: Analyze

Select: **Descriptive Statistics → Crosstabs**

Move: **Desired variable 1 into Row(s) box** **Desired variable 2 into Column(s) box**

Select: **Statistics... →** ✓ **Chi-square →** **Continue → OK**

Optional: **Statistics... →** ✓ **Contingency coefficient and/or** ✓ **Phi and Cramer's V**

Note: This procedure allows for numerous statistics to be calculated, depending on the type of data involved (e.g., nominal, ordinal).

Remember to login to the CourseMate for STAT site at www.cengagebrain.com for additional review tools and games

SPSS Instructions

Spearman Rank-Order Correlation Coefficient

Main Menu: *Analyze*

Select:	**Correlate → Bivariate**
Move:	**Desired X and Y variables into Variable(s) box**
Select:	**✓ Spearman (under Correlation Coefficients) → ✓ Two-tailed or One-tailed significance test → OK**

Notes: (1) It does not matter which variable is the X and which is the Y for this procedure as both go into the Variable(s) box.
(2) Two-tailed significance test is the default.

Mann–Whitney U Test

Main Menu: *Analyze*

Select:	**Nonparametric Tests → Legacy Dialogs → 2 Independent Samples**
Move:	**Desired dependent variable into Test Variable List box** **Independent variable into Grouping Variable box**
Select:	**Define Groups**
Enter:	**Value used for independent variable group 1 (e.g., "a" or 1)** **Value used for independent variable group 2 (e.g., "b" or 2)**
Select:	**Continue → OK**

Wilcoxon T Test

Main Menu: *Analyze*

Select:	**Nonparametric Tests → Legacy Dialogs → 2 Related Samples**
Move:	**Variable containing data for first condition of the related samples into Variable 1 of Pair 1 in the Paired Variable(s) box** **Variable containing data for second condition of the related samples into Variable 2 of Pair 1 in the Paired Variable(s) box**
Select:	**✓ Wilcoxon (under Test Type) → OK**

Kruskal–Wallis H Test

Main Menu: *Analyze*

Select:	**Nonparametric Tests → Legacy Dialogs → K Independent Samples**
Move:	**Desired dependent variable into Test Variable List box** **Independent variable into Grouping Variable box**
Select:	**Define Range**
Enter:	**Value used for independent variable group 1 (e.g., 1)** **Value used for independent variable group k (e.g., 4)**
Select:	**Continue → ✓ Kruskal–Wallis H (under Test Type) → OK**

Friedman χ^2 Test

Main Menu: *Analyze*

Select:	**Nonparametric Tests → Legacy Dialogs → K Related Samples**
Move:	**Each variable containing a condition of the related samples into Test Variables box**
Select:	**✓ Friedman (under Test Type) → OK**

Application Questions

15. A study determines the frequency of the different political party affiliations for male and female senior citizens. The following data are obtained:

		Affiliation		
		Republican	Democrat	Other
Gender	Male	18	43	14
	Female	39	23	18

(a) What are H_0 and H_a? (b) What is f_e in each cell? (c) Compute χ^2_{obt}. (d) With $\alpha = .05$, what do you conclude about gender and party affiliation in the population of senior citizens? (e) How consistent is this relationship?

16. Select the statistical procedure to use in each of the following: (a) In a study of the effects of a new pain reliever on rankings of the emotional content of words describing pain, a randomly selected group of people is tested before and after administration of the drug. (b) In a study of the effects of eight different colors of spaghetti sauce on tastiness scores, a different random sample of people tastes each color of sauce, and then the tastiness scores are ranked. (c) In a study of the effects of increasing amounts of alcohol consumption on reaction-time scores, the scores are ranked, and the same group of participants is tested after 1, 3, and 5 drinks. (d) In a study of family income, a wealthy and poor sample indicates the percentage of family income that was spent on clothing last year. Scores are then rank-ordered. (e) A study of marathon runners is conducted to describe the relationship between their finishing position in Race A and their finishing position in Race B.

17. What is the basic logic of H_0 and H_a in all nonparametric procedures for ranked data?

18. After testing 40 participants, our χ^2_{obt} was 13.31. With $\alpha = .05$, and $df = 2$, report this significant result.

19. A research article indicates that the Wilcoxon T test was significant ($p < .05$). (a) What does this test indicate about the design of the study? (b) What does it indicate about the nature of the scores? (c) What will you conclude about the relationship here?

20. A research article indicates that the Mann–Whitney U test was significant ($p < .01$). (a) What does this test indicate about the design of the study? (b) What does it indicate about the scale of the dependent scores? (c) What will you conclude about the relationship?